Fundamentals of the Petrophysics of Oil and Gas Reservoirs

Scrivener Publishing
100 Cummings Center, Suite 41J
Beverly, MA 01915-6106

Publishers at Scrivener
Martin Scrivener (martin@scrivenerpublishing.com)
Phillip Carmical (pcarmical@scrivenerpublishing.com)

Fundamentals of the Petrophysics of Oil and Gas Reservoirs

Leonid Buryakovsky, Ph.D.

Russian Academy of Natural Sciences, U.S.A. Section

George V. Chilingar, Ph.D.

Emeritus Professor of Civil, Environmental and Petroleum Engineering,
University of Southern California, Los Angeles, CA

Herman H. Rieke, Ph.D.

Professor of Petroleum Engineering, University of Louisiana at Lafayette, Lafayette, LA

and

Sanghee Shin, Ph.D.

Research Associate, Rudolf W. Gunnerman, Energy and Environmental Laboratory,
University of Southern California, Los Angeles, CA

Scrivener

WILEY

Co-published by John Wiley & Sons, Inc. Hoboken, New Jersey, and Scrivener Publishing LLC, Salem, Massachusetts.
Published simultaneously in Canada.

Cover design by Kris Hackerott.

Library of Congress Cataloging-in-Publication Data:

ISBN 978-1-118-34447-7

Printed in the United States of America

10 9 8 7 6 5 4 3 2 1

This volume is dedicated to Dr. Chengyu Fu
for his important contributions
to World Petroleum Industry
and World Economy

Contents

Preface

Petrophysics (rock physics) is a branch of applied geology relating to the study of reservoir and caprock properties and their interactions with fluids (gases, hydrocarbons, and aqueous solutions) based on fundamental methods of physics, chemistry, and mathematics. The geologic material forming a reservoir for the accumulation of fluids (oil, gas and water) in the subsurface must contain a three-dimensional network of interconnected voids (pores, vugs and/or fractures) to store the formation fluids and allow their movement within the reservoir during hydrocarbon recovery. Petrophysical applications are widely used in petroleum geology, economic geology, seismic interpretations, hydrocarbon reserve estimation, reservoir description and simulation, field development planning, and reservoir production management.

The goal of petroleum geology is to perform the exploration and to provide, if a discovery is made of a commercial oil and gas accumulation, a geological/geophysical description of the reservoir. This includes preparing an estimate of the initial reserve hydrocarbon volume. The application of petrophysics in both hydrocarbon exploration and recovery is to minimize financial risk.

The goal of economic geology is the study and analysis of geologic bodies and materials that can be utilized profitably, including carbon fuels, metals, nonmetallic minerals, and water. The application of geoscience knowledge and theory is foremost for understanding of the origin of deposits and, most importantly, how to exploit them.

The goal of reservoir engineering is to produce an integrated reservoir study in order to support a computer model of the reservoir that can implement the integration of the total reservoir database. The model must include: (1) production forecasts; (2) results of operational consequences based on management decisions, and (3) how to maintain a current reservoir model by using newly-acquired performance and field data.

Theoretical and applied petroleum geology encompasses the exploration, discovery, and integration of information to be applied to the appraisal of oil and gas basins, provinces, regions, areas, and fields that are considered as integral geologic systems. Owing to the absence of distinct boundaries, geologic systems are mostly the so-called open systems, the geologic properties of which are modified over time and space. Considering these comments, a geologic system can be defined as follows:

> *A geologic system is an organized natural assembly of interconnected and interacting elements of lithosphere having common development history and comprising a single geologic body with properties that are not inherent in its individual elements.*

In this regard the petrophysical system may be defined as a:

> *Petrophysical system is the well-organized natural assembly of interacting solid, liquid, and gaseous elements having common development history and a distinguishing set of physical and chemical properties, which manifest themselves both individually and jointly.*

In addition to the above-mentioned statement, we have to indicate also the basic principle of geological investigations, which states, "the present is the key to the past". This concept means that processes, which acted on the Earth in the past, are very similar to or are the same as those operating today. That is why the petrophysical study of reservoir and sealing-rock (caprock) properties by laboratory core analysis and/or well logging and well testing is very important to understand the origin, composition, and behavior of oil and gas reservoirs.

The study of fluid flow through porous rocks as well as rock properties themselves had begun by Austrian scientist Kozeny (1927). He solved the Navier-Stokes equations for fluid flow by considering the porous medium as an assembly of capillary tubes (pores) of the same length. Kozeny obtained relationship among permeability, porosity, and specific surface area of porous media (Kozeny, 1927). At about the same time, the Schlumberger brothers in France introduced the first well logs. These early developments led to rapid improvements of equipment, production operations, formation evaluation, and hydrocarbon recovery efficiency (Schlumberger, 1972, 1987). Therefore, in the decades following, the study of rock properties and fluid flow was intensified and became a part of the research endeavors of petroleum institutes and major oil companies. Today, most of the oil and gas companies rely on research and the application of the obtained results to the field by service companies.

A first experimental study of petrophysical properties using rock samples was by Bridgman (1918). He conducted the stress – strain testing under atmospheric pressure and at room temperature. Comparison of his experimental results with well-logging data showed a discrepancy between the two owing to the influence of formation pressure and temperature on the petrophysical properties in-situ. Bridgman was the first investigator who established deviation of physical parameters of sandstones determined at room temperature and atmospheric pressure from those obtained under elevated pressures and temperatures (Bridgman, 1936; Bridgman *et al.*, 1966).

In 1942, G.E. Archie discussed the relationship between electrical resistance of fluids in porous media and porosity and proposed an equation relating porosity and electrical resistivity. He reviewed and discussed the relationships among the types of rocks, sedimentary environments, and petrophysical properties, and suggested that specialized studies of reservoir properties of reservoir rocks and caprocks should be recognized as a separate geologic discipline called *petrophysics* (Archie, 1950, 1952).

The influence of overburden pressure on porosity was studied by Archie (1950), who established that porosity of argillaceous sediments at the Earth's surface is about 50%, whereas at a depth of 2000 m it is ten times lower. Krumbein and Sloss (1951) showed that the porosity of shale and sandstone is a function of burial depth, which influences porosity of shale more than that of sandstone. Fatt (1953, 1957a,b) was the first investigator who suggested that the physical properties of rocks are affected by the difference between the total overburden pressure and reservoir pore pressure, i.e., the net overburden (grain-to-grain) pressure.

A major contribution to petrophysical studies was made by Hedberg (1926, 1936); Athy (1930); Carman (1937, 1938, 1939); Carpenter and Spencer (1940); Klinkenberg (1941, 1951); Trask (1942); Taylor (1950); Wyllie *et al.* (1950, 1956, 1958a); Winsauer *et al.* (1952, 1953); Griffith (1952); Brooks and Purcell (1952); Fatt (1953, 1957a,b); Hall (1953); Krumbein (1955b); Weller (1959); von Engelhardt (1960); Chilingar *et al.* (1963); Chilingar (1964, 2005); Donaldson *et al.* (1969); Rieke and Chilingarian (1974), Eremenko and Chilingar (1996), Rebesco and Camerlenchi (2008), and van den Berg and Nio (2010).

Important petrophysical studies were carried out in the former USSR, e.g., Avdusin and Tsvetkova (1938); Volarovich (1940, 1960); Trebin (1945); Kotyakhov (1949, 1956); Samedov and Buryakovsky (1957); Kusakov and Gudok (1958); Buryakovsky (1960, 1970, 1977); Buryakovsky *et al.* (1961, 1975, 1982); Vassoevich and Bronovitskiy (1962); Dobrynin (1962); Teodorovich (1965); Parkhomenko (1965); Vendelshtein (1966); Khanin (1966, 1969); Dakhkilgov (1967); Shreiner *et al.* (1968); Petkevich and Verbitskiy (1970); Avchan (1972); Ellanskiy (1972); Bagrintseva (1977, 1982); Chernikov and Kurenkov (1977); Marmorshtein (1975, 1985); Morozovich (1967); Proshlyakov (1974); and Dzhevanshir *et al.* (1986).

Generalized discussion of petrophysics were published by Krumbein and Sloss (1951); Pirson (1950, 1963); Scheidegger (1957); Dakhnov and Dolina (1959); von Engelhardt (1960); Kobranova (1962, 1986); Parkhomenko (1965); Khanin (1966, 1969, 1976); Avchan *et al.* (1966, 1979); Vendelshtein (1966); Romm (1966, 1985); Griffith (1967); Gudok (1970); Dobrynin (1970); Lomtadze (1972); Volarovich (1974); Pavlova (1975); Chilingar *et al.* (1975, 1976, 1979, 1992); Kotyakhov (1977); Buryakovsky (1977, 1985a); Buryakovsky *et al.* (1961, 1982, 1985b, 1990a, 2001); Marmorshtein (1975, 1985); Bagrintseva (1977, 1982); Magara (1978); Ellanskiy (1978); Dakhnov (1982, 1985); Proshlyakov *et al.* (1987); and Tiab and Donaldson (1996).

Of major importance in petrophysical studies is the construction and investigation of petrophysical relationships. Among a large amount of contributions on the use of mathematical methods and techniques in petrophysics one should mention the following: Krumbein (1955a, 1955b); Miller and Kahn (1962); Stetyukha (1964); Krumbein *et al.* (1965, 1969); Sharapov (1965); Griffith (1967); Vistelius (1967); Harbaugh *et al.* (1970, 1977); Buryakovsky (1968, 1974b, 1982, 1985a, 1992); Buryakovsky *et al.* (1974a, 1979, 1980, 1981, 1982, 1990a, 1991); Ellanskiy (1978); Romm (1985); Abasov *et al.* (1987, 1989); Lucia (1999); Chilingar *et al.* (2005); and Cosentino (2006).

Seismic fluid detection, reservoir delineation, and rock physics is in the realm of the geophysicists. Because of the growing complexity of recently discovered oil and gas fields more reliance is being placed on seismic delineation of the properties of reservoir rocks, (such as porosity and permeability), fluid movement in time, fracture detection, pore pressure, mineralogy and saturation components in the formation. Well test data; well logs, and core data are of a scale that does not match seismic spatial detail of the variability in reservoir petrophysical properties. Some important contributors to this science are: Fertl *et al.* (1976); Gregory (1976); Nobes *et al.* (1986); Batzle and Wang (1992); Berryman (1992); Guéguen and Palciauskas (1994); Mavko *et al.* (1998); and Cohen (2007).

Various oil and gas reservoirs in clastic, carbonate and volcanic rocks are descried in this book, taking into consideration their depositional environments and depth of occurrence. Core analysis and well-logging techniques, used for the determination of such essential reservoir-rock properties as porosity (total and effective), permeability (absolute and relative to air, water, oil, and gas), oil/gas/water saturation, and wettability are described in detail. Well-logging section includes electrical, radioactive, acoustic and other tools used for subsurface investigation. Well-log analysis and interpretation includes formation evaluation based on core and log data and relationships between them. Today, the mathematical simulation of petrophysical properties and relationships including core-to core and log-to-core, and seismic-to-core-to-well-log correlation is a common industry practice. One must be aware that the scales of petrophysical properties in these correlations are of different magnitudes, ranging from 10^{-6} to 10^{6} m, which covers the microscopic to the gigascopic properties (Chilingarian *et al.*, 1996).

This book is an essential summary of theoretical studies and their practical applications in the field of petrophysics and some interdisciplinary sciences, activities conducted by the authors for more than 50 years. It represents the physical and geological background of petrophysical investigations of subsurface formations.

Leonid A. Buryakovsky
George V. Chilingar
Herman H. Rieke
Sanghee Shin

Hydrocarbon exploration and exploitation technologies have made tremendous technical progress during the past 25 years. One of the technologies that improved success is the ability to integrate reservoir information into a virtual three-dimensional reservoir. Although, such spatial computer models only represent an approximation of the real hydrocarbon reservoir, simulation has facilitated our knowledge and limitations owing to the scarcity of available data. One should consider the fact that the model is only as good as the available data, which is basically petrophysical and fluid properties of the producing formation. This book is about the background and value of having knowledge of the petrophysical properties and geological data to help maximize the hydrocarbon recovery.

The authors would also like to recommend the classical book on "Petrophysics" by D.Tiab and E.C. Donaldson, 2004, Gulf Professional Publishing, as a reference book.

Leonid Khilyuk

List of Contributors

Co-Authors
Leonid Buryakovsky, Ph.D.
Russian Academy of Natural Sciences, U.S.A. Section

George V. Chilingar, Ph.D.
Emeritus Professor of Civil, Environmental and Petroleum Engineering, University of Southern California, Los Angeles, CA

Herman H. Rieke, Ph.D.
Professor of Petroleum Engineering, University of Louisiana at Lafayette, Lafayette, LA

Sanghee Shin, Ph.D.
Research Associate, Rudolf W. Gunnerman, Energy and Environmental Laboratory, University of Southern California, Los Angeles, CA

With contributions by
Carl Richter, Ph.D.
University of Louisiana at Lafayette, Lafayette, LA

Donald G. Hill, Ph.D.
Petroleum Engineering Program, Viterbi School of Engineering, University of Southern California, Los Angeles, CA

Acknowledgement

The help extended by Dr. Henry Chuang, President of Willie International Holdings Limited of Hong Kong, China, a Petroleum Engineer, was invaluable in publishing this book. He deserves the highest of praise.

1 Introduction

1.1 Characterization of Hydrocarbon Reservoirs

Hydrocarbon accumulation requires the presence of a natural trap consisting of reservoir rocks, sealing or caprocks, and three-dimensional four-way closure. The description of reservoir rocks should include the following elements:

1. Presence of reservoir rocks
   - Depositional model (sequence stratigraphy framework)
   - Lithology
   - Structural characteristics
   - Lateral and vertical distribution
2. Quality of reservoir rocks
   - Lateral continuity and extension
   - Thickness and vertical lithological cyclicity
   - Relative heterogeneity of rock properties
   - Pore systems ranges and types
   - Transmissibility of fluids
   - Hydrocarbon potential and preservation
   - Diagenetic characteristics

The reservoir rocks are mainly sedimentary rocks, which are deposited as sediments by water, wind, or ice and made up of clastic material, chemical precipitates, and organic or biogenic debris. Sedimentary rocks have formed from sediments or debris by any of the following processes: (1) *compaction*, (2) *cementation*, and (3) *crystallization*. A simplified classification of sedimentary rocks is presented in Table 1.1.

Clastic rocks are the consolidated sedimentary rocks consisting principally of the debris of preexisting rocks (of any origin) or the solid products formed during chemical weathering of such rocks, transported mechanically (by such agents as water, wind, ice, and

Leonid Buryakovsky, George V. Chilingar, Herman H. Rieke and Sanghee Shin
Petrophysics, (1–56) © Scrivener Publishing LLC

Table 1.1 Simplified classification of sedimentary rocks.

Clastic Rocks	Carbonates	Evaporites	Organic Rocks	Other
Conglomerate	Limestone	Gypsum	Peat	Chert
Breccia	Chalk	Anhydrite	Coal	
Sandstone	Dolomite	Rock salt	Diatomite	
Siltstone	Marl	Potash	Limestone	
Mudstone				
Shale				
Limestone				

gravity) to their places of deposition. Clastic non-carbonate rocks, which are almost exclusively silicon-bearing (either as quartz or silicates) are called the *siliciclastic rocks*. Siliciclastic rocks consist of sand-, silt- and clay-size particles and their combinations.

Carbonate rocks are the rocks consisting chiefly of carbonate minerals formed by the organic or inorganic precipitation from aqueous solution of carbonates of calcium (limestone), calcium plus magnesium (dolomite), or iron (siderite). These rocks may consist also of the debris of preexisting carbonate rocks (of any origin), which have been transported mechanically to their places of deposition.

Chilingar and Yen (1982) pointed out that carbonate rocks constitute only 15 to 30% of the total volume of sedimentary rocks, whereas about 65% of the total oil and gas reserves in the World reside in carbonate reservoirs. The behavior of carbonate reservoirs differs in many respects from sandstone reservoir, mainly due to the very complex pore structure of carbonate rocks. However, the percentage of in-situ oil recovered from these reservoir rocks is often very low ($\leq$ 20%). Their origin, composition, and the diagenetic and catagenetic processes in large measure determine the petrophysical properties of carbonates and behavior of carbonate hydrocarbon reservoirs (refer to Chilingarian *et al.*, 1996).

Chert is a hard, extremely dense or compact cryptocrystalline sedimentary rock, consisting dominantly of cryptocrystalline silica (chiefly fibrous chalcedony) with lesser amount of micro- or cryptocrystalline quartz and amorphous silica (opal).

Figure 1.1 illustrates a classification based on a tetrahedron at the corners of which are placed carbonate, clay (shale), sandstone (quartz) and chert. This figure also depicts one side of this

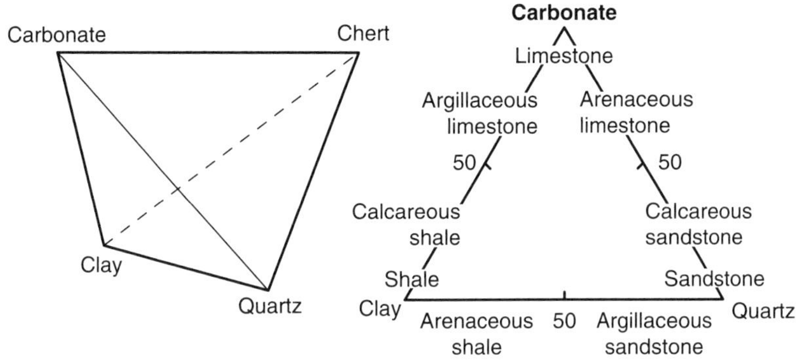

Figure 1.1 Fundamental tetrahedron for classifying sedimentary rocks.

tetrahedron so that some of the variations between contents of shale, sandstone and limestone can be seen. For example, starting from shale and going toward limestone, increasing amounts of lime will produce calcareous shale, grading into argillaceous (shaley) limestone, then to pure limestone. Similarly, on the other two edges, it is shown how the changes occur from shale to sandstone and from sandstone to limestone. The other three sides show similar variations with chert replacing one of the other constituents. Other valuable classifications of sandstones based on composition were presented by Teodorovich (1965). (e.g., Figure 1.2a,b.)

One of the most important aspects in reservoir characterization is an understanding of depositional environments in the area under study. Depositional environments and facies relationships, diagenesis (physical and chemical changes in sediments up to and *through*

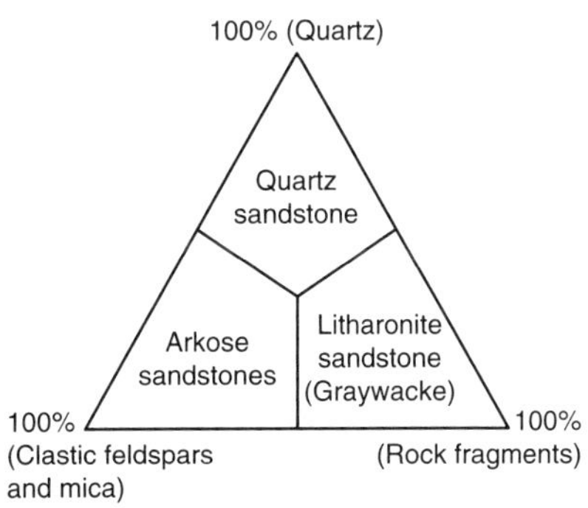

Figure 1.2a The major classification of sandstones, based on composition. (After Teodorovich, 1965.)

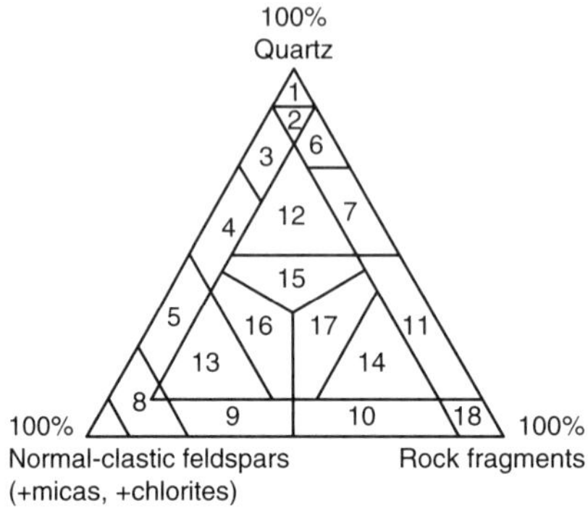

Figure 1.2b Classification of sandstones on the basis of mineralogical composition. (A) Family of quartz sandstones (1, 2, 3); (B) arkoses (4, 5); (C) two-component mineral-petroclastic sandstones (6, 7, 8, 9, 10, 11) – the latter family can be subdivided into (a) predominantly quartz sandstones (6, 7); (b) predominantly feldspathic sandstones (8, 9), and (c) predominantly petroclastic sandstones (10, 11): (D) three-component mineral – petroclastic sandstones with absolute predominance of one of the components (12, 13, 14); (E) three-component mineral – petroclastic sandstones with relative predominance of one of the components (15, 16, 17); (F) ultrapetroclastic sandstones (18). (After Teodorovich, 1967, Figure 1, p. 76.)

lithification) and catagenesis (physical and chemical changes in the lithified rock) strongly affect the size, shape, pore-space geometry, porosity, permeability, and location of clastic deposits.

Any sediment has originally a terrestrial *source* (place of origin), created by the life cycles of plants or animals (e.g., shells, leaves, logs, and organic sediments), or by weathering (chemical disintegration and physical breakdown) of parent rocks. Each sediment has a *provenance*, which is the particular area from which its components were derived and transported by water, ice, or wind into the place of *deposition*. Sediments are deposited under a variety of conditions or environments, both on land and at sea. Each environment is characterized by specific physical processes and has the particular plants and animals living within it, which contribute such fossils as bones, shells, and plant fragments. Simplified classification of environments of sediment deposition is shown in Table 1.2.

As an example, the depositional environment and stratigraphy of the South Caspian Basin are presented here owing to the geological complexity of the basin and familiarity by the authors with the area.

Table 1.2 Depositional environments.

<table>
<tr><th></th><th colspan="2">Delta Group</th><th colspan="2">Intradelta Group</th></tr>
<tr><td rowspan="3">Continental</td><td colspan="2">Aeolian deposits</td><td colspan="2">Aeolian deposits</td></tr>
<tr><td colspan="2">Alluvial deposits</td><td colspan="2" rowspan="2">Alluvial deposits</td></tr>
<tr><td rowspan="4">Deltaic deposits</td><td rowspan="2">Delta-plain deposits</td></tr>
<tr><td rowspan="2">Transitional</td><td colspan="2" rowspan="2">Coastal Intradelta-marine deposits</td></tr>
<tr><td rowspan="2">Prodelta-plain deposits</td></tr>
<tr><td rowspan="3">Marine</td><td rowspan="3">Normal marine deposits</td><td>Shelf deposits</td></tr>
<tr><td rowspan="2">Normal marine deposits</td><td>Slope deposits</td><td>Slope deposits</td></tr>
<tr><td>Deep marine deposits</td><td>Deep marine deposits</td></tr>
</table>

1.1.1 Geographical and Geological Background of the South Caspian Basin

The South Caspian Basin encompasses water areas of the South Caspian Sea, and together with land areas of Eastern Azerbaijan and Western Turkmenistan constitutes the southern portion of the Caspian Sea region (Figure 1.3a). The South Caspian Basin is separated from the Middle Caspian Basin by the Absheron-Prebalkhan zone of uplifts, which extends NW-SE connecting the Absheron and Cheleken peninsulas, forming a narrow submarine ridge (Buryakovsky, 1974c, 1993a, 1993b; Buryakovsky *et al.*, 2001) (Figures 1.3b and 1.3c).

Azerbaijan borders Russia to the north, Georgia and Armenia to the west, Turkey and Iran to the south, and the Caspian Sea to the east. Aerially, it encompasses the southeastern spurs of the

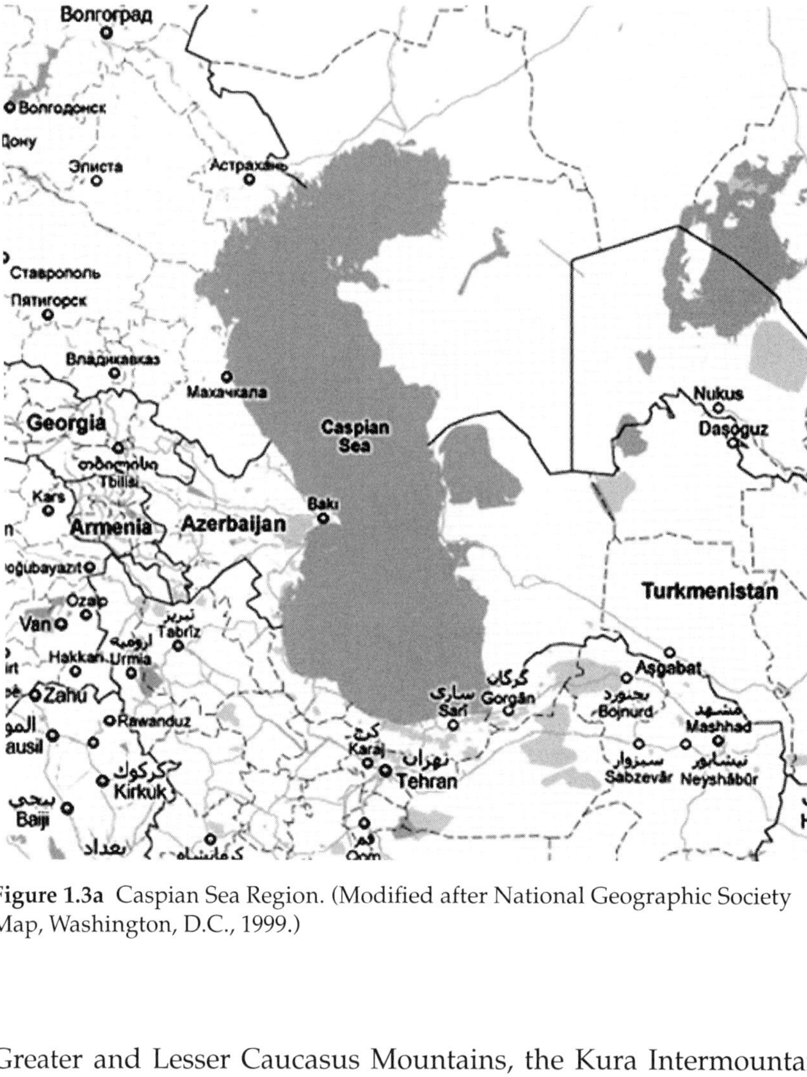

Figure 1.3a Caspian Sea Region. (Modified after National Geographic Society Map, Washington, D.C., 1999.)

Greater and Lesser Caucasus Mountains, the Kura Intermountain Depression, and Talysh Mountains (Figure 1.4 and 1.5). Azerbaijan is one of the oldest oil and gas provinces in the world. For more than 140 years oil and gas has been commercially produced onshore in Azerbaijan. The offshore production at the Caspian Sea began about 100 years ago.

Turkmenistan borders Kazakhstan to the north, Uzbekistan to the northeast and east, Afghanistan and Iran to the south, and the South Caspian Sea to the west. The onshore portion of Western Turkmenistan includes the Cheleken Peninsula on the northwest and is bordered by the Kopet-Dagh Mountains to the south.

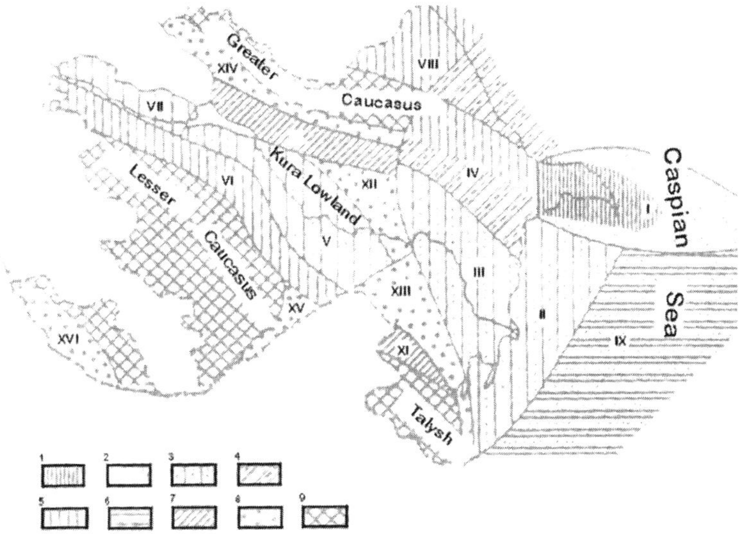

Figure 1.3b Oil and gas regional distribution, and fields and prospects of Azerbaijan and the South Caspian Basin. (Modified after the Excursion Guide-Book for Azerbaijan SSR, Vol. 1, 1984.) 1–Boundaries between oil- and gas-bearing regions, 2–boundaries between oil- and gas-bearing areas, 3–oil fields, 4–gas and gas-condensate fields; oil- and gas-bearing areas: 5–high oil and gas content, 6–moderate oil and gas content, 7–potential structure, 8–structure with low potential. Oil and gas-bearing regions and areas (areas are shown in circles): 1–Apsheron-Gobustan region (areas: 1–Aspheron; 2–Shemakha-Gobustan); II–Pre-Caspian-Kuba region; III–Kura region (areas: 3–Lower Kura, 4–Kyurdamir, 5–Gyandzha, 6–Adzhinour, 7–Kura-lori interfluve, 8–Alazan-Agrichai, 9–Dzhalilabad, 10–Baku Archipelago); IV–Araks area. Fields: 1–Balakhany-Sabunchi-Ramany, 2–Surakhany, 3–Karachukhur-Zykh, 4–Gum Deniz, 5–Goudsny, 6–Kala, 7–Buzovny-Mashtagi, 8–Darvin Bank, 9–Pirallaghi Adasi, 10–Gyurgyan Deniz, 11–Chalov Adasi, 12–Azi Aslanov, 13–Palchygh Pilpilasi-Neft Dashlary, 14–Dzanub, 15–Bakhar, 16–Binagady-Chakhnaglyar, 17–Sulutepe, 18–Yasamaly Valley, 19–Bibielbat, 20–Puta-Lokbatan, 21–Kyorgyoz-Kzyltepe, 22–Karadag, 23–Shongar, 24–Umbaki, 25–Duvanny, 26–Dashgil, 27–Chondagar-Zorat, 28–Siazan-Nardaran, 29–Saadan, 30–Amirkhanly, 31–Eastern Zagly, 32–Zagly-Tengialty, 33–Kyurovdag, 34–Karabagly, 35–Khillin, 36–Neftechala, 37–Kyursangya, 38–Mishovdag, 39–Kalmas, 40–Pirsagat, 41–Malyl Kharami, 42–Kalamadyn, 43–Muradkanly, 44–Kazanbulag, 45–Adzhidere, 46–Naftalan, 47–Mirbashir, 48–Sangachal, 49–Duvanny Deniz, 50–Khara Zyrya, 51–Bulla Deniz, and 52–Garasu.

The offshore region includes the eastern portion of the Absheron-Prebalkhan anticlinal trend and the Chikishlyar-Okarem zone (Turkmenian structural terrace); to the south it is bordered by the Alborz Range in Iran.

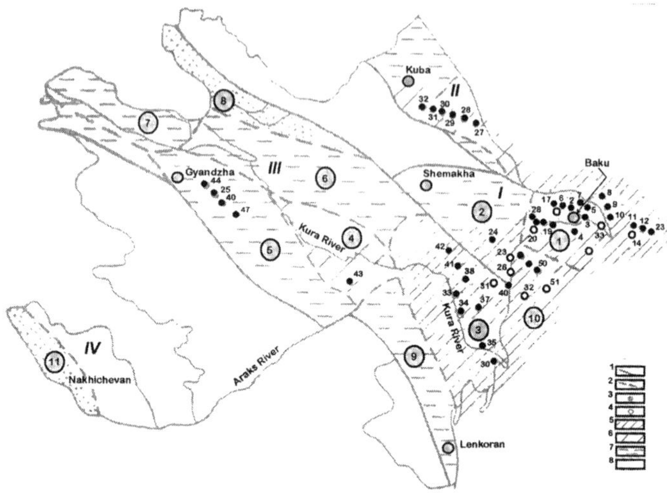

Figure 1.3c Regional distribution of oil and gas in Azerbaijan and the South Caspian Basin. (Modified after Aliyev *et al.*, 1985.) Regions: 1–with significant, proved, initial potential resources; 2–highly favorable (offshore); categories of favorability: 3–first, 4–second, 5–third; 6–areas favorable for oil and gas; 7–areas possibly favorable; 8–areas with unclarified prospects; 9–areas with no prospects. Oil- and gas-bearing areas; I–Apsheron, II–Baku Archipelago, III–Lower Kura, IV–Schemakha-Gobustan; V–Yevlakh-Agdzhabedy; VI–Gyandzha, VII–Kura-lori interfluve, VIII–Pre-Caspian-Kuba, IX–deep-water parts of the South Caspian Basin probably favorable areas: X–Adzhinour, XI–Dzhalilabad; areas with uncertain potential: XII–Dzharly-Saatly, XIII–Mil-Mugan, XIV–Alazan-Agrichai, XV–Araks, and XVI–Nakhichevan.

There are several oil and gas regions within onshore Azerbaijan. The main regions are (from north to south):

1. Pre-Caspian – Kuba Region
2. Absheron Peninsula
3. Kobystan anticlinal belt
4. Kura Intermountain Depression.

There are four offshore oil and gas regions that are within the Azerbaijan sector of the Caspian Sea (from north to south):

1. Absheron Archipelago
2. Western portion of the Absheron-Prebalkhan zone of uplifts
3. South Absheron offshore zone
4. Baku Archipelago.

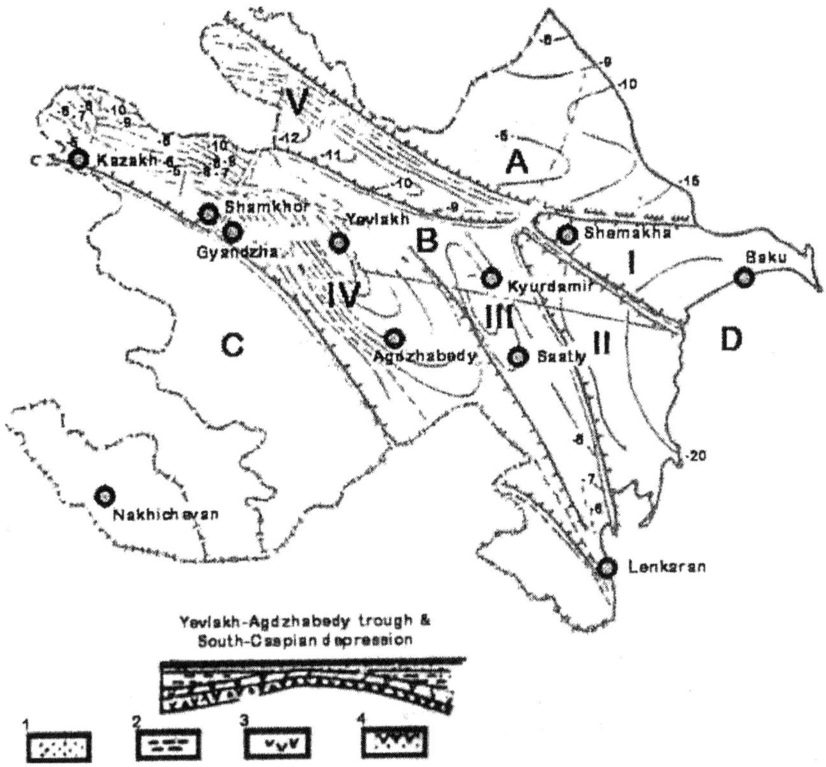

Figure 1.4 Structural pattern of Azerbaijan. (Modified after the Excursion Guide Book for Azerbaijan SSR, Vol. 11, 1984.) A–Greater Caucasus Anticlinorium; B–Kura Intermontane Depression; C–Lesser Caucasus Anticlinorium; D–South Caspian Basin: I–Gobustan-Apsheron Trough; II–Lower Kura Trough; III–Geokchai-Saatly Anticlinal Trend; IV–Yevlakh-Agdzhabedy Trough; V–Lori-Adzhinour Trough. 1–Quaternary, 2–Miocene-Paleogene, 3–Mesozoic, and 4–consolidated crust.

Offshore fields are located over the plunges of the Absheron Peninsula and the southeast Kobystan anticlinal belt.

1.1.2 Sedimentary Features of Productive Horizons in the South Caspian Basin

The main Middle Pliocene productive unit in Azerbaijan and the South Caspian Basin is called the Productive Series (sand, silt and shale interbedding). Sand-silt reservoirs contain argillaceous

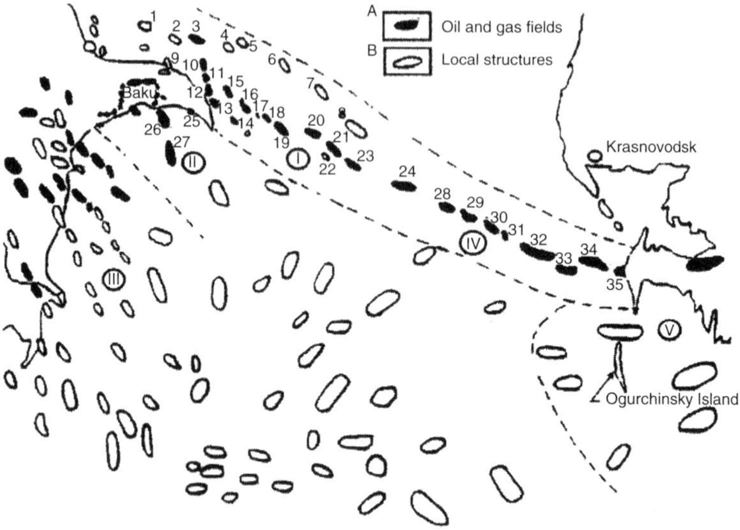

Figure 1.5 Location of structures on the Absheron Threshold. (Modified after Bagir-zadeh *et al.*, 1974.) A–Oil and gas fields; B–prospects: 1–Goshadash, 2–Apsheron Bank, 3–Agburun Deniz, 4–Gilavar, 5–East Gilavar, 6–Danulduzu, 7–Ashrafi, 8–Agburun, 9–Mardakyan Deniz, 10–Darvin Bank, 11–Pirallaghi Adasi (Northern Fold), 12–Pirallaghi Adasi (Southern Fold), 13–Gyurgyan Deniz, 14–Dzhanub, 15–Khali, 16–Chalov Adasi, 17–Azi Aslanov, 18–Palchygh Pilpilasi, 19–Neft Dashlary, 20–Gyuneshli, 21–Chyragh, 22–Ushakov, 23–Azeri, 24–Kyapaz, 25–Shakh Deniz, 26–Gum Deniz, 27–Bakhar, 28–Livanov-West, 29–Livanov-Center, 30 Livanov-East, 31– Barinov, 32–Gubkin (Western, Central, Eastern), 33–Zhdanov (Western, Eastern, Pre-Cheleken Dome), 34–LAM, 35–Cheleken.

material, whereas the shale contains sand and silt components. Stratigraphic section and typical well logs (resistivity and SP) of the Productive Series, Azerbaijan, are given in Figure 1.6 and 1.7. The thickness of the Productive Series increases in the direction of the central part of the South Caspian Basin from 1500 m within the Absheron Peninsula to 3150 m within the Absheron Archipelago, to 4150 m within the South Absheron Offshore Zone, and to 4400 m within the Baku Archipelago.

The Productive Series includes the following formations:

1. Upper division
   a. Surakhany
   b. Sabunchi
   c. Balakhany
   d. Fasila

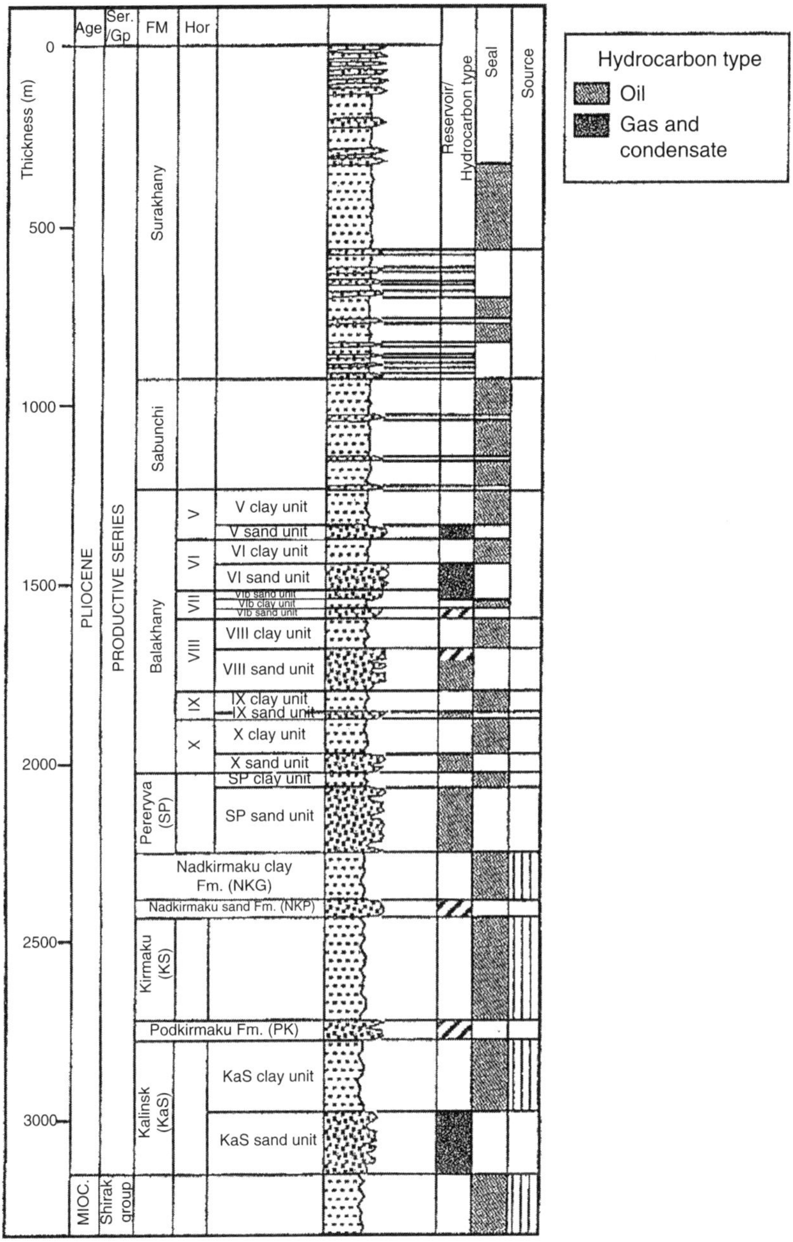

Figure 1.6 Productive series stratigraphic/lithologic column.

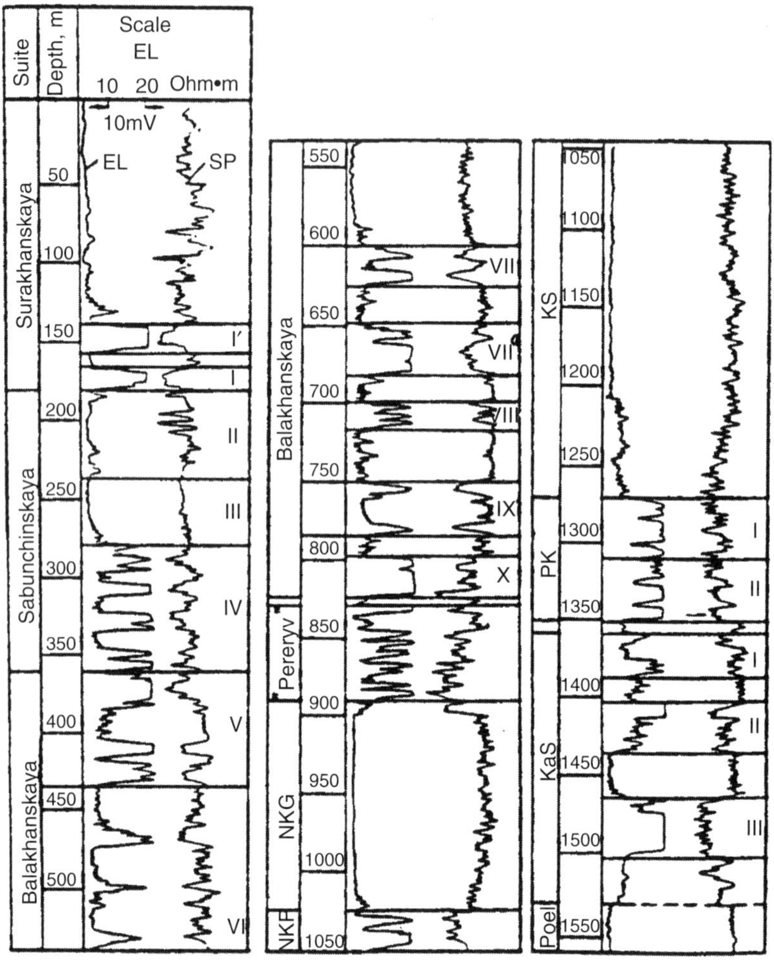

Figure 1.7 Typical resistivity logs of the productive series.

2. Lower division
    a. Nadkirmaku Shale
    b. Nadkirmaku Sand
    c. Kirmaku
    d. Podkirmaku
    e. Kala

The terrigenous (siliciclastic) rocks of Productive Series have gray color in the lower division, whereas above they are grayish brown. The rocks within the Absheron Peninsula are composed of quartz and feldspar, whereas within the Absheron and Baku archipelagoes they become polimictic-arkose, arkosic-graywacke and graywacke. The cement is usually composed of clay and calcite with a significant predominance of clay. Sorting of the siliciclastics improves noticeably upward in each sedimentary sequence.

1.1.3 Depositional Environment of Productive Series, Azerbaijan

Core data, paleontological and log analyses suggest that the Productive Series sediments were deposited in a relatively shallow-water, fluvial-deltaic environment as is evident from the paleogeographic and subsidence curves (Figure 1.8). The large volume of clastic rocks, forming the Productive Series, indicates the proximity of sediment sources. The Russian Platform, Kilyazi-Krasnovodsk anticlinal trend, islands existing north of the Absheron Peninsula and Absheron Archipelago, and the southeastern slope of the Greater Caucasus, served as primary sources for clastic material (Absheron type lithofacies) for Absheron Peninsula and the adjacent Caspian Sea offshore. Weathering of older Mesozoic-Paleogene volcanic and

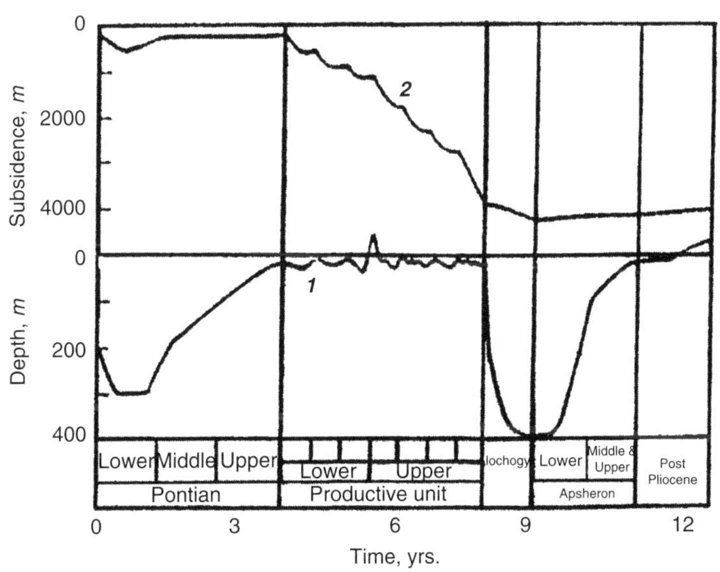

Figure 1.8 Pliocene paleogeographic (paleodepth) curve (1) and subsidence curve (2) for the Absheron Peninsula and adjacent offshore areas. (Modified after Buryakovsky *et al.*, 1982.)

sedimentary rocks from the Greater and Lesser Caucasus and Talysh Mountains, served as primary sources (Gobustan type of lithofacies) of sediments for the Lower Kura Region and the Baku Archipelago. The clastics were transported and deposited by Paleo-Volga, Paleo-Ural, Paleo-Kura and other paleo-rivers.

The major distribution pattern for reservoir rocks in the Absheron Oil and Gas Region as a whole, and within individual areas in particular, is a systematic change in mineral composition and decrease in grain size with increasing distance from the provenance (Buryakovsky, 1970, 1974c). With increasing distance to the south and southeast from the paleo-shoreline of the North Caspian Sea, depth to the productive reservoirs increases, sand content decreases, and shale and silt contents increase. More drastic changes occur in the transitional zone from the Absheron Peninsula, Absheron Archipelago and South Absheron Offshore Zone to the northern Baku Archipelago, where Absheron-type lithofacies, although preserving their main features, include more Gobustan-type lithofacies. The main changes are the following:

a. Shale content increases from 15% to 40%.
b. Sand content decreases from 40% to 15%.
c. Silt content changes in the range of 40% to 62%.
d. Grain size decreases from 0.08 to 0.02 mm.
e. Sorting is practically constant.

The Productive Series is divided into seven sedimentary sequences according to the transgressive/regressive cycles during development of the sedimentary basin. Upward through the section, they are:

1. Kala Formation (KaS)
2. Podkirmaku and Kirmaku formations (PK + KS)
3. Nadkirmaku Sand and Nadkirmaku Shale formations (NKP + NKG)
4. Fasila Formation plus X and IX Balakhany units
5. VIII, VII and VI Balakhany units
6. V Balakhany unit and IV, III and II Sabunchi units
7. I, I', A, B, C, D Surakhany units.

Each sequence displays fining upward, from coarse-grained sands at the base to the finer sands, siltstones and shales at the top. Furthermore, in each sequence, the shale content increases and the sand content decreases up the stratigraphic column. For instance, within the fifth sequence at the Bakhar Field, shale content increases upward from 17.8% in Unit VIII to 29.9% in Unit VI; silt content changes, respectively, from 49.2% to 69.1%, and sand content decreases from 33.0 to 2.0%. This depositional pattern is dependent on the tectonic regime and depositional environment of the South Caspian Basin.

Shallow-marine fossils, fresh-water ostracods, and glauconite in the core samples indicate a mingling of marine and continental environments, especially at the base of each transgressive/regressive cycle. Individual layers in the suites of the Productive Series have been identified as stream-mouth bar deposits, distributary channel-fill sands, point-bar sands, crevasse sands, or transgressive-sheet deposits. Stream-mouth bar and point-bar deposits often occur as a deltaic couplet with point-bar sands of the delta plain prograding across underlying stream-mouth bars of the delta front. These delta-plain deposits either cut into or are slightly separated from the underlying delta-front deposits. Such a deltaic couplet is often found throughout the Absheron Peninsula and Absheron Archipelago at the base of Fasila Suite (the first break in deposition). More distinct rocks, however, characterize the upper intervals of each transgressive/regressive cycle. This portion of upper parts of transgressive/regressive cycles appears to indicate the migration of delta or distributary-channel system, such that the delta began to build elsewhere. Many of these rocks appear to be crevasse sands formed as the distributary reached the flood stage and broke through a levee into adjacent interdistributary bay areas.

Due to increase in shale content upward for each cycle throughout the stratigraphic section, the reservoir thickness diminishes to the upper part of each cycle and clearly affects log responses (for example, average resistivity decreases toward the top of each cycle). Principles of cyclic sedimentation were applied for subdividing the sedimentary section and for selecting intervals for reserve estimation.

To analyze the sequences of the Productive Series sedimentary section, the authors used a special parameter, which demonstrates relative sand content within an individual transgressive/regressive cycle. The individual cycle consists of two layers, i.e., sand/silt (reservoir rock) and shale (non-reservoir rock). Ratios of sand-silt to shale layers within each individual transgressive/regressive cycle allow plotting the curve of sand content variation in the entire sedimentary sequence. When the individual cycles are combined to constitute a sequence of higher order, there is an overall decrease in the sand, and the shale content increases toward the sequence top. On this basis, the authors have defined the following levels of cyclic sequences:

1. Unit (with two layers).
2. Pack (with 4 to 6 layers).
3. Group (with 8 to 12 layers).
4. Formation (with 12 to 30 layers or more).

Individual layers can be defined as the group of layers where one observes a large increase in the shale content toward the sequence top. This provides a systematic correlation within the area and indicates the oil and gas contents. The formation level applies to thick sequences (suites), which have been identified at the Absheron Peninsula by a number of scientists on the basis of grain-size distribution and mineral composition of sedimentary rocks (Potapov, 1954, 1964). This procedure is used for well log stratigraphic correlations.

1.2 Reservoir Lithologies

The most common reservoir rocks are sands, sandstones, and carbonates including limestone and dolomite (Pustovalov, 1940; Pettijohn, 1957). Sometimes the weathered and fractured igneous and volcanic rocks may serve as the oil and gas reservoirs. To be commercially productive, the reservoir rocks must have sufficient thickness, areal extent, and pore space to contain an appreciable volume of hydrocarbons, and must yield the contained fluids at a satisfactory rate when the reservoir is penetrated by a well.

1.2.1 Clastic Rocks

Clastic rocks (mainly siliciclastics) are the good reservoir rocks, which are made-up from granular rocks, such as sands, sandstones, siltstones and sand-silt varieties. The key characteristics of clastic rocks are: (1) grain-size distribution, (2) grain sorting and rounding, (3) cement type and distribution, (4) structure and texture of a rock, (5) geometry of pore space and grain packing system, and (6) porosity and permeability.

1.2.1.1 Grain-size Characteristics of Clastic Rocks

The granular rocks are characterized by the *grain* or *particle size*, which ranges from colloidal particles up to pebbles and boulders.

Other characteristics of grains are their *sorting* and *roundness*. Poorly sorted sediments are composed of many different sizes and/or densities of grains mixed together. Well-sorted sediments, however, are composed of grains that are of similar size and/or density. Well-sorted sediments are usually composed of well-rounded grains, because the grains have been abraded and rounded during transportation. Conversely, poorly sorted sediments are usually angular, because of the lack of abrasion during transportation. The sharpness of corners on grains of sediment, viewed in profile (side view), is a measure of roundness. The well-rounded, subrounded, subangular, and angular grains are distinguished.

To understand grain-size distribution, as well as sorting and roundness of grains in a given rock, a grain-size analysis is applied along with the following procedures.

1. Direct measurement and observation of individual fragments of pebbles, cobbles, and bounders.
2. Sieving to separate pebbles, sand, and coarse silt.
3. Settling velocity for measuring the size of silt and clay particles.
4. Microscopic observation of sand, silt, and clay particles.
5. Scanning electron microscopy for studying of very small sedimentary features.

The smaller particles are defined by their volumetric diameters, i.e., the diameter of a sphere with the sane volume as the particle. The statistical proportions or distribution of particles of defined size fraction of sediment or rock is determined from the particle-size analysis.

There are numerous grain-size classifications. Examples include: Udden grade scale, Wentworth grade scale, Atterberg grade scale, Tyler standard grade scale, and Alling grade scale. Each of these scales is logarithmic. In American practices, the grain size is measured by the logarithmic grade scale devised by Udden in 1898 (Udden, 1914). This scale uses 1 mm as the reference point and progresses by the fixed ratio of ½ in the direction of decreasing size and of 2 in the direction of increasing size, such as 0.25, 0.5, 1, 2, 4. The extended version of the Udden grade scale was proposed by Wentworth (1922, 1924), who modified the size limits for the common grade terms, but retained the geometric interval or constant ratio of ½. The scale ranges from clay particles (diameter less than 1/256 mm) to boulders (diameter greater than 256 mm). The Wentworth grade scale was modified by Krumbein in 1934, who proposed a logarithmic transformation of the scale, in which the negative logarithm to the base 2 of the particle diameter (in millimeters) is substituted for the diameter value, i.e., $\varphi = -\log_2 d$, where φ is the phi value and d is the particle diameter. The integers for the class limit range from –5 for 32 mm to +10 for 1/1024 mm (refer to Table 1.3). The classification of the grain size of sedimentary rocks is shown in Table 1.4 and Table 1.5.

Table 1.3 Pore/grain size diameter conversion table (millimeter to phi units).

mm	ϕ	mm	ϕ
4.00	–2.00	0.044	4.50
3.36	–1.75	0.037	4.75
2.83	–1.50	0.031	5.00
2.38	–1.25	0.026	5.25
2.00	–1.00	0.022	5.50
1.68	–0.75	0.019	5.75
1.41	–0.50	0.016	6.00
1.19	–0.25	0.013	6.25
		0.011	6.50
1.00	0.00	0.0093	6.75
		0.0078	7.00
0.841	0.25	0.0066	7.25
0.707	0.50	0.0055	7.50
0.500	1.00	0.0047	7.75
0.420	1.25	0.0039	8.00
0.351	1.50	0.0033	8.25
0.297	1.75	0.0028	8.50
0.250	2.00	0.0023	8.75
0.210	2.25	0.0020	9.00
0.177	2.50	0.0016	9.25
0.149	2.75	0.0014	9.50
0.125	3.00	0.0012	9.75
0.105	3.25	0.00098	10.00
0.088	3.50	0.0005	11.00
0.074	3.75	0.0003	12.00
0.062	4.00		
0.052	4.25		

Table 1.4 Classification of sedimentary rocks (siliciclastics and carbonates) based on grain size.

<table>
<tr><th rowspan="2">Size, mm</th><th rowspan="2" colspan="2">Siliciclastics</th><th colspan="4">Carbonates</th><th rowspan="2">Size, mm</th></tr>
<tr><th colspan="2">Grains</th><th colspan="2">Crystals</th></tr>
<tr><td>256.0</td><td>Boulders</td><td rowspan="4">Conglomerate</td><td rowspan="4">Calcirudite*</td><td rowspan="10">Grain-supported</td><td rowspan="3">Coarsely megacrystalline</td><td rowspan="4">Megacrystalline</td><td>256.0</td></tr>
<tr><td>64.0</td><td>Cobbles</td><td>64.0</td></tr>
<tr><td>4.0</td><td>Pebbles</td><td>4.0</td></tr>
<tr><td>2.0</td><td>Granule</td><td>Finely megacrystalline</td><td>2.0</td></tr>
<tr><td>1.0</td><td>Very coarse sand</td><td rowspan="5">Sand</td><td rowspan="4">Calcarenite*</td><td>Very coarsely crystalline</td><td rowspan="5">Crystaline</td><td>1.0</td></tr>
<tr><td>0.50</td><td>Coarse sand</td><td>Coarsely crystalline</td><td>0.50</td></tr>
<tr><td>0.25</td><td>Medium sand</td><td>Medium crystalline</td><td>0.25</td></tr>
<tr><td>0.125</td><td>Fine sand</td><td>Finely crystalline</td><td>0.125</td></tr>
<tr><td>0.062</td><td>Very fine sand</td><td rowspan="2">Calcarenite* (very fine grained)</td><td>Very finely crystalline</td><td>0.062</td></tr>
<tr><td>0.031</td><td>Coarse silt</td><td rowspan="4">Silt</td><td>Coarsely microcrystalline</td><td rowspan="5">Crypto-microcrystalline</td><td>0.031</td></tr>
<tr><td>0.016</td><td>Medium silt</td><td rowspan="4">Calcilutite*</td><td rowspan="4">Mud-supported</td><td>Medium microcrystalline</td><td>0.016</td></tr>
<tr><td>0.008</td><td>Fine silt</td><td>Finely microcrystalline</td><td>0.008</td></tr>
<tr><td>0.004</td><td>Very fine silt</td><td>Very finely microcrystalline</td><td>0.004</td></tr>
<tr><td></td><td>Clay</td><td>Clay</td><td>Cryptocrystalline</td><td></td></tr>
</table>

In Russia, Pustovalov proposed the simplest and most convenient decimal classification of sedimentary rocks in 1940 (Table 1.6).

The main parameters, which characterize the pore space of clastic rocks, are the grain-size distribution coefficients. These coefficients are widely used for developing depositional models and clastic rock classifications. These coefficients are determined

Table 1.5 Particle size scale. (After Chilingar *et al.*, 1967, p. 102.)

Size, mm	Reference										Size, mm
	Krynine (1948)	**Howell (1922)**	**DeFord (1946)**		**Pettijohn (1949)**	**Mollazal (1961)**		**Folk (1959, 1962)**	**The present authors**		
4.0	Coarsely crystalline (grained)	Macro-crystalline	Mega-grained	Phaneric	Calcirudite	Macro-crystalline	Coarse Medium Fine	Extremely coarsely crystalline	Macrocystalline (macrograined)	Very coarse	4.0
2.0	Medium crystalline (grained)				Calcarenite	Meso-crystalline	Coarse	Very coarsely crystalline		Coarse	2.0
1.0	Finely crystalline (grained)	Meso-crystalline	Meso-grained					Coarsely crystalline	Mesocrystalline (mesograined)	Coarse	1.0
0.5							Medium			Medium	0.5
0.25								Medium crystalline		Fine	0.25
0.1							Fine			Very fine	0.1
0.062					Calcilutite	Finely-crystalline		Finely crystalline			0.062
0.05									Finely crystalline (finely grained)		0.05
0.025											0.025
0.016								Very finely crystalline			0.016
0.01		Crypto-crystalline							Aphanic	Micro-crystalline	0.01
0.004			Micro-grained	Aphanic		Aphanic	Micro-crystalline	Aphano-crystalline			0.004
0.0025											0.0025
0.001			Crypto-grained				Crypto-crystalline			Crypto-crystalline	0.001

Table 1.6 Grain size classification of sedimentary rocks.

Rock Structure (size range)	Grain/Particle-Size Classification	Grain/Particle Size, mm
Psammitic (1.0–0.1 mm)	Coarse sand	1.0–0.5
	Medium sand	0.5–0.25
	Fine sand	0.25–0.1
Aleuritic (0.1–0.01 mm)	Coarse silt	0.1–0.05
	Medium silt	0.05–0.025
	Fine silt	0.025–0.01
Pelitic (0.01–0.0001 mm)	Coarse clay	0.01–0.001
	Fine clay	0.001–0.0001

from the cumulative frequency distribution or cumulative distribution function and include upper, middle (median), and lower quartiles of the grain-size distribution. Upper, middle (median) and lower quartiles are the arguments of the cumulative grain-size distribution function corresponding to a probability of 0.75, 0.50, and 0.25, respectively, and are designated as Q_{75}, Q_{50} and Q_{25} (Table 1.7).

The common grain-size distribution is the lognormal one, logarithm of which follows a normal or Gaussian distribution. The plot of lognormal distribution is a continuous, infinite, bell-shaped curve that is symmetrical about its geometric mean or median of a grain-size distribution. To obtain the models of the grain-size probability density curves and cumulative distribution curves, the writers used all possible combinations of three main grain-size fractions consisting of sand (1.0–0.1 mm), silt (0.1–0.01 mm), and clay (0.01–0.001 mm) particles (Figure 1.9). These models allow one to calculate the three above-mentioned quartiles or grain-size distribution coefficients.

Another two grain-size distribution coefficients (Trask, 1942) include the grain sorting index and the asymmetry of grain-size probability distribution.

Sorting index, S_0, is the probable deviation from the median grain size calculated from the cumulative grain-size distribution function:

Table 1.7 Summary of grain-pore-size and pore-throat-size measures used in conjunction with graphic analysis. (After Chilingar *et al.*, 1972, p. 353.)

Measure of central tendency:
1. *Median* (D_{50}) is the diameter which is larger than half of the pores in the distribution and smaller than the half (i.e., the middlemost member of the distribution). It reflects the overall pre size as influenced by the chemical or physical origin of the rock and any subsequent alteration. It may be a very misleading value, however.
2. *Mean* (D_M) is the measure of the overall average pore size: $D_M = (D_5 + D_{15} + D_{25} \ldots D_{85} + D_{95})/10$ or $D_M = (D_{16} + D_{50} + D_{84})/3$
3. *Mode* (D_m) is the most frequently occurring pore diameter (peak of frequency curve). If two dominant pore sizes are present which could result when there is a mixture of two or more different porosity types (vugy, oolitic, fracture, intergranular, etc.,), then the frequency curve is bimodal.
Measure of dispersion:
Pore sorting (S_p) is a standard deviation measure of the pore sizes in a sample (Folk and Ward, 1957) $S_p = (D_{84} - D_{16})/4 + (D_{95} - D_5)/6.6$
Measure of asymmetry:
Skewness (Sk_p) measures the non-normality of a pore-size distribution: $Sk_p = (D_{84} + D_{16} - 2D_{50})/2\,(D_{84} - D_{16}) + (D_{95} + D_5 - 2D_{50})/2(D_{95} - D_5)$ A symmetrical curve has a Sk_p value of 0; limits in which Sk_p varies are as follows: $-1 \leq Sk_p \leq 1$. Positive values indicate that the curve has a tail in the small pores. Negative values indicate that the curve is skewed toward the larger pores.
Measure of peakedness:
Kurtosis (K_p) is a measure of the degree of peakedness, that is, the ratio between the spread of the pore diameters in the tails and the spread of the pore diameters in the central portion of the distribution: $K_p = (D_{95} - D_5)/2.44(D_{75} - D_{25})$ Normal curves have a K_p of 1, whereas platykurtic (bimodal) distributions may have a K_p value as low as 0.6. A curve represented by a high narrow peak (very leptokurtic) may have K_p values ranging from 1.5 to 3. D_n is the pore diameter in phi units as the nth percentile. The methodology of pore- and pore-throat-size analysis (see Folk and Ward, 1957 and McCammon, 1962)

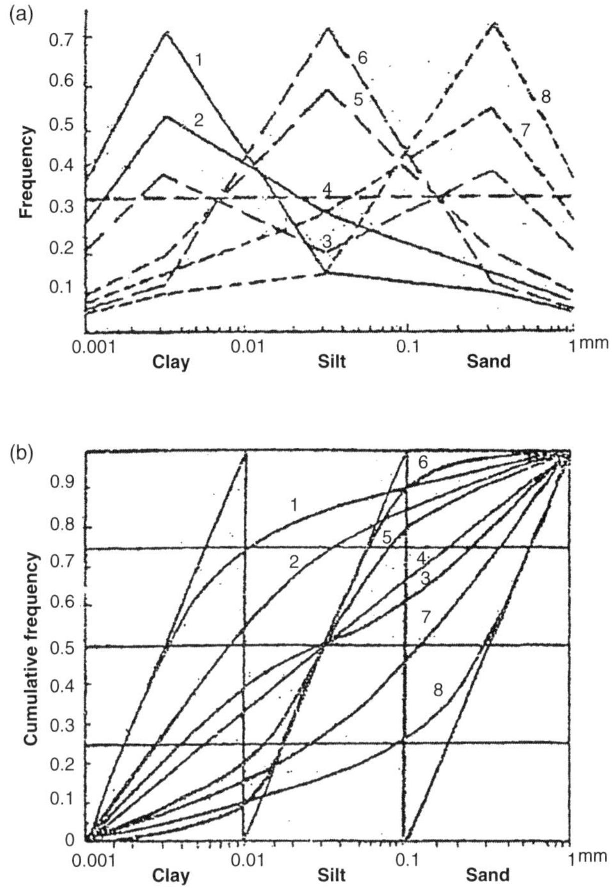

Figure 1.9 Grain-size distribution models: (a) frequency distribution, (b) cumulative frequency distribution. (Modified after Buryakovsky, 1985.). Lithology: 1–clay, 2–loam, 3–sandy-clayey poorly sorted rock, 4–unsorted rock, 5–silt, 6–sandy-clayey silt, 7–loamy sand, and 8–sand.

$$\lg S_0 = \frac{\lg Q_{75} - \lg Q_{25}}{2} \tag{1.1}$$

or

$$S_0 = \sqrt{\frac{Q_{75}}{Q_{25}}} \tag{1.2}$$

A sediment-sorting scale (Khanin, 1966) includes four classes: well-sorted sediments, S_0 = 1–1.78; medium-sorted sediments, S_0 = 1.78–2.32; poorly sorted sediments, S_0 = 2.32–2.86; and unsorted sediments, S_0 >2.86.

Asymmetry index, S_a, is calculated from the following formulae:

$$\lg S_a = \lg Q_{75} - 2\lg Q_{50} + \lg Q_{25} \tag{1.3}$$

or

$$S_a = \frac{Q_{75}Q_{25}}{Q_{50}^2} \tag{1.4}$$

When $S_a > 1$, the fine-grained material prevails, whereas when $S_a < 1$, the coarse-grained material is prevalent.

1.2.2 Pore Throat Distribution in Carbonate Rocks

Pore systems in carbonate rocks usually contain both pores and inter-connections between these pores (pore throats). Isolated vugs, however, are common in some carbonates. The pore system may have two extremes: (1) the size of the pores approaches that of the interconnecting pore throats, and (2) the size difference between the two is very large. In the mercury injection test, the size distribution between pores and pore throats is an artificial one. Mercury injection pressure is indicative of the pore throat sizes. It is usually assumed, therefore, that the pore throat sizes control the injection.

Aschenbrenner and Achauer (1960) found that both pore and pore throat sizes were essentially log-normally distributed (straight-line relationship between logarithm of pore size or pore throat size and cumulative percentage on probability paper) in Paleozoic carbonates of the Williston Basin and in the Rocky Mountains. Inasmuch as most pore- and pore-throat size distributions tend to be log-normal, this provides a method for estimating their size

Based on extensive data, Figure 1.10a illustrates a simple frequency curve showing a normal distribution with approximately 68% of the pore throat diameters occurring between D_{16} and D_{84} [one standard deviation (1.3ϕ) on either side of the mean (6.1ϕ); $\sigma_\phi = (D_{84} - D_{16})/2 = 1.3$]. Figure 1.10b illustrates the data from Figure 1.10a as a cumulative frequency curve, whereas in Figure 1.10c the cumulative frequency of pore throat sizes is plotted on Cartesian probability paper. The probability scale is designed in such a manner that a symmetrical cumulative pore-size frequency curve plots as a straight line on the graph indicating a unimodal relationship.

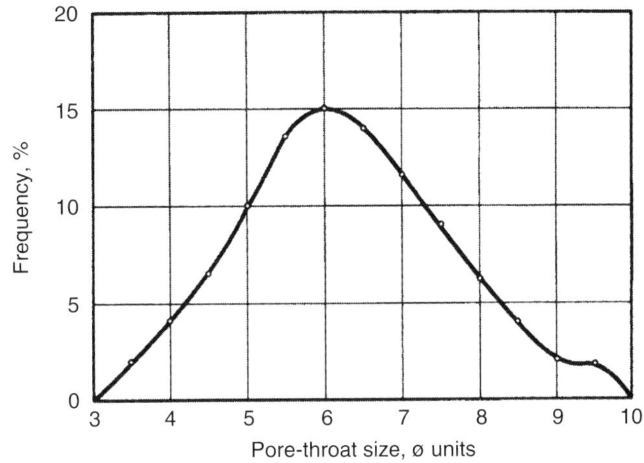

Figure 1.10a Frequency distribution of pore throat sizes in a carbonate rock.

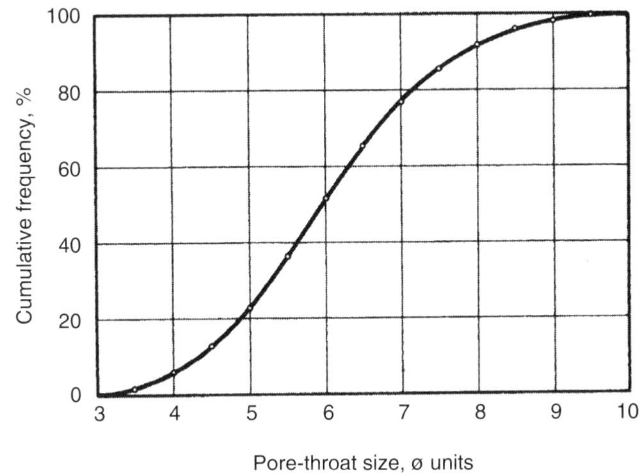

Figure 1.10b Cumulative frequency distribution of pore throat sizes (data from Figure 10a).

1.2.2.1 Cementation of Clastic Rocks

Another important characteristic of clastic rocks is their cementation while sediments become lithified or consolidated into hard, compact rocks through the deposition or precipitation of minerals in the spaces among the individual grains of the sediment. Cementation may occur simultaneously with sedimentation, or the cement may

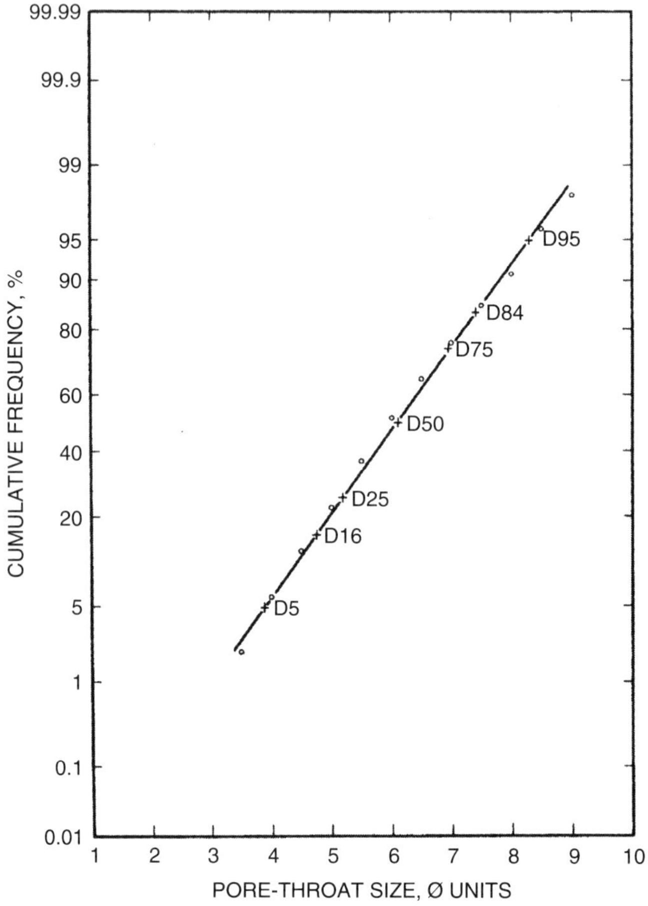

Figure 1.10c Cumulative lognormal curve on Cartesian probability paper, showing positions of various percentile ordinates (data from Figure 10a).

be introduced at a later time. The most common types of cement are silica (quartz, opal, and chalcedony), and carbonates (calcite, dolomite, and siderite). Other cements include barite, gypsum, anhydrite, and pyrite. Clay minerals constitute argillaceous cement. There are several types of cementation depending on the type of sediment supporting material: (1) mud matrix within the grain-supported sediment, including (*a*) film-like, (*b*) meniscus-like, and (*c*) pore-filling cementation; and (2) grains in a mud-supported sediment (basal cementation). Table 1.8 shows a simplified classification of clastic rock cementation.

Table 1.8 Classification of cements of clastic rocks.

Type of Cement	Content of Cement, %
Film-type	3–10
Meniscus-type	5–10
Partly pore-filling	10–20
Pore-filling	20–30
Abundant pore-filling	30–40
Basal	>40

Texture and structure of clastic rocks are the major morphologic features of a rock. Both terms are used to describe the physical appearance or geometric aspects of a rock. The term *texture* is generally used for the smaller features or particles composing a rock, whereas *structure* is used for those features that indicate the way the rock is organized or made up of its components.

Grain shape and roundness (Figures 1.11 and 1.12), grain size and sorting, grain orientation and packing, and chemical composition determine the texture of sedimentary rocks. A specific combination of these variables may reveal information about diagenetic and catagenetic processes and mechanisms acting during transportation, deposition, and compaction and deformation of sediments

Roundness is a measure of the sharpness of the particle edges, regardless of shape. One accepted method for determining roundness is to view the particles two-dimensionally, and determine the ratio of the average radius of curvature of the particle's corners to the radius of the largest circle that can be inscribed in that particle. The general method for estimating roundness is microscopic measurement of a number of grains and visual comparison to a standard chart (Figure 1.11) such as those introduced by Griffith (1967).

The degree of roundness commonly varies with size. Larger-diameter sand or gravel particles are usually more rounded than the smaller ones. Maturity and degree of weathering affect this relationship. Freshly broken fragments, which tend to be angular near the source, assume a greater degree of roundness as a result of weathering and abrasion during transportation.

Sphericity is some times confused with roundness. Although they are related in a certain degree, roundness is primarily a measurement of the angularity of a particle's corners; whereas sphericity is

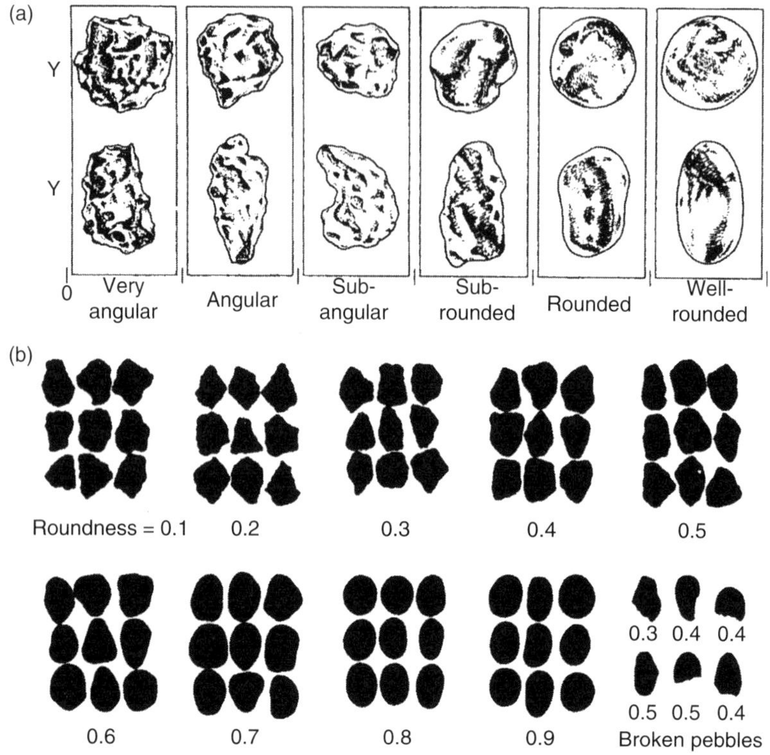

Figure 1.11 (a) *Roundness* images and classes. Columns show grains of similar roundness but different sphericity. (After Powers, 1953, modified by Pettijohn *et al.*, 1972, in Chilingarian and Wolf, 1975, p. 15, Figure 1.11.). (b) Images for estimating visual roundness. (After Krumbein, 1941, in: Chilingarian and Wolf, 1975, p.15, Figure 1.12.)

a measure of the degree the shape of the particle approaches that of a sphere. Images for estimating visual sphericity visually are given in Figure 1.12. True sphericity as defined by Wadell (1934) as the surface area of a sphere of the same volume as the particle divided by the actual surface area of the solid.

A capsule shaped object could have a roundness factor of unity, whereas if its surface area were compared to that of a sphere of the same volume, using Wadell's definition, sphericity, the ratio would be far less than unity. A more practical formula for sphericity, also introduced by Wadell (1934), is to divide the nominal diameter of the particle (the diameter of a sphere of the same volume as the particle) by the diameter by the circumscribing sphere. Krumbein

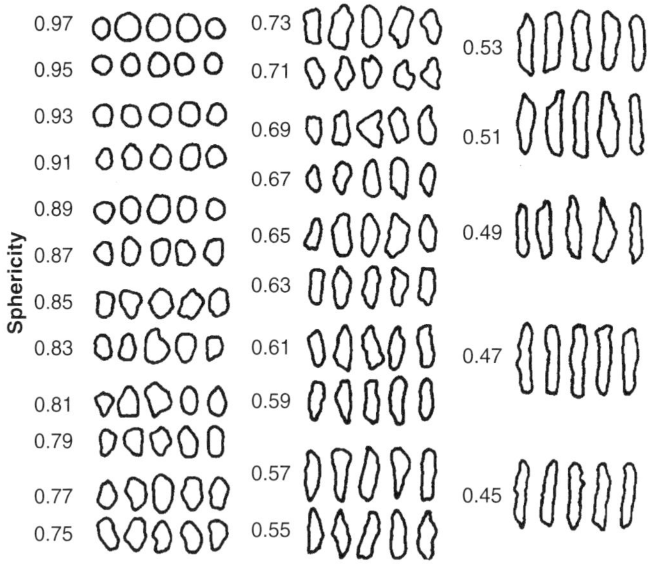

Figure 1.12 Images for estimating visual sphericity. (After Rittenhouse, 1943, in: Chilingarian and Wolf, 1975, p. 16, Figure 1.13.)

(1941) introduced a definition of sphericity based on volumes: he defined sphericity as the cube root of the volume of the particle divided by the volume of the circumscribing sphere. Factors that control the shape and roundness of particles include:

a. the original shape of the fragment,
b. durability of the fragment (hard mineral grains such as quartz and zircon are rounded less during transport than soft grains of feldspar and pyroxenes),
c. structure of the fragment (cleavage or bedding),
d. nature of the geologic agent (wind is more effective in rounding grains than water),
e. nature of the action to which the fragment is subjected and rigor of the action, and
f. residence time and distance imposed by the action.

The methods of determining parameters of rock texture and structure are the following:

1. Rock outcrop observation.
2. Core sample study.

3. Study of thin-sections under the optical microscope.
4. Study of core chips under the scanning electron microscope.
5. X-ray diffraction analysis.

A variety of structures exist in sedimentary rocks. Some of these structures formed at the time of sediment transportation and deposition. These structures are referred to as *primary sedimentary structures*. The most obvious of these is *stratification* (layering of sediments). Most layers of sediments (*strata*) accumulate in nearly horizontal sheets. Strata less than 1 cm in thickness are called *laminations*; whereas strata 1 cm or more in thickness are called *beds*. Surfaces between strata are called *bedding planes*, which represent surfaces of exposure that existed between sedimentary depositional events. Some stratification is inclined, and is referred to as *cross-stratification*.

Individual strata may also be *graded*. Normally, graded beds are sorted (becoming finer upward), a feature caused when (1) sediment-laden currents suddenly slow down as they enter a standing body of water, (2) current flow terminates, or (3) a depth of depositional basin gradually increases. In these cases, each stratum is internally graded from coarse sediments on bottom to fine sediments on top.

Many sedimentary rocks contain structures that formed after deposition. For example, desiccation cracks often form while wet deposits of mud shrink on drying. Such structures are referred to as *secondary sedimentary structures*.

Geometry of grain packing is a quantitative and qualitative presentation of the grain-packing system. Geometry of grain-packing system is very complex and depends on the specific features of grain packing and cementing. The most important geometrical parameters are the *proximity of grains*, *density of grains*, and *density of cement* of the system (Winsauer and Gaither, 1953; Kahnn, 1956).

The proximity of grains P_p is determined from the following formula:

$$P_p = q/n \cdot 100\% \tag{1.5}$$

where q is the number of grain contacts crossed by micron-scale ruler and n is the total number of grains crossed by the ruler.

Maximum proximity value may reach 100%, when all the grains are in contact with each other. Minimum (zero) value occurs when no one-grain is in contact with the others.

Generally: $0 \leq q \leq (n-1)$.

The density of grain compaction $P_{d \cdot g}$ is determined from the following formula:

$$P_{d \cdot g} = m \sum_{i=1}^{n} g_i / t \cdot 100\% \tag{1.6}$$

where m is the magnification of microscope; g_i is the number of micron-scale ruler points crossing a single grain; t is the total length of the ruler; and n is the total number of grains at all positions of the ruler.

Maximum value of the density of grains may reach 100% in the case when all ruler crossings are occupied by grains. Practically, this case is impossible for the granular (clastic) rocks.

The density of cementation $P_{d.c}$ is the relative content of cement in the rock. This parameter is calculated from the following formula:

$$P_{d \cdot c} = \left[100 - \sum_{i=1}^{k} c_i / k \right] (\%) \tag{1.7}$$

where c_i is the number of micron-scale ruler points covering i-th site of cement and k is the number of observed micron-scale ruler positions.

Except for the above-mentioned parameters, the other important parameters are (Chernikov and Kurenkov, 1977):

1. Ratio of packing proximity to the density of grains: $P_{c \cdot g} = P_p / P_{d \cdot g}$ *(relative compaction of grains)*.
2. Ratio of packing proximity to the density of cement: $P_{c \cdot c} = P_p / P_{d \cdot c}$ *(relative compaction of cement)*.

These parameters account for both mutual relations between individual grains and grain proportions in the rock.

Figure 1.13 illustrates one of the positions of the micron-scale ruler crossing the thin-section points on a grain, crossed by micron-scale ruler l – number of points on the cement matrix, crossed by micron-scale ruler area. Grains are counted from one side of the micron-scale ruler.

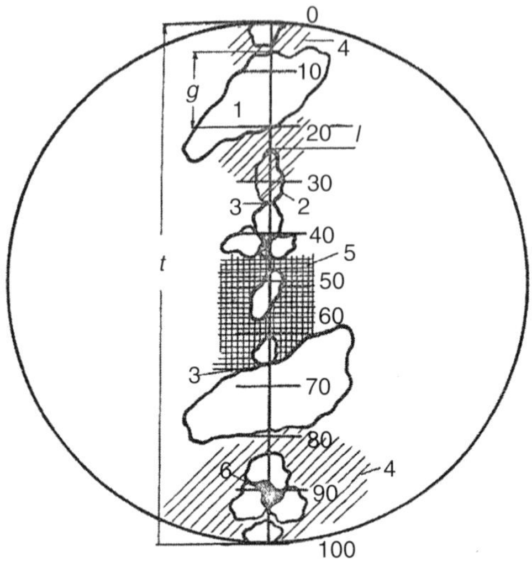

Figure 1.13 Example of microscopic image of a rock thin-section. Microscopic features: 1–quartz grain, 2–feldspar grain, 3–contact between grains, 4–argillaceous cement, 5–calcareous cement, and 6–pore space.

The grain-packing system is characterized by the following types of intragrain contacts (Taylor, 1950):

1. Point or tangent contacts – adjacent grains touch one another in a single point.
2. Linear contacts – grains connect one another through an interface.
3. Convex-concave contacts – grains are connected along the relatively smooth curve.
4. Suture contacts – grains are connected along the irregular spike-like curve.

Two other important pore-space geometrical parameters are *tortuosity* and *clearance* between the grains or in the rock matrix.

The tortuosity τ is the ratio of the effective length L_e (path of fluid flow) to the overall direction of flow, L (length of the rock sample):

$$\tau = L_e / L \tag{1.8}$$

The clearance φ is the ratio of area A_1 of openings between grains or matrix components (as are seen on the thin-section visible under the microscope) to the total area A_2 of the thin-section:

$$\varphi = A_1/A_2 \tag{1.9}$$

Sometimes, clearance is referred to as the surface porosity. It is believed that in the absence of isolated pores, not effective for fluid flow, the product of tortuosity by clearance equals to the porosity of granular rock, i.e.,

$$\tau\varphi = \phi \tag{1.10}$$

In the presence of isolated pores, this product should be less than porosity and may be somewhat similar to the effective porosity.

1.2.2.2 Porosity and Permeability of Clastic Rocks

The porosity and permeability of the reservoir rocks are the most fundamental physical properties with respect to storage and transmission of fluids. Porosity of clastic rocks is controlled primarily by grain sorting (i.e., by the extent of mixing of grains of various sizes), cementation, and by the way the grains are packed together.

Porosity is at a maximum when grains are spherical and all of one size. However, porosity becomes progressively lower as the grains become more angular and pack together more closely. Artificially mixed clean sand has measured porosity of about 43% for extremely well-sorted sands, almost irrespective of grain size, decreasing to about 25% for very poorly-sorted medium-to-coarse sands; whereas the very fine-grained sands have over 30% porosity. The total porosity and bulk density of some sediments and sedimentary rocks are presented in Table 1.9.

There are four principal definitions of porosity:

a. Absolute (total) porosity, ϕ_a, ratio of the pore (void volume), V_p, to the bulk volume of sample, V_b.
b. "Effective" porosity, ϕ_{eff}, ratio of the interconnected pore volume, $V_{inter.p}$, to the bulk volume of sample, V_b. The writers prefer to call it open porosity.
c. Void ratio, e, ratio of the pore (void volume), V_p, to the grain/solids volume, V_{gr}.

Table 1.9 Total porosity and bulk density of some sediments and sedimentary rocks.

Lithology	**Total Porosity, %**	**Bulk Density, g/cm^3**
	Primary sediments	
Silty mud	60–70	–
Sandy mud	30–70	1.27–1.94
Lime mud	65–87	–
Clay	10–75	1.20–3.18
Clayey silt	50–60	0.80–1.80
Diatomite	30–80	0.40–1.57
Loess	30–60	1.14–1.93
	Sedimentary rocks	
Sand	4–40	1.3–2.3
Sandstone	1–30	1.3–3.6
Siltstone	1–40	1.5–3.2
Shale	1–35	1.3–3.2
Sandy shale	1–25	1.8–2.9
Claystone	1–25	1.6–3.3
Marl	1–35	2.0–3.1
Chalk	10–50	1.8–2.6
Limestone	0.5–40	1.3–3.5
Dolomite	0.1–40	1.9–3.5
Anhydrite	0.2–15	2.3–3.0
Rock salt	0–5	2.1–2.3

d. Effective porosity, as proposed by the writers, is equal to the "effective" porosity as defined above (*b*) minus the irreducible fluid saturation.

Relation between the void ratio and the absolute porosity is as follows:

$$e = \phi/(1 - \phi) \tag{1.11}$$

The ease with which fluids move through the interconnected pore spaces of a reservoir rock is called permeability. Numerical expressions of permeability are measured in Darcies (D) after Henry d'Arcy, a French engineer, who in 1856 devised a means of measuring the permeability of porous rocks. A rock has a permeability of one Darcy (1 D) when 1 cm^3 of a fluid with a viscosity of 1 cP (centipoise) flows through a 1 cm^2 of cross section of rock in 1 s under a pressure gradient of 1 atm/cm. Because most reservoir rocks have an average permeability considerably less than one Darcy, the usual measurement is in millidarcies (mD), i.e., one thousandth of a Darcy.

The magnitude of permeability depends on wettability, i.e., on whether (1) the fluid does not wet the solid surfaces of the rock and, therefore, occupies the central parts of the pores, or (2) the fluid wets the solid surfaces and thus tends to concentrate next to the rock surfaces and in smaller pores. The nature, distribution, and amount of immobile phase affect the effective permeability. The effective permeability as defined by the writers is the permeability of a core containing an irreducible fluid.

The relative permeability to a fluid is defined as the ratio of effective permeability at a given saturation of that fluid to the absolute permeability at 100% saturation. The terms $k_{ro}(k_o/k)$, $k_{rg}(k_g/k)$, and $k_{rw}(k_w/k)$ denote the relative permeability to oil, to gas, and to water, respectively (k is the absolute permeability, often a single-phase liquid permeability). The relative permeability is expressed in percent or as a fraction.

In waterflooding projects or in natural water-drive pools, the relative permeability to oil and to water is of great importance. Where water and oil flow together, the relative permeability is affected by many factors, which include (1) relative dispersion of one phase in the other, (2) time of contact with pore walls, (3) amount of polar substances in the oil, (4) degree of hardness of water, (5) relative amount of carbonate material in porous medium, and (6) temperature knowledge of the distribution of porosity and permeability is required for the efficient development, management, and prediction of future performance of an oilfield.

1.2.3 Carbonate Rocks

Carbonate rocks represent a complex group, which is difficult to study. The carbonate rocks include limestones composed mostly of calcite ($CaCO_3$) and dolomites, containing both calcium and magnesium [$CaMgCO_3)_2$].

Limestone is composed of more than 50% carbonate minerals; of these, 50% or more consist of calcite and/or aragonite. A small admixture of clay particles or organic matter imparts a gray color to limestones, which may be white, gray, dark gray, yellowish, greenish, or blue in color; some are even black. Dolomites are rocks, which contain more than 50% of the minerals dolomite and calcite (plus aragonite), with dolomite being more dominant. The pure dolomite mineral is composed of 45.7% $MgCO_3$ and 54.3% $CaCO_3$, by weight; or 47.8% CO_2, 21.8% MgO, and 30.4% CaO. Dolomites are quite similar to limestone in appearance and, therefore, it is difficult to distinguish between the two with the naked eye. On the basis of CaO/MgO ratios, Frolova (1959) proposed the classification presented in Table 1.10. The origin, occurrence, classification, and physical and chemical aspects of carbonate rocks are presented in detail by Chilingar *et al.* (1967a,b).

It is very important to evaluate as correctly as possible various properties of carbonate rocks. A good case in point is the Fullerton Clearfork dolomitic limestone reservoir in the Permian Basin. Bulnes and Fitting (1945) reported that 82% of the core samples had permeability of less than 1 mD. The problem of what to use for minimum productive permeability becomes very acute in such instances. According to Bulnes and Fitting, if 1 mD were used as the minimum productive permeability instead of the actual value of 0.1 mD, the resulting estimated ultimate recovery would be 70% in error. The core analysis of some carbonate rocks is complicated by the presence of fractures and solution cavities. In order to analyze such rocks, "whole" or "large" core analysis, whereby the entire core is analyzed instead of small plugs, was developed.

Dunham proposed an excellent classification of limestones in 1962 on the basis of their texture and mud content. Limestones which are composed of particles of less than 2 mm in size and which retained their original, depositional texture can be classified as follows:

1. Lime mudstone with less than 10% grains in a mud-supported sediment.
2. Lime wackestone with more than 10% grains in a mud-supported sediment.
3. Lime packstone with mud matrix within the grain-supported sediment.

Table 1.10 Frolova's classification of dolomite-magnesite-calcite series. (After Frolova, 1956, p. 35.)

Name	Content, %			CaO/MgO Ratio
	Dolomite	Calcite	Magnesite	
Limestone	5–0	95–100	...	>50.1
Slightly dolomitic limestone	25–5	75–95	...	9.1–50.1
Dolomitic limestone	50–25	50–75	...	4.0–9.1
Calcitic dolomite	75–50	25–50	...	2.2–4.0
Slightly calcitic dolomite	95–75	5–25	...	1.5–2.2
Dolomite	100–95	0–5	...	1.4–1.5
Very slightly magnesian dolomite	100–95	...	0–5	1.25–1.4
Slightly magnesian dolomite	95–75	...	5–25	0.80–1.25
Magnesian dolomite	75–50	...	25–50	0.44–0.80
Dolomite magnesite	50–25	...	50–75	0.18–0.44
Slightly dolomitic magnesite	25–5	...	75–95	0.03–0.18
Magnesite	5–0	...	95–100	0.00–0.03

4. Lime grainstone with no mud matrix within the grain-supported sediment.
5. Lime boundstone in which original components are bound together.

The following two types of limestones are distinguished, if they are composed of particles of less than 2 mm in size and in which the depositional texture has been destroyed by recrystallization,

1. Crystalline limestone with a fine texture.
2. Sucrosic ("sugary") limestone with a coarse texture.

Embry and Clovan (1971) expanded Dunham's classification to include limestones containing more than 10% of the clasts larger than 2 mm in size (coarse clasts):

1. Floatstone with coarse clasts in the matrix-supported sediment.
2. Rudstone with coarse clasts in the clast-supported sediment.

Carbonates are quite different from siliciclastic rocks, especially because of their susceptibility to post-depositional changes, particularly dolomitization involving the action of magnesium-bearing water (seawater or percolating meteoric water). Chemical equation explaining the molecular replacement of limestone by dolomite was proposed by Elie de Beaumont in 1836 as follows:

$$2CaCO_3 + MgCl_2 \rightarrow CaMg(CO_3)_2 + CaCl_2$$

Chilingar and Terry (1954) showed that a definite relationship exists between porosity and degree of dolomitization as exhibited in the Asmari Limestone in Iran (Chilingar *et al.*, 1972; Sarkisyan *et al.*, 1973).

T. F. Gaskell of the British Petroleum Co. Ltd. (personal communication, 1963) determined the porosity and density of carbonate reservoir rocks in Southwestern Iran. The average density values for the different oilfields, grouped in ranges of porosity of 0–4.0%, 4.1–8.0%, 8.1–12.0%, and >12.1% are presented in Table 1.11. The mean values were weighted according to the number of observations for each oilfield. A certain amount of the density scatter may be due to impurities in the limestones, variation in the degree of secondary cementation subsequent to dolomitization, and so on. The gradual trend of density from 2.70 g/cm^3 at low porosity to 2.80 g/cm^3 for the high-porosity group indicates that dolomitization gives rise to porosity (Figure 1.14). Inasmuch as at 20°C the density of calcite is 2.71 g/cm^3 and that of dolomite is 2.87 g/cm^3, the average values given in Table 1.11 correspond to the percentage of dolomitization given in Table 1.12. These results are in close accord with those obtained by Chilingar and Terry (1954). This relationship also presents the possibility of determining porosity from matrix density of drill chips and grains.

Table 1.11 Relationship between porosity and specific gravity of Iranian carbonate rocks (Asmari Limestone).

Oil Field	Porosity Range*(%)			
	0–4.0	4.1–8.0	8.1–12.0	≥ 12.1
Haft Gel	2.68 ± 0.04 (14)	2.73 ± 0.09 (9)	2.75 ± 0.14 (8)	2.68 ± 0.12 (20)
Naft Khaneh	2.62 ± (1)	2.77 ± 0.08 (3)	2.81 ± 0.05 (9)	2.83 ± 0.05 (24)
Gach Saran	2.71 ± 0.09 (7)	2.77 ± 0.08 (10)	2.78 ± 0.08 (11)	2.79 ± 0.08 (8)
Agha Jari	2.74 ± 0.09 (7)	2.74 ± 0.09 (10)	2.73 ± 0.06 (5)	2.81 ± 0.09 (9)
Naft Sefid	2.67 ± 0.04 (3)	2.73 ± 0.04 (3)	2.76 ± 0.06 (9)	2.76 ± 0.09 (12)
Lali	2.74 ± 0.08 (6)	2.74 ± 0.03 (6)	2.79 ± 0.04 (3)	2.79 ± 0.03 (4)
M-i-S	2.67 ± 0.12 (11)	2.71 ± 0.11 (8)	2.73 ± 0.12 (9)	2.80 ± 0.06 (12)
Mean	2.70 (49)	2.74 (49)	2.76 (53)	2.80 (89)

*The ± figures are mean square errors of the average values, and the figures in parentheses are the numbers of observations.

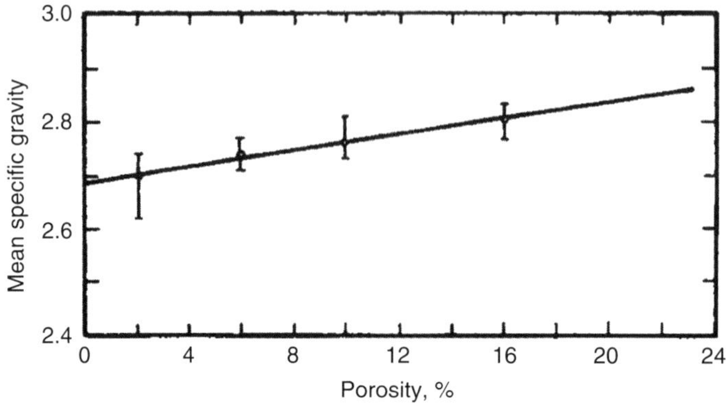

Figure 1.14 Relationship between specific gravity and porosity of Iranian carbonate rocks (Asmari Limestone). (After T.F. Gaskell, British Petroleum Co, Ltd. Personal communication, 1963.)

Table 1.12 Relationship between porosity and density of Asmari Limestone in Iran.

Porosity (%)	Specific Gravity	Dolomitization (%)
0–4.1	2.70	0
4.1–8.0	2.74	20
8.1–12.0	2.76	32
≥ 12.1	2.80	58
	2.84	82

As in the case of siliciclastic rocks, carbonate rocks that had higher initial porosity, underwent the most extensive diagenetic changes. It should be noted that lithification of carbonates rocks takes place much faster than that of sandstones and siltstones. This results in an earlier completion of the process of mechanical compaction.

More than thirty different natural processes, which are controlled by local and regional factors, occur during the diagenesis and catagenesis of carbonates (Chilingar *et al.*, 1979; Larsen and Chilingar, 1983). Lithification of carbonate sediments is of biochemical, physicochemical, and mechanical nature. To some extent, these processes occur simultaneously and change both the composition and the pore geometry of sediments and rocks. With time, their rates are reduced.

An essential difference between mechanical and biochemical – physicochemical processes is that the former acts in one direction with results being mostly irreversible. Biochemical and physicochemical processes, on the other hand, can take place in different directions; thus, increase and decrease in secondary porosity of carbonate rocks can occur periodically depending on the environmental conditions. Inasmuch as the mechanical processes are unidirectional and usually irreversible, possibly they play a major role in changing the original (primary) porosity of carbonate rocks. Thus, there is similarity with compaction of terrigenous (siliciclastic) rocks.

Degree of consolidation, dissolution and cementation under the overburden pressure is important. Increase in overburden load as a result of subsidence of sediments leads to the solution of crystals under pressure, i.e., differential solution takes place in more strained parts of grains with a subsequent deposition of material on the surfaces having lower potential energy. In addition, grains (and crystals) may get flatter parallel to the surface of stratification. These processes decrease the initial porosity of carbonate rocks.

Carbonates can be also extensively fractured. In this situation, even without porosity and permeability in the main body of the formation, commercial amount of oil can exist.. On the basis of fracture width, fractures can be classified into:

1. *supercapillary* (width greater than 0.26 mm),
2. *capillary* (width from 0.26 to 0.0001 mm), and
3. *subcapillary* (width less than 0.0001 mm).

Mar'enko (1978) proposed another classification of fractures, but the writers prefer the following classification:

1. fine macrofractures (width = 1 – 10 mm),
2. fine fractures (width = 0.1 – 1 mm),
3. very fine fractures (width = 0.01 – 0.1 mm),
4. hair-thin fractures (width = 0.001 – 0.01 mm), and
5. microfractures (width = 0.0001 – 0.001 mm).

Subsequent dissolution may enlarge initial fractures and, thus, increase fracture porosity as shown in Figure 1.15. Often, one can observe vugs along the extent of fractures. In carbonates, the porosity, permeability, and pore space distribution are related to both

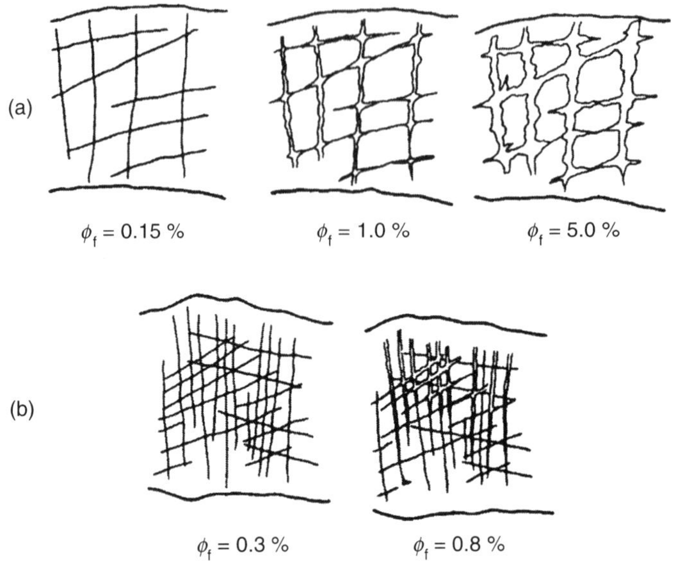

Figure 1.15 Development of fracture porosity in carbonate reservoir rocks that have (a) low insoluble residue (IR) content and (b) high IR content. (After Tkhostov *et al.*, 1970, Figure 11.)

the depositional environment of the sediment and the changes that have taken place after deposition. When volume loss occurs due to solution and recrystallization, irregular voids are formed called vugs (vuggy porosity).

The initial porosity of carbonates often approaches that of sandstones in that their structure consists of aggregates of oolites, grains, and crystals. The initial (primary) porosity of carbonates depends on their genetic type to a great extent: it is the largest in biogenic and clastic (detrital) varieties, whereas it is considerably lower in cloddy and chemogenic ones (excluding chemogenic oolitic limestones). According to Aksenov *et al.* (1986), values of maximum porosity of carbonate rocks considering their structural-genetic types are: biogenic – 24%, biogenic-detrital – 24%, clotted-cloddy – 13%, crystalline-granular – 4%, pelitomorphic – 2%, and oolitic and pisolitic – 24%.

Permeability is controlled by the size of the passages (pore throats) between the much larger pores and vugs. Mercury injection into the rock pore space in the laboratory measures the size of pore throats rather than those of the void space. Consequently, a highly porous rock may have little or no permeability if these interconnections are very narrow or absent. On the other hand, some very fine-grained carbonate rocks have an extensive network of interconnected pore space with enough permeability to be able to yield commercial volumes of oil. Intercrystalline pores tend to be interconnected, and rocks with high intercrystalline porosity are normally permeable as found in many highly-productive dolomite reservoir rocks.

Studies are continuing on the effects of post-sedimentation processes on the properties of carbonate reservoir rocks (e.g., see Sarkisyan *et al.*, 1973). Many good bioclastic carbonater reservoir rocks (i.e., high porosity and permeability) originate in the shallow parts of the basins. Sulfatization, calcification, and silicification affect adversely the reservoir-rock properties. The secondary mineralization processes, however, indirectly improve the flow capacity of rocks (permeability) by creating heterogeneity, which favors the subsequent formation of fractures and solution cavities. Dolomitization, in general, either creates or increases porosity.

Bioherms. Considerable attention is paid to the origin of bioherms and reefs and the properties of composing rocks, mainly because of practical considerations. Many carbonate reservoirs are present in reefs and bioherms.

Korolyuk and Mikhaylova (1970) presented an elaborate classification of bioherms and reefs. They defined *organic structure* as a geologic body formed as a result of growth on each other of attached or colonial organisms together with the complex of associated rocks (Figure 1.16). During certain periods of their growth, various organic structures could have been "wave breakers". Biorhythmites are characterized by repeated occurrence of bioherms, biostromes, and other related bodies in the sequence of bedded rocks.

The classification of Korolyuk and Mikhaylova (1970) is based on lithologic-morphologic principles rather than paleogeographic ones (Figure 1.16, 1.17, 1.18, and 1.19). Korolyuk and Mikhaylova (1970) recognized the following three types of reef complexes: I – Reef complex that is composed of organic core and greatly subordinate, low-volume, flanking organo-detrital deposits. II – Reef complex that includes massive organic structure with adjoining, detrital, flanking deposits; the volume of flanking deposits (fore-reef and back-reef) is smaller than that of the core. Among detrital rocks, coarse-detrital varieties play a major role. III – Reef complex that includes small organic core and much more massive (voluminous) flanking deposits (various rocks).

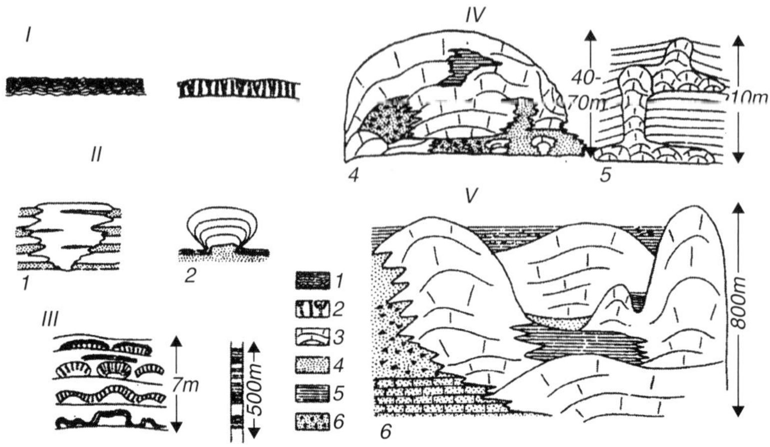

Figure 1.16 Types of organic structures. (After Korolyuk and Mikhaylova, 1970, Figure 1, p. 230.) I – Biostromes, 0.5 – 5 m thick and 10 – 100 m long; II – bioherms, oncoids, 1 – 10 m; III – biorhythmites; IV – biohermal massive, 10 – 100 m; and V – reef massive, hundreds of meters 1 – Archaeocyathidal bioherms, Cambrian, western Siberia (after I.T. Zhuravleva) (arrested growth); 2 – stromatolitic bioherms. Cambrian, western Siberia (free-growing); 3 – stromatolitic biorhythmites, Cambrian, western Siberia; 4 – biohermal massive, Jurassic, western Crimea; 5 – biohermal massive, Sarmatian, Moldavia; and 6 – reef massive , Jurassic, Caucasus Lithology; 1 – bedded stromatolites; 2 – columnar stromatolites; 3 – biohermal massive limestones; 4 – organo-clastic limestones; 5 – chemical carbonates; and 6 – coarse detrital limestones.

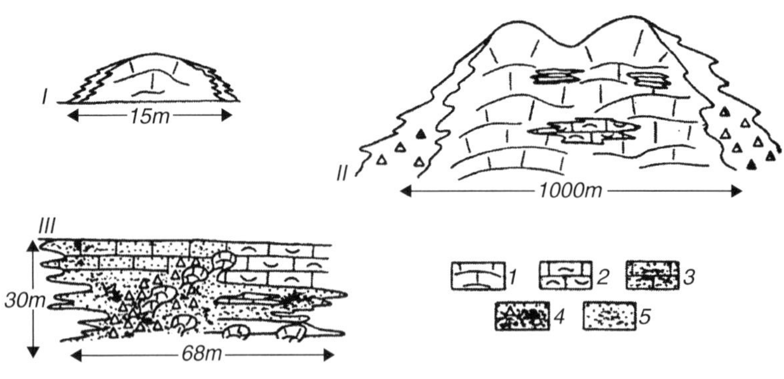

Figure 1.17 Types of reef complexes. (After Korolyuk and Mikhaylova, 1970. Figure 2, p. 231.) I – Stromatolitic-bryozoan bioherm, Ordovician, Baltic region; II – reef massive, Upper Jurassic, Crimea; and III – reef complex, Upper Jurassic, western Crimea. Lithology: 1 – biohermal limestone; 2 – organic-detrital limestone; 3 – fine-detrital limestone; 4 – limestone breccia; and 5 – limestone gravels and sandstones.

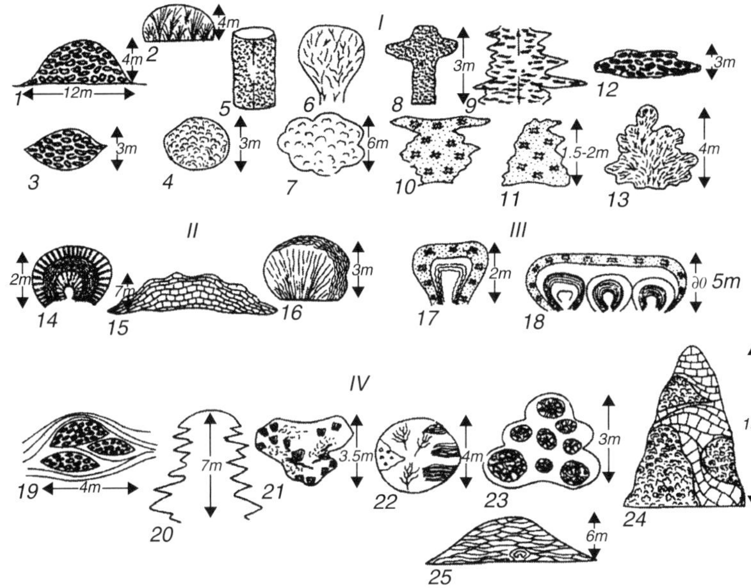

Figure 1.18 Types of bioherms. (After Korolyuk and Mikhaylova, 1970. Figure 3, p. 231.) 1 – Simple homogeneous; II – simple zonal; III – complex zonal; and IV – complex spotty. 1 – Archaeocyathidal, Cambrian, western Siberia (after I.T. Zhuravleva): 2 – coralline, Devonian, northern Caucasus; 3 – Archaeocyathidal-algal, Cambrian, Tuva; 4 – renaltsician, Cambrian, western Siberia (after I.T. Zhuravleva); 5 and 8 – bryozoan-nubecularian, Sarmatian, Moldavia (after V.S. Sayanov); 6 – coralline, Jurassic, western Crimea; 9 – algal, Jurassic, Crimea; 10 and 11 – bryozoan, Neogene, Ukraine (after L.A. Belokrys); 12 – Archaeocyathidal-agal, Cambrian, Altay (after V.D. Fomin); 13 – bryozoan, Neogene, Kerch; 14 – stromatolitic, Cambrian, western Siberia; 15 – *Rhodophyceae* (algal), Jurassic, Crimea; 16 – bryozoan, Neogene, Crimea, ; 17 and 18 – bryozoan algal (Rhodophyceae), Neogene, Crimea (after L.A. Belokrys); 19 – Archaeocyathidal, Cambrian, Siberia (after I.T. Zhuravleva); 20 – coralline-stomatoporan, Devonian, Novaya Zemlya (after Patrunov); 22 – coralline-algal, Jurassic, Crimea; 23 – serpulo-bryozoan, Neogene, western Crimea, and 24 – coralline-bryozoan, Jurassic, Crimea.

Bioherms were divided into four types: I – Homogenous bodies constructed by one or two types of organisms (e.g., coralgal and bryalgal), with minor admixture of other organic limestones. II – Zonal bioherms composed of one or two frame-building organisms, systematically close to each other. III – Zonal bioherms composed of several remotely related frame-building organisms, but occurring in regular layers. IV – Complex, spotty bioherms formed by several frame-builders, distributed in clusters and, as a rule, accompanied

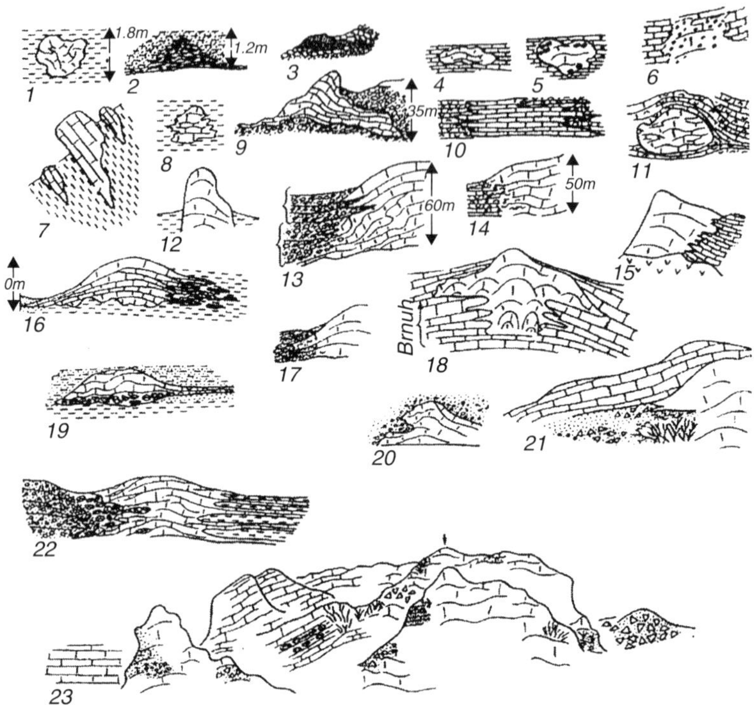

Figure 1.19 Types of contacts of organic structures. (After Korolyuk and Mikhaylova, 1970, Figure 4., p. 233.) 1-11 – *Small forms*: 1 – Upper Oxfordian, Sudak area, Crimea; in clays, butt-joint; 2 – Upper Oxfordian; Lysaya (Bold) Mountain, Crimea; in clay, butt-joint; 3 – Oxfordian, Panagiya landmark, Crimea; in sands and conglomerates, growth; 4 – Upper Devonian, Frontal Ridge of Northern Caucasus (after S.M. Kropachev and I.V. Krutu); in limestones, butt-joint; 5 – Neocomian, Crimea (after E.I. Kuz'micheva); in limestones; 6 – Silurian, Vaazalemma, Estonia; in limestones, intergrowth; 7 and 8 – Permian, Darvaz (after M.A. Kalmykova) in clays, wedging-in; 9 – Oxfordian, Panagiya landmark, Crimea, in sands and conflomerates, butt-joint; 10 – Kimmeridgian, Karaba-Yayla, Crimea, in limestones, wedging-in; and 11 – Oxfordian, Delyamet-Kaya, Crimea, in limestones, enveloping. Twelve to twenty two – *biohermal massives*: 12 – Upper Oxfordian, Sakharnaya Golovka (Sugar Head), Crimea; in clays; 13 – Oxfordian, Khart-Kaya, Crimea: in sandstones and conglomerates; complex, upper part – wedging-in, lower part –intergrowth; 14 – Oxfordian-Kimmeridgian, Redant, northern Caucasus; in limestones, gradual; 15 – Oxfordian- Kimmeridgian, Georgia (after N.S. Bendukidze); in limestones, wedging-in; 16 – Upper Oxfordian, Lysaya (bold) Mountain, Crimea; in clays, complex and gradual wedging-in (right side); 17 – Upper Oxfordian, Fiagdon River, Northern Caucasus; in sandstones and conglomerates, complex envelopment and intergrowth; 18 – Samatian, Pogornichany, Moldavia (after V.S. Sayanov); in limestones, upper part – complex envelopment, lower – wedging-in; 19 – Upper Oxfordian, Likon Mountain, Crimea, in clays, left side – lens-like wedging-in, right side – butt-joint and gradual; 20 – Upper Permian, Abago Range, Greater Caucasus (after A.A. Belov; in sandstones and conglomerates, complex, erosion and envelopment; 21 – Oxfordian- Kimmeridgian, Gizeldon River, Northern Caucasus; in limestones; 22 – Kimmeridgian, Demirdzhi Yayla Mountain (center), Crimea; in clays; and 23 – Oxfordian-Kimmeridgian reef massive, Oshtein Mountain (arrow); Glavnyy (main) Caucasus Range.

by a rich variety of other organisms. The role of non-biohermal deposits (detrital, etc.) is significant.

1.2.4 Carbonate versus Sandstone Reservoirs

Around sixty five percent of World oil reserves reside in carbonates. Unfortunately, the recovery factor from carbonates is much lower than those from sandstone reservoirs. Thus, an all-out effort should be made to correct this unfortunate situation. Some of the reasons for this sad situation can be summarized as follows:

1. Greater heterogeneity of carbonate reservoir rocks compared with sandstones. Often, heterogeneity can be observed even on a thin-section scale.
2. Carbonate rocks appear to be more oil-wet than water-wet compared with sandstones.
3. The presence of fractures adds new dimension to be fluid flow problems.
4. The presences of double porosity system [i.e., low-permeability matrix and fractures (including vugs)] results in very low recoveries. The greatest challenge is to how to move the oil from the low-permeability matrix into the fractures. Most of the flow occurs in fractures.
5. In carbonates that have intergranular porosity, permeabilities parallel and perpendicular to the bedding are about equal, whereas in sandstones the horizontal permeability is much greater than the vertical permeability.
6. Tectonically caused overpressures in lithified carbonates defy the existing predictive techniques.

1.2.5 Volcanic/Igneous Rocks

Fractured volcanic rocks occasionally play an important role in creating reservoirs and traps for hydrocarbon accumulation. Reserve estimation in such traps requires sophisticated methods of studying reservoir rock properties, such as density of fractures, specific surface area, width of fractures, irreducible fluid saturation, pore space structure, porosity, and permeability (Kondrushkin and Buryakovsky: 1987; Abasov *et al.*, 1997).

As an example, the productive volcanic Muradkhanly Oilfield (Figure 1.20) in the center of the Kura Depression, Azerbaijan, is described here. Reservoirs have been formed in the weathered volcanic rocks of the upper portion of Upper Cretaceous section. Oil traps here were formed by transgressive overlapping by Maikop shales in the shallowest part, and by Eocene terrigenous-carbonate rocks on the western flank. Commercial oil reserves are associated with the fractured Upper Cretaceous volcanic rocks. Productivity of the Eocene terrigenous, carbonate, pyroclastic rocks, and the Middle Miocene terrigenous-carbonate rocks encountered in this field is lower than that of the Upper Cretaceous volcanic rocks.

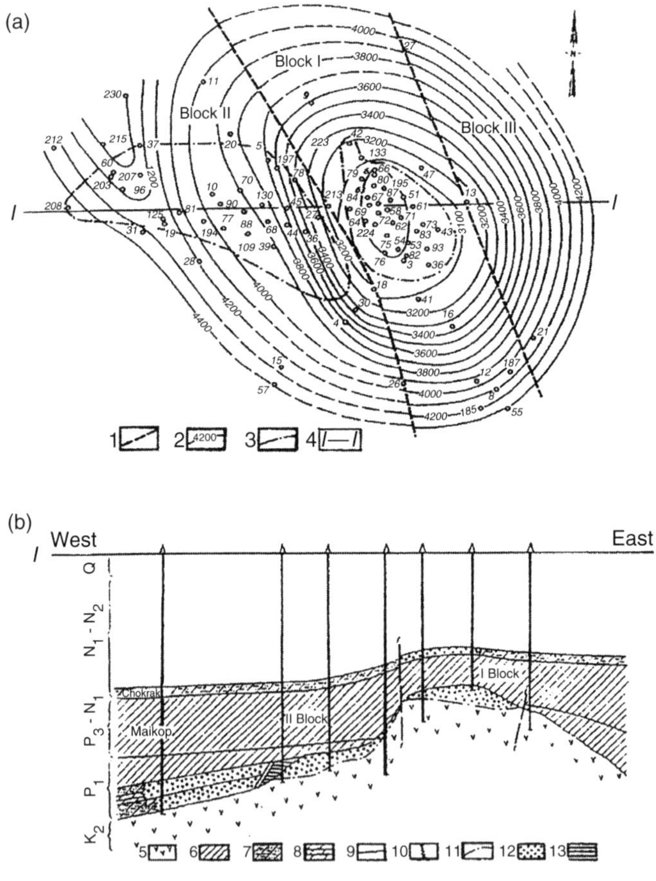

Figure 1.20 Muradkhanly Oilfield: (a) structural map on the top of volcanic rocks; (b) cross-section along the line west to east I – I. (Modified after Buryakovsky *et al.*, 2001.) 1–Faults; 2–Contour lines on top of volcanic rocks; 3–initial OWC; 5–volcanic rocks; 6–clay/shale; 7–alternation of sand, silt and shale; 12–oil reservoir.

Logs from the Muradkhanly Oilfield indicate that an anticline is present above the volcanic rocks at a minimum depth of 3000 m. Within the 4200-m contour line, the overall field size is 15 × 11 km. The dips vary from 10 to 20°. The structure is cut by two faults and, hence, is divided into three separate blocks (Figure 1.20). Oil reserves are concentrated in the crestal area (Block I) and at the western flank of the structure (Block II).

The Upper Cretaceous includes undisturbed volcanic rocks: pyroxene-andesite; biotite-, hornblende, and pyroxene-trachyandesite; porphyry and amygdaloidal basalts; and products of alteration due to weathering of volcanic rocks with admixture of clastic material (tuff-sandstones, tuff-breccia, and tuff-gritstone; also see Figure 1.21). Penetrated thickness of sedimentary and volcanic rocks ranges from 3 to 1952 m.

The porosity of volcanic rocks is of fracture-vuggy and intergranular type. Large intergranular pores, vugs and fractures are present in the core samples. Large pores are 1 mm (average) in diameter, whereas vugs have diameters of 2 cm (average). Microfractures, which contain mainly calcite and argillaceous cement, have widths of ≥ 0.1 mm. Oil is present in large intergranular pores, vugs, and fractures. During drilling, lost circulation (up to 100 m^3/d) and high oil flows (up to 500 metric tons per day) in several wells suggest that there are long and wide fractures in the volcanic rocks.

Microfractures have been studied in 4 × 5 cm thin-sections. Microscopic fracture porosity ranges from 0.04 to 0.004%, fracture

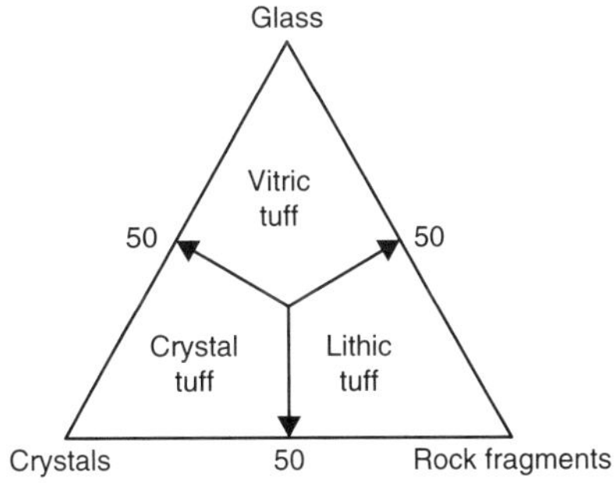

Figure 1.21 Triangular diagram and end members of tuffs. (After Pettijohn *et al.*, 1972, Figure 7-9, p. 269; see O'Brien, 1963; Fisher, 1961; courtesy of Springer-Verlag, New York).

permeability varies from 0.16 to 6.90 mD, and average fracture density (total length of fractures per unit area) is 0.30 cm/cm^2.

Scanning Electron Microscope (SEM) micrographs show that the volcanic rock texture depends on the original properties of the unweathered rocks with subsequent imprint of weathering. Oil flows (up to 500 t/d) in several wells suggest that there are long and wide fractures in the volcanic rocks.

Petrographic studies show that reservoir properties depend on the degree of weathering of volcanic rocks. The formation of large pores and vugs is due to the plagioclase dissolution. Sometimes, when plagioclase and other minerals are dissolved, micro-caverns are formed.

Microfractures have been studied in 4 × 5 cm thin-sections. Micro-fracture porosity ranges from 0.04 to 0.004%, fracture permeability varies from 0.16 to 6.90 mD, and average fracture density (total length of fractures per unit area) is 0.30 cm/cm^2.

Scanning Electron Microscope (SEM) micrographs show that the volcanic rock texture depends on the original properties of the unweathered rocks with subsequent imprint of weathering and alteration. Alteration of ash resulted in the formation of such clay minerals as smectite (montmorillonite), chlorite and biotite during diagenesis and catagenesis. Secondary matrix pores vary in size from 1 to 200 μm. Pores are often connected by irregularly curved fractures, 10–600 μm long and 0.5–10 μm wide.

Mercury injection studies show that the volcanic rock matrix within the unproductive and/or low-productive sections contains up to 60–75% of small pores with radii less than 0.1 μm, i.e., subcapillary pores not involved in fluid migration. Diameters of pore throats, which are important for fluid movement, range from 0.25 to 6.3 μm. A power-law correlation between the pore throat diameter and matrix (intergranular) permeability is as follows:

$$k = 0.0525 d_{ch}^{2.85} \tag{1.12}$$

were k is the permeability in mD and d_{ch} is the pore throat diameter in μm.

The porosity of volcanic rocks studied in core samples by the saturation method varies within a wide range (0.6 to 28%), with an average value of 13%. The intergranular permeability is low; it varies from 0 to 10 mD, with an average value of 1 mD. The unusual combination of high porosity and very low permeability is explained by the complex and non-uniform structure of the porous space. Finely-porous rocks have complex pore structure and curved channels. The 0.1- μm subcapillary pores are not involved in fluid migration. The secondary matrix porosity includes pores (0.25 μm up to 1 mm in size) and vugs (larger than 1 mm in size). Commonly, these pores and vugs are partly filled with kaolinite, illite, smectite (montmorillonite), ferro-oxides, and zeolites; some clays are dispersed and highly swelling. Clay mineral content (mainly authigenic) in rocks is variable and can reach 40% or more. The petrophysical study shows that if the content of highly-dispersed clay is more than 40%, then the water saturation of rocks is almost 70% and even higher. Under these conditions, rocks cannot be considered productive.

Oil is present both in the rock matrix (pores and vugs) and in the micro- and macro-fractures. The intergranular matrix permeability is very low, and the oil saturation of reservoir rocks is distributed unevenly. Oil is produced mainly from zones that have hydrodynamic connections with the fracture systems. For quantitative evaluation of volcanic reservoirs, core samples from inside-perimeter wells with oil production and outside-perimeter wells without fluid flow were analyzed. The two statistical distributions of porosity were compared, and the average porosity values were determined. The secondary porosity (ϕ_1) (vugs and fractures) can be determined using the following formula:

$$\phi_1 = (\phi_2 - \phi_3)/(1 - \phi_3) \tag{1.13}$$

where ϕ_2 is the porosity within the productive zones in the inside-perimeter wells and ϕ_3 is the porosity within the unproductive zones in the outside-perimeter wells.

The average secondary porosity is 1.8%. Depth intervals with high porosity (the secondary pores, vugs, and fractures) were determined using log data (electrical, radioactive, sonic, and caliper) and well test data. Thickness of these intervals can be considered as the effective (oil-bearing) reservoir thickness (net pay). These intervals have been identified using porosity determined from log data. Two porosity cut-off limits were identified:

1. Lower limit: for impermeable, unproductive rocks, porosity is less than 7–8%.
2. Upper limit: for water-bearing intervals with the content of highly-dispersed clay minerals of 40%, porosity exceeds 20%.

Electrical logs were used to estimate the intergranular porosity and initial oil saturation. Based on the log analysis, the oil saturation in fractures is about 100%, whereas the oil saturation in the matrix is about 50%. Weighted average oil saturation of the whole formation (including the secondary pores, vugs, and fractures) is about 90%.

Porosity and permeability were measured at a depth of 450 to 500 m from the top of volcanic rocks. Deeper intervals, i.e., from 1000 to 2000 m, are dry or showed insignificant flow of water. The most productive zone is the upper section of volcanic rocks, 25–30 m thick. Here, one can observe uniform and intensive secondary rock alterations and strong oil flow in most of the wells. The oil-saturated intervals are distributed from the top of volcanic rocks to a depth of 10–50 m in some wells, and to a depth of 100 m and deeper in others. As shown in Figure 1.20b, the bottom of oil accumulation is located at different depths in different volcanic rocks. This means that there is no continuous and flat oil-water contact; instead, it has a wave-shaped configuration. The real oil-reservoir boundaries intersect the contour lines on the top of volcanic rocks. Oil is present in the secondary porosity of these rocks.

The reservoirs are characterized by non-uniform oil content, both in lateral and vertical directions. Consequently, the initial oil production rates vary within the following wide limits:

1. 1 to 30 t/d (7 to 220 bbl/d) in 48% of wells.
2. 30 to 100 t/d (220 to 750 bbl/d) in 35% of wells.
3. >100 t/d (>750 bbl/d) in 17% of wells.

The maximum initial water production in most wells (58%) is 10 m^3/d. Initial reservoir pressure and temperature are 55 MPa and 125°C, respectively. The initial reservoir pressure is higher than the bubble-point pressure by 40 MPa and higher than the normal hydrostatic pressure by 20 MPa. Gas/oil ratio is equal to 30 m^3/t and the average density of oil is 0.880 g/cm^3 at standard conditions. The oil is paraffinic, with low sulfur content.

1.2.6 Classification of Hydrocarbon Accumulations Based on the Type of Traps

Hydrocarbon accumulations can be classified into three major types and subtypes (Eremenko and Chilingar, 1996):

I. *Traps formed by folding (with or without faults).* Accumulations formed as a result of folding are usually associated with the bedded reservoirs. The complexity of structure (sometimes even isometric), size, and especially heights are caused by the trap and reservoir position in the sedimentary basin. Over the central areas of tectonic plates, the traps are gentle and sometimes very large. Over the plate margins, transition zones and, especially, collision zones, the folds are higher, steeper and with a clearly expressed trend. The accumulations may be classified using some other parameters, too. In particular, oil-water contours in such accumulations are closed and, in plan view, have oval or more intricate shapes, and form rings.

II. *Traps formed within various buildups*. Accumulations formed within various buildups are usually associated with the massive-type reservoirs. Most common are accumulations in biogenic buildups (reefs and bioherms). Sometimes, biostromes are mistakenly attributed to the same class. Included here are large accumulations with huge flow rates due to the presence of fractures and vugs in carbonates. Some investigators also include in this group the erosional projections of the metamorphic and volcanic rocks (fault-bounded or bounded by erosional surfaces), which may contain accumulations, e.g., White Tiger Field in Vietnam.

III. *Traps that are limited by the depositionally imposed facies changes*. Lithologic and stratigraphic traps of Group III include facies pinch-outs, stratigraphic unconformities, and contact of the reservoir with the impermeable rock up-section. Such traps may be associated with the bedded reservoirs on the monoclines or on the flanks of anticlines. These traps may contain rather large accumulations. They may be associated with bedded reservoirs confined on every side. In such a case, they form large accumulations. Water saturation contours impinge on the trap (impermeable barrier). This type of accumulations is very common: about 50% of all known accumulations.

Accumulations of Types I, II, and III are formed in accordance with the gravitational ("anticlinal") theory. By far, not all known accumulations, however, belong in the described three types or combinations thereof. Also, not all of them formed in accordance with the gravitational theory. These unconventional accumulations are discussed below (Types IV through VII).

IV. *Dominance of capillary forces over the gravity force*. Oil or gas found in hydrophilic rocks occupies coarser-grained reservoir rocks, which are sealed by water-saturated fine-grained reservoir rocks. Examples of such accumulations associated with relatively coarse-grained sandstone lenses (e.g., 100-ft sandstone in Appalachian Oil and Gas Province, USA) were presented by Brod (1957). The authors of this book reviewed large number of commercial accumulations all over the world and have not been able to find another such clear-cut example. Although the appearance of capillary forces is frequently observed, the formation and preservation of the accumulations cannot be attributed to these forces. It should be kept in mind that water and gas lenses exist within oil accumulations; water is sometimes encountered updip in pinched-out reservoirs (e.g., Productive Series of the Absheron Peninsula in Azerbaijan, and Maykopian sandstones in the Northwestern Caucasus).

V. *Dominance of hydraulic forces*. The hydraulic forces (Figure 1.22) can cause a tilt in the oil- water interface. To determine that tilt, a trigonometric function or the Savchenko's (1977) equation could be used.

$$\Delta h = \Delta p_{norm} / g(\rho_{water} - \rho_{oil}) \quad (1.14)$$

where Δh is the amount of shift at the edge of accumulation, Δp_{norm} is the difference in normalized pressures; ρ_{water} and ρ_{oil} are the density of water and oil, respectively; and g is the gravitational acceleration.

A barrier (facies change, stratigraphic unconformity, and a fault) often turns out to be a barrier due to the presence of pressure difference across the barrier, rather than because of the appearance of an impermeable barrier in the way of fluid movement.

Figure 1.22 shows possible relations of the position of the piezometric surface (normalized pressure head) and the position of the oil-water contact. The necessary condition for the pore preservation of hydraulically-trapped accumulation next to a fault is a higher potential head of the water next to the fault zone than that of the productive formation (the surplus pressure is included). This condition may exist if, for instance, there is a communication along the fault between the accumulation and the reservoir with AHFP (overpressure).

In monoclines, the accumulations can be preserved (Figure 1.22c, e and f) if the potential head decreases down dip in locations where the dip increases (Figure 1.22e) or the dip of the piezometric surface

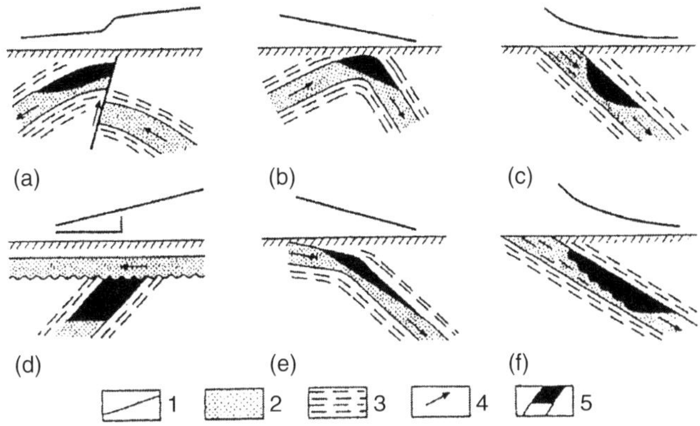

Figure 1.22 Conceptual cross-sections showing hydraulically trapped oil and gas accumulations. (a) Next to conductive faults. (b) At the anticlinal crest. (c) In monoclines within the areas of changing reservoir properties. (d) In monoclines underneath the stratigraphic unconformities. (e) At structural noses of monoclines. (f) Near the reservoir shale-out boundaries. 1 – Piezometric surface; 2 – reservoir; 3 – shales; 4 – direction of water movement; and 5 – oil and gas accumulations.

decreases (Figure 1.22f). The latter is possible when the reservoir-rock properties change (i.e., capillary forces enter into play). The oil-water contours can close onto themselves (but crossing the structural contour lines on top of the reservoir) or can abut on the trapping barrier. (The contributions of Plotnikov, 1976; Gattenberger, 1984; and Mikhaylov, 1984, on hydrodynamic traps are noteworthy.)

Neither of the described types, however, owes its existence to the hydraulic forces exclusively. They can exist only under condition of the combined interaction of several different forces: (1) hydraulic and gravity, (2) hydraulic and capillary, or (3) hydraulic + capillary + gravity forces. The effect of hydraulic forces is commensurate with that of gravity and capillary forces

VI. *Gas accumulations in synclines or in monoclines devoid of structural highs*. Examples of such accumulations have been presented by Masters (1979) and Perrodon (1984). There is a gas accumulation in the Deep Basin Monocline, in Alberta, Canada. The latter accumulation resides in the Mesozoic sandstone, which is more than 3 km high (the thickness of individual gas intervals is 10–150 m). The sandstone is water-saturated updip the gas accumulation, with an improvement in petrophysical properties. The gas reserves are nearly 11.3 TCM. The gas accumulation of Milk River Field (Canada), with 250 BCM of reserves, is another similar example. The gas accumulation of San Juan Field (USA) resides in the Mesozoic sandstone in the synclinal part of the structure, with reserves of 700 BCM. The sandstone is water-saturated over the flanks. The porosity and permeability within the gas-saturated portion are 14% and 1 mD, respectively, whereas in the water-saturated portion, ϕ = 25% and k = 100 mD. For further discussion, see Chilingar *et al.* (2005).

2

Characterization of Hydrocarbon Reservoirs

2.1 Petrophysical Parameters

The reliable geological interpretation of log analysis results requires a reliable definition of relationship among petrophysical and reservoir parameters of oil-gas-water-bearing rocks. Petrophysical relationships are based on the laboratory analyses of core samples saturated with formation fluids. Core analyses are conducted under surface (ambient) and subsurface (in-place or reservoir or downhole) conditions.

The basic petrophysical parameters needed to evaluate a petroleum reservoir are its porosity, permeability, fluid saturation, areal extent, and formation thickness. These parameters can be estimated from three common sources: core, well logging, and pressure test analyses. In this chapter, the writers describe the various petrophysical parameters.

2.2 Porosity, Void Ratio, and Density

Virtually all detrital rocks are porous (i.e., contain void space) to some extent. The voids in sand are particularly important in the study of compaction; because compaction is associated with reduction in pore space. Under extremely high pressures, there is also a reduction in volume of solids; however, in most studies of compaction, reduction in solids volume has been ignored.

The relative volumes of voids and solids can be expressed in terms of (1) porosity and (2) void ratio. With few exceptions, geologists and petroleum engineers prefer the term porosity, whereas soils and civil engineers use the term void ratio. It should be pointed out here that the various disciplines in geology and engineering all have distinct sets of nomenclature and symbols for rock parameters.

Leonid Buryakovsky, George V. Chilingar, Herman H. Rieke and Sanghee Shin
Petrophysics, (57–150) © Scrivener Publishing LLC

Both porosity and void ratio are related to bulk volume of a rock. Bulk volume V_b is defined as the sum of the volumes of the voids or pores, V_v or V_p, and the solids, V_s:

$$V_b = V_p + V_s \tag{2.1}$$

Porosity ϕ is the ratio of the void space to the bulk volume and is usually expressed in percent:

$$\phi = (V_p / V_b) \times 100 \tag{2.2}$$

When used in an equation, however, the decimal equivalent is usually used, e.g., 0.5 instead of 50%. Comparison charts to aid in visual estimation of porosity are presented in Figure 2.1.

Void ratio, e, is extremely important in compaction studies and is defined as the ratio of the voids volume to the solids volume:

$$e = V_p / \mathbf{V}_s \tag{2.3}$$

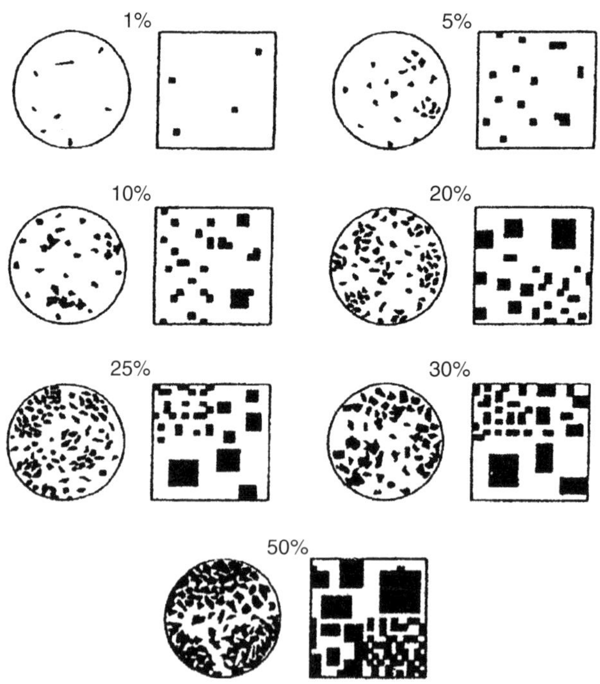

Figure 2.1 Comparison charts to aid in visual saturation of percent of porosity. Circles after Terry and Chilingar (1955); squares after Folk (1951a.)

also:

$$e = \phi/(1 - \phi) \tag{2.4}$$

or:

$$\phi = e/(1 + e) \tag{2.5}$$

Obviously, porosity can never exceed a value of 100% or 1.0 (fractional porosity), whereas void ratio often exceeds unity in fine-grained sediments and clays. Figure 2.2 shows the relationship between the void ratio and porosity in the commonly-occurring range.

Porosity is a measure of the volume of voids related to the bulk volume, which changes when compaction takes place. The volume of solids remains essentially constant under compression, while the bulk volume decreases. Subtracting one porosity value

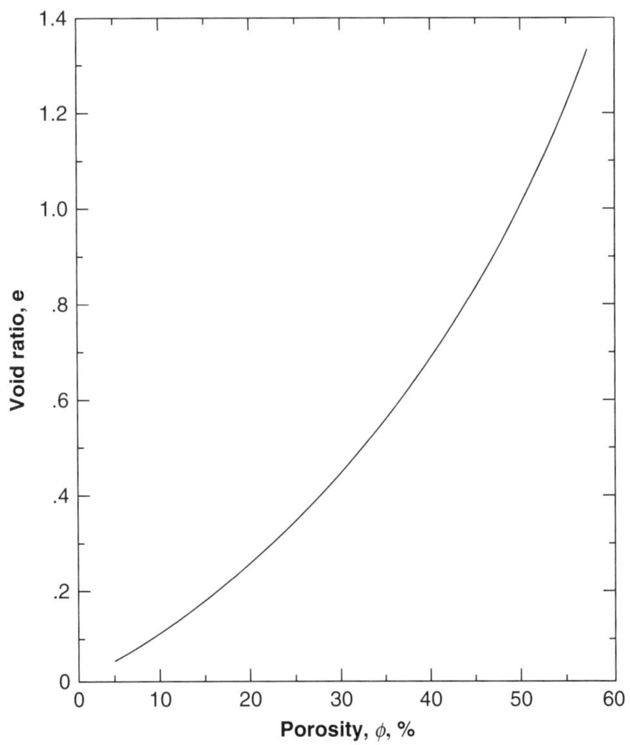

Figure 2.2 Relationship between void ratio and porosity.

from another, therefore, cannot represent the total volume change, because the porosities are expressed in reference to different bulk volumes.

Equations 2.6 and 2.7 can be used in calculations because they relate the change in bulk and pore volumes to the volume of the solids, which is considered to be constant. A change in the bulk volume V_b can be expressed in terms of porosity:

$$\delta V_b = 1 - (1 - \phi_1)/(1 - \phi_2) \tag{2.6}$$

where δV_b = change in bulk volume, ϕ_1 = porosity at time 1, and ϕ_2 = porosity at time 2 (after compaction). Inasmuch as the volume of the solids remains constant, δV_b is also equal to δV_p.

The decrease in porosity per unit of the original bulk volume can be expressed as:

$$\delta\phi = \delta e / (1 + e_0) \tag{2.7}$$

where $\delta\phi$ = change in porosity related to the original V_b, e_0 = original void ratio, and δe = change in void ratio. Equation 2.7 is preferred when void ratios are used, because calculations are simplified.

Specific weight, γ, is often used in conjunction with porosity and void ratio. It is defined as the weight per unit volume, whereas density, ρ, is the mass per unit volume and is equal to γ/g where g is the gravitational acceleration. The term "density", however, is often used to designate specific weight, which often results in erroneous calculations. Mass ρ is attracted by the earth with a force $\gamma = (\rho \times g)$. For example, if the specific weight of water is equal to 62.4 lb/ft^3, then the density expressed in terms of slugs/ft^3 is equal to 1.94 (= 62.4 (lb/ft^3)/32.17(ft/sec/sec)).

Bulk specific weight can be either "dry" or "wet" depending upon the nature of the fluid in the pore spaces. The unit weight of dry sand (only air is present in pore spaces) is equal to:

$$\gamma_{db} = (1 - \phi)\gamma_s \tag{2.8}$$

where γ_{db} = weight per unit of dry bulk volume, and γ_s = specific weight of solids (grains).

The unit weight of wet sand is expressed as:

$$\gamma_{wb} = (1 - \phi)\gamma_s + \phi\gamma_f \tag{2.9}$$

or:

$$\gamma_{wb} = \gamma_s - \phi(\gamma_s - \gamma_f) \tag{2.10}$$

where γ_{wb} = weight per unit of wet bulk volume, γ_s = specific weight of solids (grains), and γ_f = specific weight of fluid in the pores.

Quartz sands have an average specific gravity of 2.65 with reference to water (specific gravity = specific weight of a material at 60°F: (specific weight of water at 60°F)*. If the weight of one cubic foot of fresh water is assumed to be equal to 62.4 lb, then one cubic foot of solid silica having a specific gravity of 2.65 would weigh 165.4 lb. If one cubic foot of dry sand weighs 137.3 lb, then its dry, bulk specific gravity would be equal to 2.2 (= 137.3/62.4). The porosity of this dry sand can be calculated using Eq. 2.8: $\gamma_{db} = (1 - \phi)\gamma_s$ or $\phi = (\gamma_s - \gamma_{db})/\gamma_s = (2.65 - 2.2)/2.65 = 0.17$ or 17%. If the sand was saturated with water, its bulk specific weight would be equal to the weight of the solids (137.3 lb) plus the weight of the water (0.17 × 62.4 lb) or 147.91 lb/ft^3.

It can be easily illustrated on using idealized spheres that porosity is dependent upon the method of packing. If packed cubically, then spheres of equal size would have a maximum possible void space of about 47.6% (Slichter, 1897–1898). If packed rhombohedrally, the porosity is reduced to a minimum of about 26% (Figure 2.3). It is obvious that sphere size does not change porosity when unit volumes with sides at least 2 radii in length are examined. In nature, owing to variation in size of grains and their angularity, usually the porosity of a sand or sandstone will be less than the values specified for spherical grains. It has also been demonstrated that in a mixture of spherical particles having different diameters, stacking arrangement does affect porosity. The introduction of a second set of spheres, small enough to fit in the pore space between the larger set, can reduce porosity to about 13%.

Usually, finer-grained sediments exhibit greater porosity when deposited than coarse-grained ones. A well-sorted, well-rounded, loosely compacted medium to coarse-grained sand may have a porosity of about 37%, whereas poorly–sorted fine-grained sand with irregularly-shaped grains may have a porosity in excess of 50%. An admixture of irregular-shaped, tabular and bladed

---

*(°C × 1.8) + 32 = °F; [(°F + 40)/1.8] − 40 = °C

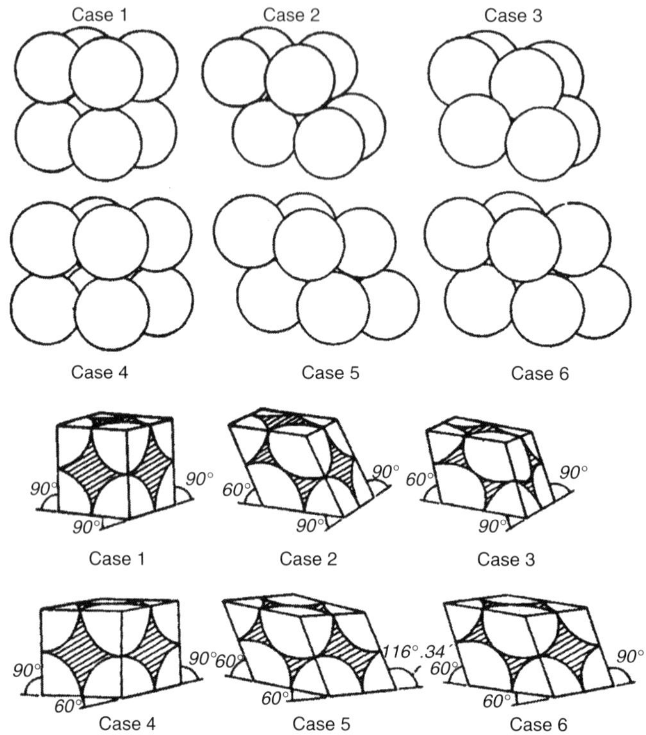

Figure 2.3 Several different ways of packing spheres. (After Graton and Frazer, 1933.)

particles usually gives rise to a higher porosity values because of particle bridging. The wide variability of porosity owing to depositional environment is best illustrated by the greywackes, which may have either a high or a very low porosity value, depending on the amount of fine–grained material filling the pores. Clays and silts may have porosities as high as 50–80% when freshly deposited.

The terms "effective porosity" and "total porosity" are often used in petroleum geology and reservoir engineering studies. These terms differentiate between the interconnected pores through which fluids can move and the total pore space, regardless of its ability to transmit fluid.

In the opinion of the authors, the term "effective" (open, intercommunicating) porosity as used in the U.S. should be abandoned. Instead, effective porosity should be defined as the open porosity minus the irreducible fluid saturation.

2.2.1 Quantitative Evaluation of Porosity in Argillaceous Sediments

The relationship between porosity of shales and clays and their burial depth has been studied by numerous investigators (refer to Rieke and Chilingarian, 1974). Figure 2.4 shows the variation in porosity values with depth from one area to another. This is due to the fact that porosity of argillaceous sediments is a complex function of numerous natural factors, often superimposed on each other (Dzevanshir *et al.*, 1986). These factors include: (1) geologic age; (2) effective stress; (3) lithology; (4) mineralogy; (5) tectonic stresses; (6) depositional rate; (7) thickness of the formations; (8) grain sorting: (9) grain orientation; (10) geothermal temperature; (11) hydrocarbon saturation; (12) amount and type of cementing

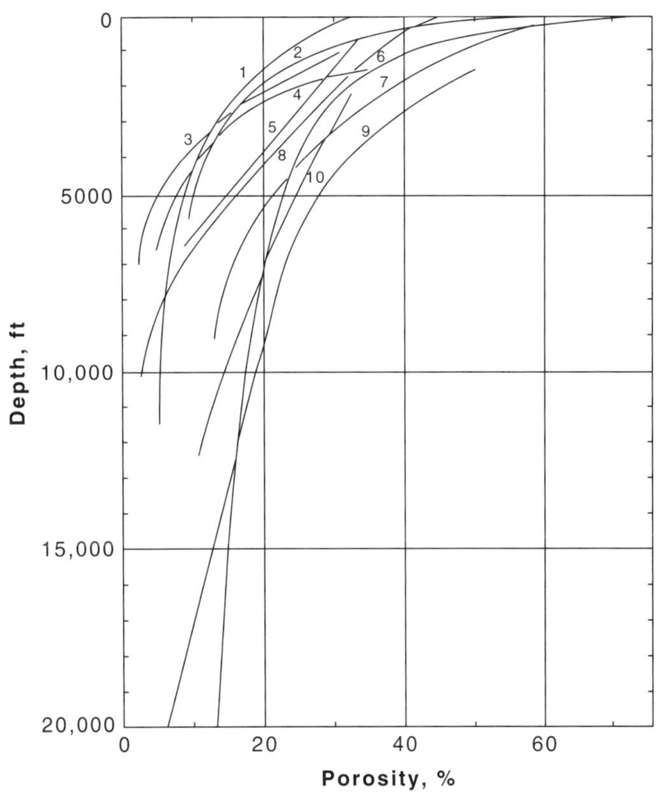

Figure 2.4 Relationship between porosity and depth of burial for shales and argillaceous sediments. 1 = Proshlyakov (1960); 2 = Meade (1966); 3 = Athy (1930); 4 = Hosoi (1963); 5 = Hedberg (1036); 6 = Dickinson (1953); 7 = Magara (1968); 8 = Weller (1959); 9 = Ham (1944); and 10 = Foster and Whalen (1966).

material, and (13) chemistry of the interstitial solutions. The magnitude of the above variables complicates any quantitative assessment of the impact of these individual parameters on the porosity of argillaceous sediments. Dzevanshir *et al.* (1986) proposed one method of solving this problem. The solution is to establish dependence of porosity of argillaceous sediments on the most important factors, such as the geologic age, lithology, and burial depth. The coefficient of irreversible compaction is related to the geologic age and the lithology. These prominent parameters either over-shadow or incorporate the influence of other factors of lesser importance.

The following formulas were derived in order to quantitatively evaluate the role played by various parameters in maintaining porosity in clayey sediments. The coefficient of irreversible compaction, β (MPa^{-1}), was defined by Athy (Buryakovskiy *et al.*, 1986, pp. 54 and 97) as:

$$\phi_D = \phi_s e^{-\beta pe} \tag{2.11}$$

where ϕ_D is the fractional porosity at burial depth, D, in m; ϕ_s = fractional porosity at the surface, and p_e = effective pressure in MPa. Equation 2.11 can be expressed as follows on assuming $\phi_s = 0.4$:

$$\phi_D = 0.4\phi_s e^{-0.014\beta D} \tag{2.12}$$

Figure 2.5 shows a family of straight lines on semilogarithmic paper which represent the coefficient of irreversible compaction. Overprinted upon this family of straight lines are the actual compaction curves of argillaceous rocks. On knowing the coefficient of irreversible compaction for each one of the straight lines (Eq. 2.12), it is possible to determine graphically its average value for actual curves (Table 2.1).

As indicated in Eq. 2.12, with the exception of depth of burial, all other variables are included in the term β (Dzevanshir *et al.*, 1986). Correlation of this coefficient with geologic age and lithology becomes apparent when one compares the curves of different geologic age and lithology obtained by Weller (1959), Vassoevich and Bronovitskiy (1962), Dobrynin (1970), Durmish'yan (1973), and Proshlyakov (1974) with curves corresponding to sediments of the same geological age in Azerbaijan, obtained from areas having different lithologies (Figure 2.5). Scherer (1987) showed that in sedimentary basins having average geothermal gradients (<4°/100 m), the first-order parameters influencing porosity in compacting

Table 2.1 Coefficient of irresponsible compaction (β) of clays. (after Dzevanshir *et al.*, 1986, p. 172, table 1.)

Curve description	$\beta \times 10^{-3}$ (MPa^{-1})
Weller(1959)	58.5
Proshlyakov (1974) and Dobrynin 1970)	42.8
Vassoevich and Bronovitskiy (1962)	33.6
Apsheron Peninsula and Archipelago	42.1
Southwestern part of Apshron peninsula and northern part of Baku Archipelago	27.1
Southern part of Baku Archipelago	19.3

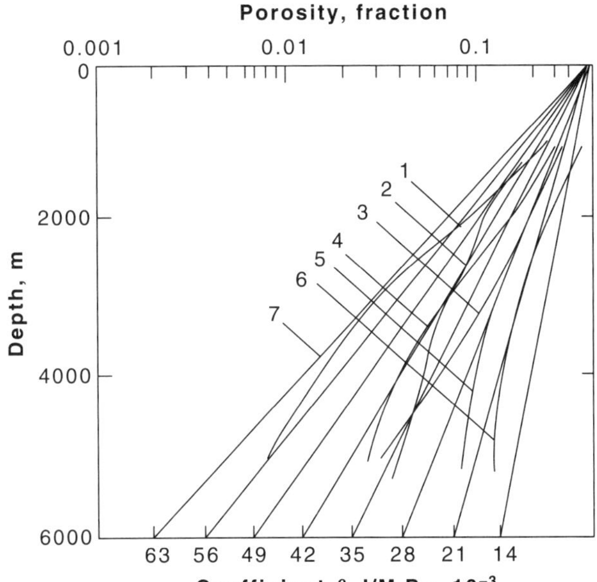

Figure 2.5 Relationship between porosity and depth of burial of clays. Cofficient of irreversible compaction β is shown. 1 = Weller (1959); Aralsorskiy Well SG-1; 3 = Vassoevich and Bronovitskiy (1962); 4 = Apsheron Peninsula and Archipelago, Azerbayjan; 5 = southwestern part of Apsheron and northern part of Baku Archipelago, Azerbayjan; 6 = southern part of Baku Archipelago and Along-Kurinskaya Depression; and 7 = family of curves calculated using Eq. 2.12. (After Dzevanshir *et al.*, 1986, p. 171, fig. 2.)

sandstones are age (time of burial),amount of detrital-quartz content, maximum depth of burial, and sorting. Overpressured sandstones seem to retain about 1.9% porosity for every 1000 psi above hydrostatic pressure.

The coefficient β depends on the duration of sample loading as shown by experimental data (Terzaghi, 1961; Dobrynin, 1970; Rieke and Chilingarian, 1974). In an argillaceous sedimentary sequence with increasing thickness of shale and decreasing number of sandstone reservoir rocks, the shales remain more porous owing to greater difficulty of pore fluid expulsion from the shaley sediments. The following Eq. 2.13 was obtained by Dzevanshir *et al.* (1986) from interrelationships among the coefficient of irreversible compaction, geologic age, and lithology. Porosity at a burial depth D (ϕ_D) can be calculated by using the following equations.

$$\beta = (26.61 \log A - 8.42) \times 10^{-3}$$
$$\beta = (140 - 166.6 \log R) \times 10^{-3}$$
$$\phi_D = \phi_0 \exp[-0.014(13.3 \log A - 83.5 \log R + 2.79) \times 10^{-3} D] \tag{2.13}$$

where ϕ_0 is the initial fractional porosity of clays; A = geologic age in millions of years, and R = ratio of thickness of clays to the total thickness of terrigenous complex. The nomogram presented in Figure 2.6 enables rapid solution of Eq. 2.13. Dzevanshir *et al.* (1986) reported that the absolute error does not exceed 3% and Eq. 2.13 gives practical usable results.

2.3 Permeability

Permeability is the measure of the ability of a porous rock to transmit a fluid under the pressure gradient (differential pressure). The **absolute permeability**, k, is the ability of a rock to conduct a single fluid (gas, oil, or water) at 100% saturation in the rock pore space with that fluid. **Effective/phase permeability** is the ability of a rock to conduct one fluid phase (gas, oil, or water) in the presence of other fluid phases. **Relative permeability** to a fluid is the ratio of effective/phase permeability at a given saturation value to the permeability at 100% saturation (the absolute permeability). The terms $k_{r0}(k_0/k)$, $k_{rg}(k_g/k)$, and $k_{rw}(k_w/k)$ denote the relative permeabilities to oil, gas, and water, respectively. The relative permeability is expressed in percent or as a fraction.

Permeability is measured by an arbitrary unit called the Darcy, D, which is named after Henry d'Arcy, a French engineer, who in 1856 devised a method of measuring the permeability of porous

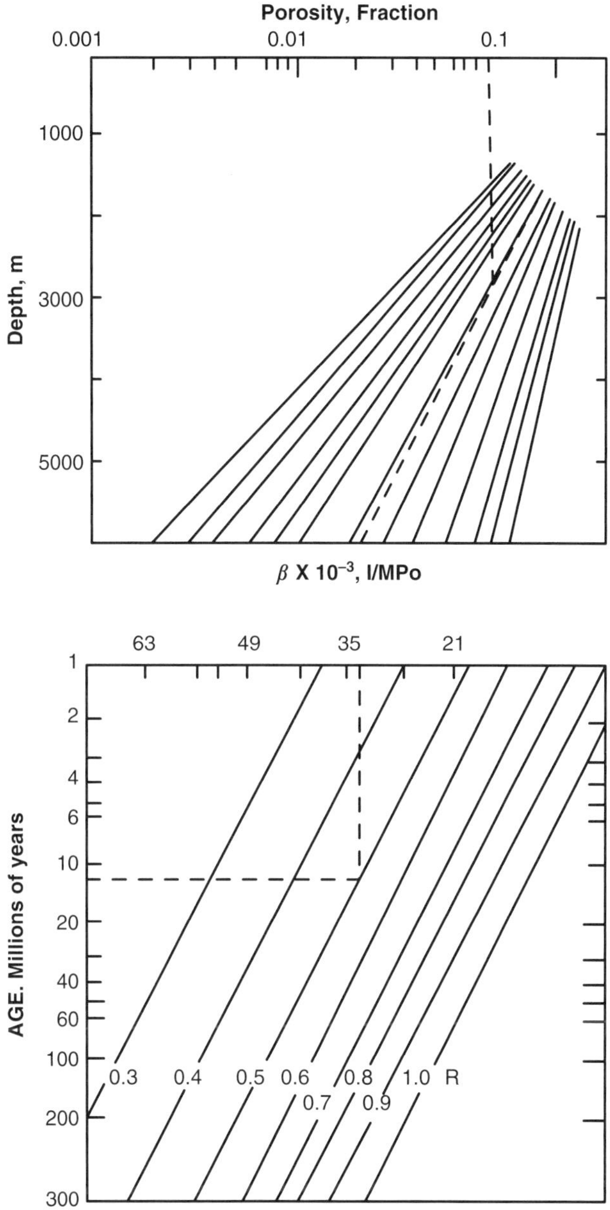

Figure 2.6 Nomogram for determining porosity at a particular depth of burial using geologic age and lithology (ratio of thickness of shales/total thickness of sediments) as controlling factors.

rocks. One Darcy is 1 cm^3 per second of a fluid having viscosity of 1 cP flowing through a 1 cm^3 cross-section of rock under a pressure gradient of 1 atm/cm. Because most reservoir rocks have an average permeability considerably <1 Darcy, the usual measurement units are in millidarcies (mD). Sometimes the term "perm" is used and is equal to 1.127 D.

In the majority of formations there is a simultaneous existence of more than one phase in the pore space (oil and gas; oil and water; gas and water; or oil, gas, and water). The concept of effective permeability implies that all, except for one phase, are immobile. Inasmuch as part of the effective pore space is occupied by another phase, correction factor must be used. The magnitude of effective permeability depends on **wettability,** i.e., on whether (1) the mobile phase does not wet the solid mineral surfaces of the rock and, therefore, occupies the central parts of the pores, or (2) the immobile phase wets the solid surfaces and thus tends to concentrate in the smaller pores. The nature, distribution, and amount of immobile phase affect the effective permeability.

The results of effective/phase permeability measurement are shown graphically in the triangular diagram, where the apex represents the 100% saturation point of the respective phase (oil, gas, water). Contour lines of equal permeability to reservoir fluids are drawn in order to evaluate the test results and characteristics of simultaneous multiphase fluid flow (Figure 2.7). Composition of three-phase flow through the porous media (after Leverett, 1941) is illustrated by Figure 2.8. The diagram of relative permeability to oil and to water versus the oil and water saturation is constructed and data points for the relative phase permeability to oil and water in

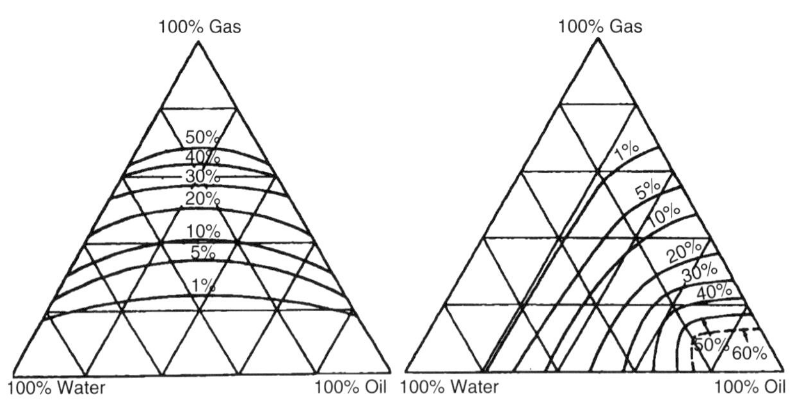

Figure 2.7 Triangle diagram of ternary mixtures (water, oil and gas) with lines of equal relative permeability (in %): (a) to gas (as a function of water and oil saturation); (b) to oil (as a function of gas and water saturation). (After Leverett, 1940; courtesy of AIME.)

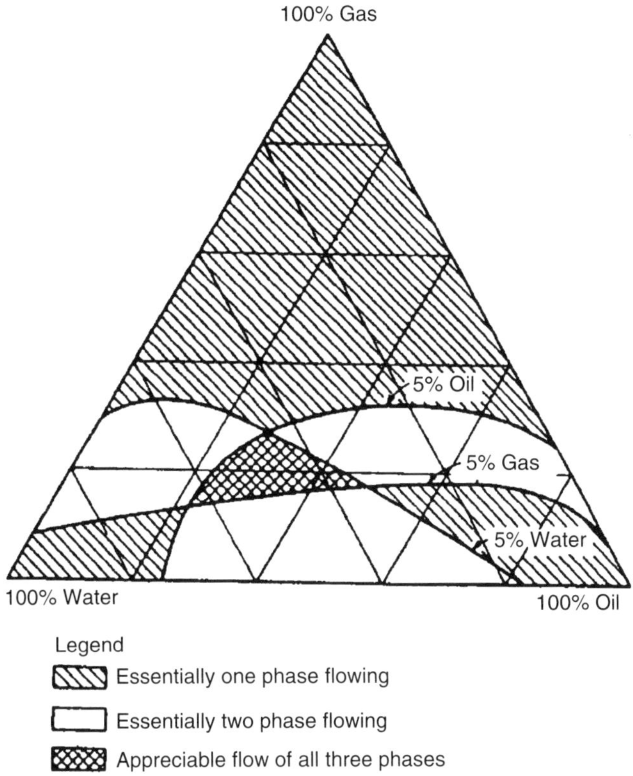

Figure 2.8 Composition of a three-phase flow through the porous media. (After Leverett and Lewis, 1940, courtesy of AIME.)

the presence of gas phase are depicted on the diagram. Examples of curves of relative permeability to oil and to water vs. water saturation are shown in Figures 2.9 to 2.11.

In waterflooding operations or in natural water-drive reservoirs, the relative permeabilities to oil and to water are of great importance. Where water and oil flow together, the relative permeabilities are affected by many factors, which include:

1. relative dispersion of one phase in the other,
2. time of contact with the pore walls,
3. amount of polar substances in the oil,
4. degree of hardness of water,
5. relative amount of carbonate material in porous medium, and
6. temperature.

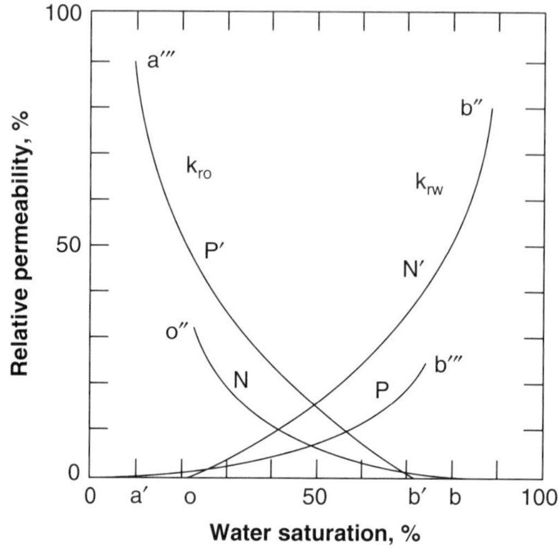

Figure 2.9 Relative permeability curves for polar and nonpolar oil. Curves P and P′ are for polar oil, whereas curves N and N′ are for non-polar oil. (After Babalyan, 1956, p. 145.)

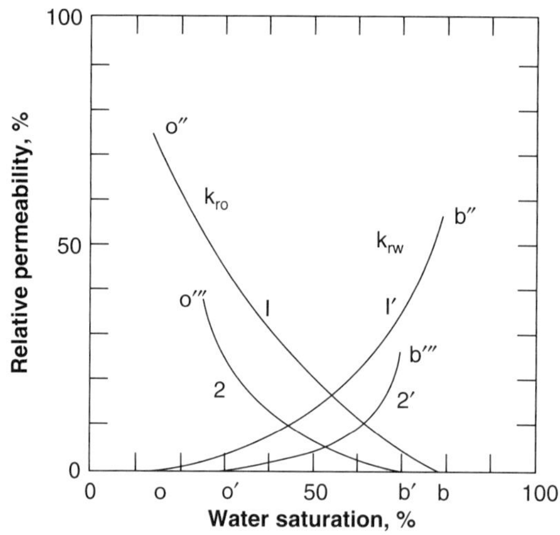

Figure 2.10 Relative permeabilities in oil and to water for polar oil + alkaline water (curves 1 and 1′) and for polar oil + hard water (curves 2 and 2′). (After Babalyan, 1956, p. 148.)

Based on the laboratory experiments by Sinnokrot and Chilingar (1961), the relative permeability to the continuous phase (dispersion medium) is greater than the relative permeability to the

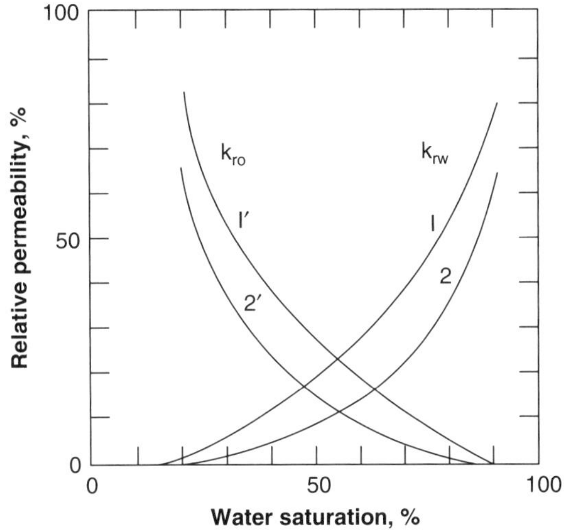

Figure 2.11 Relative permeability curves for non-polar oil with alkaline water (curves 1 and 1′) and for nonpolar oil + hard water (curves 2 and 2′). (After Babalyan, 1956, p. 145.)

discontinuous (dispersed, internal) phase. With increasing degree of dispersion, the relative permeability increases for both continuous and discontinuous phases. The degree of dispersion increases with decreasing interfacial tension and increasing time of coalescence of dispersed-phase droplets.

Sticking (attachments) of the dispersed phase to solid surface depends on (1) interfacial tension, (2) contact angle, (3) time necessary for the coalescence of droplets and lenses of the mobile portion of a dispersed phase, and (4) thickness of dispersion medium (continuous phase) layer attached to the solid surface. The relative permeability of the dispersed phase decreases if its droplets stick to the solid surfaces. The thickness of water film on solid surfaces is decreased in the presence of surface-active substances, which adsorb on the surfaces. In the case of alkaline water, which contains certain amount of salts of organic acids (soaps), the adsorbed layer (film) is thinner than in the case of hard or distilled water.

At low water saturation, water is present as a dispersed phase. The intensity of its transition from a dispersed phase into a dispersion medium (continuous medium) is determined by the coalescence intensity of water droplets and intensity of their sticking (attachment) to solid surfaces. The water saturation at which water changes from a dispersed phase into a continuous phase decreases with decreasing time of coalescence and sticking of water droplets to solid surfaces.

With increasing concentration of polar substances in the oil, the cumulative water production decreases. The change in oil production rate upon increasing the concentration of polar substance in oil is quite rapid initially, and then it slows down and eventually stabilizes when the polarity of oil reaches a certain limit. As shown in Figure 2.9, the critical saturation for water decreases (point a moves to a') and oil saturation increases (point b moves to b') with increasing concentration of polar substances in the oil. With decreasing concentration of polar substance in oil, the relative permeability to water sharply increases (point b' "moves to b''"), whereas that to oil decreases (point a' "moves to a''"). This is due to the fact that attraction of nonpolar oil to solid surfaces is negligible and that the mobile oil presents less resistance to flow of water than does the immobile oil.

According to Babalyan (1956), in the case of polar oil and water, the oil production is greater when the water is alkaline than when it is hard. This is due to the change in the critical saturation of both phases (Figure 2.10). As shown in Figure 2.10, the relative permeability curves of oil + alkaline water lie above those of oil + hard water, because the following is true in the case of alkaline water: (1) low interfacial tension between oil and water; (2) low values of contact angle; (3) slow coalescence of oil droplets in water; and (4) greater degree of dispersion of oil in water. The intensity of the transformation of oil into a dispersed phase is greater in the alkaline than in the hard waters.

In the case of nonpolar oil, the attachment of oil to solid surface is negligible in the presence of both alkaline and hard waters. When nonpolar oil flows with either alkaline or hard water, there is no change in critical saturations and hence the recovery of oil and water is the same in each case. The relative permeability curves with alkaline water, however, lie above those with hard water (Figure 2.11).

Some of the controversial aspects of the above discussion will be resolved by the future research. Honarpour *et al.* (1992) give details on the relative permeability of carbonate reservoirs.

In general, the permeability of unconsolidated sands commonly ranges from 1 to 6 *D*, whereas that in lightly-cemented consolidated, samples ranges from 0.2 to 3 D. Indurated, well-cemented rocks have a permeability range of <0.001 to 1.0 D; the permeability of a fracture or a vug can be very high.

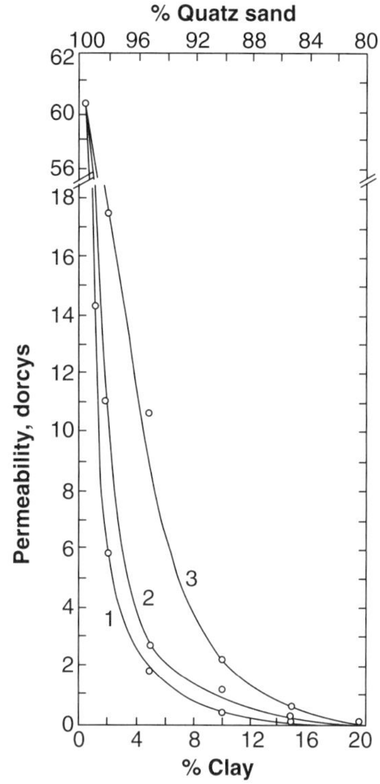

Figure 2.12 Effect of clay content on permeability of 0.35 to 0.5 mm sand fraction. 1 – smectite (montmorillonite); 2 – polymictic clay; 3 – kaolinite. (After Tsvetkova, 1954.)

Tsvetkova (1954) clearly showed that the presence of clay minerals (especially those that swell when contacted by fresh water) greatly reduces the permeability of sandstones (Figure 2.12). This is especially true for smectite clays where 2% of the clay in coarse-grained sand lowers the permeability 10-fold, whereas the presence of 5% of the clay lowers the permeability 30-fold. Sandstones are practically impermeable if they contain 6–9% smectite clay; on the other hand, sandstones containing as much as 15% of the low-swelling kaolinite clay can still remain permeable.

2.3.1 Porosity/Permeability Relationship

A general quantitative relationship between permeability and porosity is hard to achieve because the relationship depends on several independent parameters: (1) the interconnection of the pores; a rock may

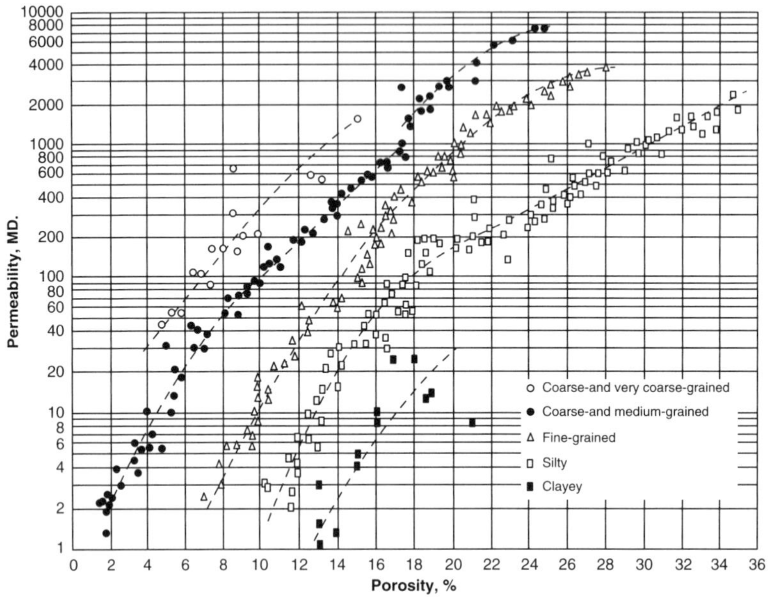

Figure 2.13 Relationship between porosity and permeability of various types of sandstones. (After Chilingar, 1964, Fig. 2, in Chilingarian and Wolf, 1975, p. 33, Fig. 1-25.) Cores contained irreducible water saturation.

be porous and exhibit little or no permeability if the pores are not interconnected, such as in the case of pumice (pyroclastic rock having unconnected pores) and in carbonate formations containing isolated vugs; (2) grain-size distribution, e.g., a rock having a narrow grain-size distribution will have a greater porosity than a rock that contains fine particles mixed with larger particles; (3) pore-size distribution; and (4) tortuosity (the flow path in the rock). Empirical relationships between permeability and porosity can be developed for specific sediments and rocks, such as those shown in Figure 2.13 (Chilingar, 1964).

2.3.1.1 *Effect of Bedding Orientation on Permeability and Porosity*

Geologists and petroleum engineers are well aware that vertical permeabilities, k_v, in carbonate reservoirs commonly exceed horizontal permeabilities, k_h, especially in the case of carbonate reefs. Such relationships in carbonates can be attributed to either acidic ground water having a dissolving effect or by compaction-derived low pH fluids moving vertically. Both cases can create solution channels, vugs, caverns, and enlarged pre-existing fractures. In contrast,

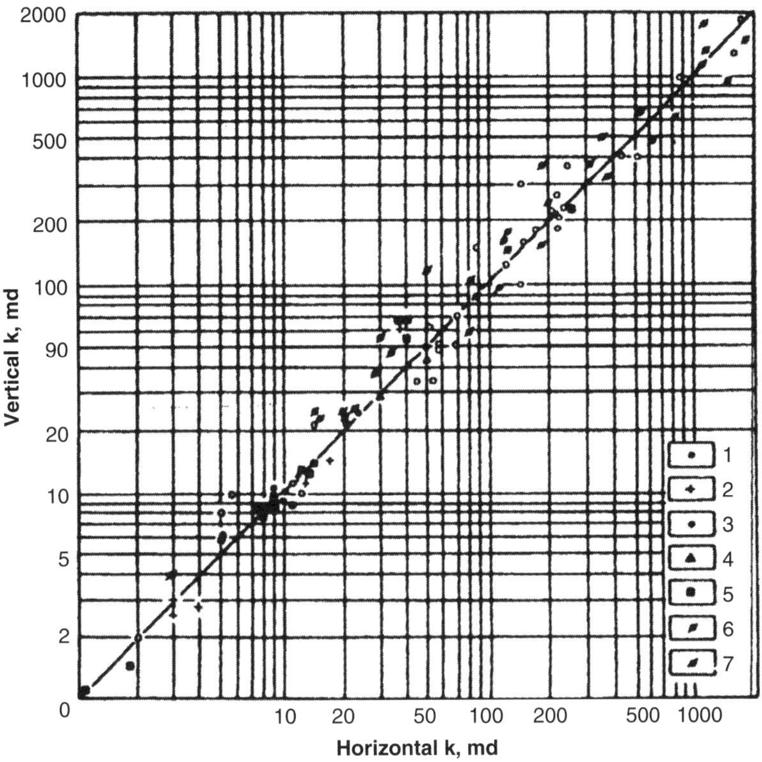

Figure 2.14 Relationship between horizontal permeability and vertical permeability in dolomites with intergranular porosity. (1) Kuybyshev Along-Volga Region; (2) Saman-Tepe; (3) Uchkyr; (4) Adam-Tash; (5) Orenburg; (6) Vuktyl; (7) Urta-Bulak. (After Bagrintseva, 1977, p. 45.)

horizontal permeability is generally greater than the vertical permeability in sandstone reservoirs. In dolomites having intergranular porosity, vertical permeabilities commonly are nearly equal to horizontal permeabilities (Figure 2.14). This may not be the case for fractured carbonate reservoirs (Figure 2.15).

2.3.1.2 *Effective and Absolute Permeability Relationships in Carbonate Rocks*

The relationship between absolute permeability, k_a, and effective permeability, k_e, to gas is presented in Figures 2.16 and 2.17. The effective permeability to gas was measured on cores containing irreducible water saturation, S_w. This is obviously more meaningful than measuring the permeability of dry cores.

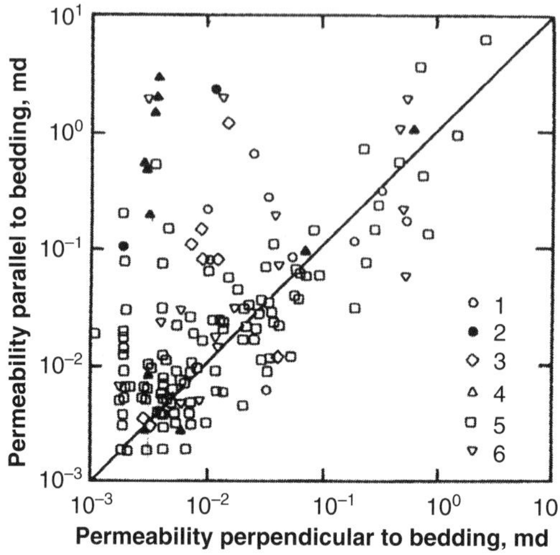

Figure 2.15 Relationship between permeability parallel to the bedding and permeably perpendicular to the bedding in fractured carbonate reservoir rocks. 1 – Kuybyshev Along-Volga Region; 2 – Volgograd Along-Volga Region; 3 – Saman-Tepe; 4 – Adam-Tash; 5 – Orenburg; 6 – Vuktyl. (After Bagrintseva, 1977, p. 47, Fig. 3.)

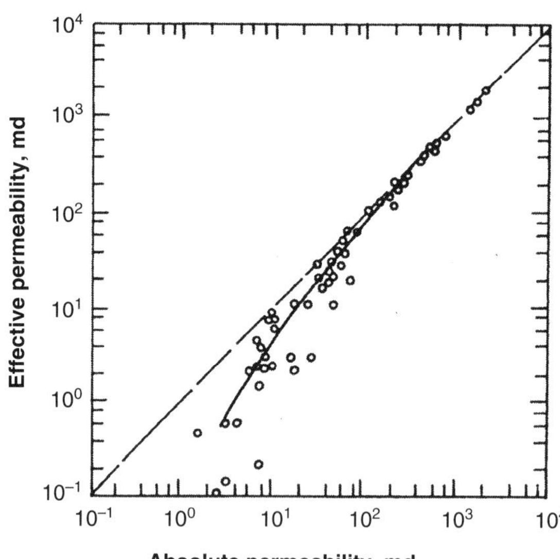

Figure 2.16 Relationship between absolute permeability, k_a, and effective permeability, k_e, to gas in the presence of irreducible fluid saturation for porous carbonate reservoir rocks of Kuybyshev Along-Volga Region, Russia. 1 – empirical curve of effective permeability to gas. (After Bagrintseva, 1977, p. 130, Fig. 43.)

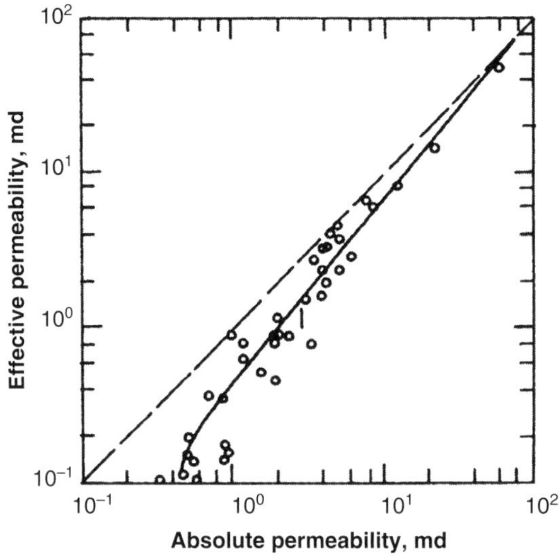

Figure 2.17 Relationship between absolute permeability, k_a, and effective permeability, k_e, to gas in the presence of irreducible fluid saturation for porous carbonate reservoir rocks of the Orenburg deposit in Russia. Intergranular type of porosity: 1 – empirical curve of effective permeability to gas, $k_e = k_a(0.566 + 0.143\log k_a)$. (After Bagrintseva, 1977, p.110, Fig. 34.)

2.3.1.3 *Permeability and Residual Water Saturation Relationships in Some Carbonate Rocks*

The relationship between permeability and residual water saturation can be expressed by the following formulas:

$$S_{wr} = 65.25 - 9.49\ \ln(k_a + 1) \quad \text{with } \sigma = 10.6 \tag{2.14}$$

$$\ln S_{wr} = 4.39 - 0.33\ \ln(k_a + 1) \quad \text{with } \sigma = 0.367 \tag{2.15}$$

$$\ln(k_a + 1) = 5.765 - 0.073 S_{wr} \quad \text{with } \sigma = 0.929 \tag{2.16}$$

Number of samples tested $n = 654$, and the correlation coefficient, $R = 0.83 \pm 0.08$. The relationship between irreducible fluid saturation and permeability of Kuybyshev Along-Volga carbonate rocks of Russia is given in Figure 2.18.

2.3.1.4 *Relationship Between Median Pore Diameter and Permeability of Carbonates*

As an example, for carbonates rocks of Kuybyshev along Volga Region, Russia, the median diameter of pores can be related to permeability (absolute, k_a, and effective, k_e) and to residual water saturation as follows (Figure 2.19):

$$\ln(k_a + 1) = 1.217 \pm 1.265 \text{ mD} \quad \sigma = 1.052; \; R = 0.73 \pm 0.18; \; n = 55 \quad (2.17)$$

$$\ln(k_e + 1) = 0.657 + 1.421 \text{ mD} \quad \sigma = 1.143; \; R = 0.74 \pm 0.13; \; n = 52 \quad (2.18)$$

$$S_{wr} = 49.23 - 11.55 \text{ mD} \quad \sigma = 7.8; \; R = 0.79 \pm 0.10; \; n = 55 \quad (2.19)$$

$$\ln(k_e + 2) = 0.238 + 1.318 \text{ mD} \quad \sigma = 1.411; \; R = 0.60 \pm 0.17; \; n = 58 \quad (2.20)$$

$$S_{wi} = 50.1 - 12.21 \text{ mD} \quad \sigma = 15.5; \; R = 0.54 \pm 0.18; \; n = 64 \quad (2.21)$$

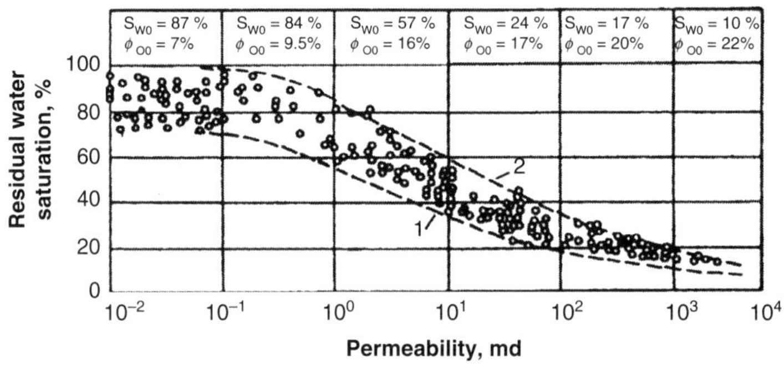

Figure 2.18 Relationship between residual water saturation, S_{wr}, and permeability, k, for carbonate rocks of Kuybyshev Along-Volga deposits, Russia. Average values for residual water saturation, S_{wra}, and average open porosity (effective porosity as used in the United States), ϕ_{0a}, are presented. (1) Lower limit of residual water content; (2) upper limit of residual water content. (After Bagrintseva, 1977, p. 129, Fig. 42.)

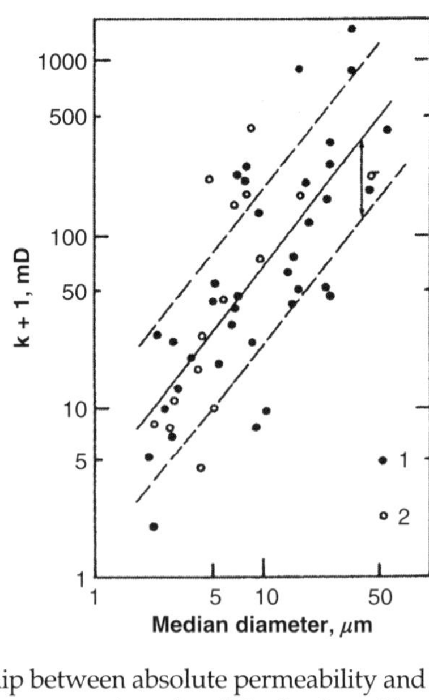

Figure 2.19 Relationship between absolute permeability and median pore size in μm ($\sigma = 1.025$). 1 – Carbonate reservoir rocks of Kuybyshev, Along-Volga Region, Russia; 2 – carbonate rocks of Central Asia. (After Bagrintseva, 1977, p. 134, Figure 46.)

2.4 Specific Surface Area

Specific surface area, s_b, is defined as the surface area of the pores per unit of bulk volume, and is dependent upon the number, shape, size, and length of the pore channels. Specific surface area may be determined by (1) laboratory analysis, (2) a theoretical approach relating surface area to porosity and permeability, or (3) a statistical method such as the one proposed by Chalkley *et al.* (1949). Another method of expressing surface area is to relate it to the pore volume rather than to the bulk volume. In this case, the notation, s_p, is used.

2.4.1 Derivation of Theoretical Equation Relating Porosity, Permeability, and Surface Area

In a reservoir modeled by a bundle of capillary tubes, the rate of flow, q, is given by the Hagen-Poiseuille equation:

$$q = N\pi r^4 \Delta p / 8\mu L \tag{2.22}$$

where q = volumetric flow rate in cm^3/s; N = the number of capillaries; r = capillary radius, cm; Δp = differential pressure across the capillaries, dynes/cm^2; μ = the fluid viscosity, Poise; and L = length of the capillaries, cm.

The Darcy equation for the rate of flow, q, is:

$$q = kA\Delta p / (\mu L) \tag{2.23}$$

where q = volumetric rate of flow, cm^3/s; k = permeability, D; A = total cross-sectional area, cm^2; Δp = differential pressure, atm; μ = fluid viscosity, cP; and L = length of the flow path, cm.

If, instead, viscosity in expressed in Poises and differential pressure in dynes/cm^2, then:

$$q = 9.869 \times 10^{-9} k\Delta p / (\mu L) \tag{2.24}$$

The porosity, ϕ_c, of this bundle of capillary tubes may be expressed as the capillary volume, V_c, per unit of bulk volume, V_b:

$$\phi_c = V_c / V_b = N\pi r^2 L / (AL) = N\pi r^2 / A \tag{2.25}$$

Thus, the total cross-sectional area, A, of the bundle of tubes is:

$$A = N\pi r^2 / \phi_c \tag{2.26}$$

The average capillary tube radius, r, maybe found by combining Eqs. 2.22, 2.23, and 2.26:

$$r = 2(2k/\phi_c)^{0.5} \quad (2.27)$$

The surface area per unit of pore volume, s_p, is given by:

$$s_p = N2\pi rL/N\pi r^2L = 2/r \quad (2.28)$$

On substituting the value of the capillary tube radius from Eq. 2.27 into Eq. 2.28, the surface area can be expressed as:

$$s_p = (\phi_c/2k)^{0.5} \quad (2.29)$$

Solving Eq. 2.29 for permeability yields:

$$k = (\phi_c/2s_p^{\ 2}) \quad (2.30)$$

Inasmuch as a porous rock is more complex than a bundle of capillary tubes, a constant, K_{ef} is introduced. Thus, the equation for permeability becomes:

$$k = \phi_c/K_{ef}s_p^{\ 2} \quad (2.31)$$

Equation 2.31 is the familiar Kozeny-Carman equation (1937). Carman (1937) has noted that the constant, K_{ef}, is actually a complex combination of two variables, *i.e*, a shape factor for the pores, s_{hf}, and a tortuosity factor, τ:

$$K_{ef} = (s_{hf})\tau \quad (2.32)$$

Tortuosity is equal to the square of the ratio of the effective length, L_e, to the length parallel to the overall direction of flow of the pore channels, L:

$$\tau = (L_e/L) \quad (2.33)$$

Thus, the Kozeny-Carman constant, K_{ef}, is a function of both the shape of each particular pore tube and the orientation of the pore tube relative to the overall direction of fluid flow.

Several theoretical relationships between tortuosity and porosity have been developed for simplified models, two of which are presented below. They are not, however, applicable to more complex media.

$$\tau = (F\phi)^2 \text{(after Wyllie and Rose, 1950)} \tag{2.34}$$
$$\tau = F\phi \quad \text{(after Cornell and Katz, 1953)} \tag{2.35}$$

where F is the formation resistivity factor which is expressed as:

$$F = R_0/R_w \tag{2.36}$$

In the above equation R_0 is the electrical resistivity of a formation 100% saturated with formation water and, R_w, is the formation water resistivity. The formation factor embodies the effect of grain size, grain shape, grain distribution, and grain packing. The effect of the porosity of formation is thus included in the formation factor.

On measuring the values for L_e and L by sonic transit time theory or by determining the tortuosity from a pore distribution concept, using a capillary pressure curve (Faris *et al.*, 1954), one can determine the value of tortuosity. Studies by Winsauer *et al.* (1952) on natural cores yielded the following equation:

$$\tau = (F\phi)^{1.2} \tag{2.37}$$

This equation has been confirmed by the independent work of several authors (Faris *et al.*, 1954; Powell, 1940).

Wyllie and Spangler (1952) determined the pore shape factor, s_{hf}, in the case of unconsolidated sphere and bead packs. They indicated that the s_{hf} "constant" ranges from 2.13 to 3.32. The basic assumption of these studies is that the length of the collective pore tubes varies considerably, but the cross-sections of this series of pore tubes are all of the same shape. For a medium of uniform grains and pores (such as sphere and bead packs) this is probably fairly accurate, but for a nonuniform medium it is not true. With the data collected by Wyllie and Spangler (1952), but using a value for tortuosity of $(F\phi)^{1.2}$, rather than $(F\phi)^2$, the calculations of s_{hf} were found not to give a constant value at all, but a variable which increases with increasing K_{ef}. Wyllie and Spangler (1952) calculated the values for K_c after measuring the porosity, permeability, and surface area by using the pindrop technique. Utilizing Eq.2.37 and data from Wyllie and Spangler (1952) and Wyllie and Gregory (1955), one can obtain a straight line on log-log paper having an equation of (Chilingar *et al.*, 1963):

$$s_{hf} = 1.55 K_c^{\,0.455} \tag{2.38}$$

Because s_{hf} and K_{ef} are interrelated by the tortuosity, the expression for the shape factor, and the overall Kozeny factor can be expressed in terms of easily obtained parameters (Chilingar *et al.*, 1963):

$$s_{hf} = 2.24 F\phi \tag{2.39}$$

and

$$K_c = 2.24 (F\phi)^{2.2}. \tag{2.40}$$

The final expression for the surface area per unit of pore volume in cm^2/cm^3 is, therefore:

$$s_p = 2.11 \times 10^5 / (F^{2.2}\phi^{1.2}k)^{0.5} \tag{2.41}$$

where the permeability, k, is expressed in millidarcys (Chilingar *et al.*, 1963).

Kotyakhov developed the following formula for the surface area per unit of bulk volume, s_b, (cm^2/cm^3 of sands (Eremenko, 1960; Kotyakhov, 1949).

$$s_b = 7000 (\phi^3/k)^{0.5} \tag{2.42}$$

where ϕ is the fractional porosity and k is the permeability expressed in Darcys. This equation is another version of the Kozeny-Carman equation:

$$k = 10^8 \phi^3 / (s_{hf})(\tau)(s_b)^2 \tag{2.43}$$

For a consolidated rock ($s_{hf} = 2$) and assuming that $\tau = 1$, which, in effect, means that the rock sample is equivalent to a bundle of capillary tubes, one obtains:

$$k = 10^8 \phi^3 / (2)(1)(s_b)^2 \qquad \text{or} \qquad s_b \approx 7000 (\phi^3/k)^{0.5}$$

It is clear that Kotyakhov's (1949) equation is an oversimplification for rocks. If values of $\tau = 1.25$ and $s_{hf} = 2.5$, representative of unconsolidated sands or calcarenites, are inserted in Eq. 2.43 the surface area may be expressed as :

$$s_b = 5650(\phi^3/k)^{0.5} \tag{2.44}$$

where s_b = surface area in cm ^{2} per unit of bulk volume (cm^3); ϕ = fractional porosity; and k = permeability in darcys.

Equation 2.44 should give a good approximation of surface area for unconsolidated sediments, but should not be used for consolidated rocks.

Pirson (1958) proposed the following formula for determining the surface area of carbonate rocks:

$$k = 10^8/2Fs_p^{\,2} \tag{2.45}$$

or

$$s_p = 10^4(1/2Fk)^{0.5} \tag{2.46}$$

where k = Klinkenberg permeability, D; s_p = surface area per unit of pore volume, cm^2/cm^3; and F = formation resistivity factor (dimensionless), which is related to the porosity by the following equation:

$$F = \phi^{-m} \tag{2.47}$$

where the cementation factor m varies from 1.3 for unconsolidated sands and oolitic limestones to 2.2 for dense limestones.

Shirkovskiy (1969) proposed the following formula for determining the specific surface area (surface area/unit bulk volume), s_b, of fragmental rocks without interstitial water in cm^2/cm^3:

$$s_b = \phi^{1.5/}(2k^{0.5}\tau^{0.5})^{0.5} \tag{2.48}$$

where ϕ = porosity, fractional; k = permeability, perm (1 D = 1.02 × 10^{-8} perm); and τ = tortuosity. The specific surface area in cm^2/cm^3 with correction for interstitial water is equal to:

$$s_{bw} = s_b(1 - S_w)^{0.75} \tag{2.49}$$

where S_w = interstitial water saturation, fraction.

Permeability and porosity measurements may be used in the Carman-Kozeny equation to determine the surface area of unconsolidated sands:

$$s_g = 31.8\rho^{-1} (\phi/\kappa k)^{0.5} \tag{2.50}$$

where s_g = specific surface area per unit weight, m^2/g: ρ^{-1} = density, g/cm^3; ϕ = porosity; κ = adjustable textural factor; and k = permeability, D.

An adjustable textural factor, κ, is equal to 5.0 for unconsolidated sands, as suggested by Carman (1937), Willey and Rose (1950), and Donaldson *et al.* (1974). The textural factor is related to the tortuosity of fluid flow path. The basic form of the equation was developed by Kozeny (1927) and critically reviewed and extended by Carman (1937) who conducted numerous experiments to verify the validity of the equation. Willey and Rose (1950) extended the equations to applications for the petroleum industry and Donaldson *et al.*, (1974) proved its validity for unconsolidated sands by comparing the results to measurements made by gas adsorption. In addition, they presented experimental data showing that the specific surface areas of the Kozeny-Carman equation ranged from 26 to 43. Thus, although the equation is accurate for unconsolidated samples, it cannot be used for consolidated porous media. The Carman-Kozeny equation provides a measure of the external surface of the solid particles contacted by the fluid moving through the porous medium. On the other hand, the gas adsorption method yields a measure of the total external areas and surface contributed by dead-end pores that hold but do not transmit fluid.

The specific surface area per unit of grain volume of rock, s_{gn}, maybe expressed through grain-size distribution. There is a simple relation between s_p and s_{gv}:

$$s_{gn} = s_p\phi(1-\phi) \tag{2.51}$$

Buryakovsky and Samedov (1961) have proposed the following formula for the specific surface area per unit of grain volume of rock, s_{gv}, consisting of equal-size spheres:

$$s_{gv} = A/V_b = N\pi d^2(1-\phi)V_{sph} = 6N\pi d^2(1-\phi)/N\pi d^3 = 6(1-\phi)/d \tag{2.52}$$

where A is the surface area, cm^2; V_b is the bulk volume of rock, cm^3; V_{sph} is the volume of spheres, cm^3; N is the number of spheres per unit volume; d is the diameter of a single sphere, cm; and ϕ is the fractional porosity.

Inasmuch as the specific surface area is influenced by the clay content, C_{clay}, the following empirical equation was developed by Buryakovsky *et al.* (1990a; 2001):

$$s_{gv} = 75(1 - \phi)C_{clay} + 532(C_{clay} + 7.1)(1 - \phi) \quad (2.53)$$

Assuming that $\phi_{ave} = 0.25$, the correlation between s_{gv} and C_{clay} can be simplified as follows:

$$s_{gv} = 56.3C_{clay} + 400. \quad (2.54)$$

Numerous experimental studies were conducted using core samples from the South Caspian oil and gas fields (Buryakovsky, 1985b). The statistical formula, similar to the Kozeny-Carman equation, was proposed for these cores in the following form:

$$k = \alpha(\phi^3/s_{gv}), \quad (2.55)$$

where a is the coefficient accounting for the influence of grain-size distribution and tortuosity of pore channels (somewhat similar to the Kozeny constant, K_c.). Its numerical value varies for different oilfields and formations. If porosity is expressed in percent, the permeability in mD, and specific surface area per unit of grain volume in cm^2/cm^3, the average value of coefficient α is equal to 64×10^6, based on the analyses of 517 cores.

2.4.2 Relationship Between Specific Surface Area (Area Per Unit of Pore Volume) and Permeability of Carbonate Rocks

The relationship between specific surface area, s_p, and permeability of carbonate rocks from various oilfields in Russia is presented in Figure 2.20.

2.4.3 Relationship Between Specific Surface Area and Residual Water Saturation of Carbonate Rocks

The relationship between the specific surface area, s_p, and residual water saturation from various oilfields in Russia is presented in Figures 2.21–2.23.

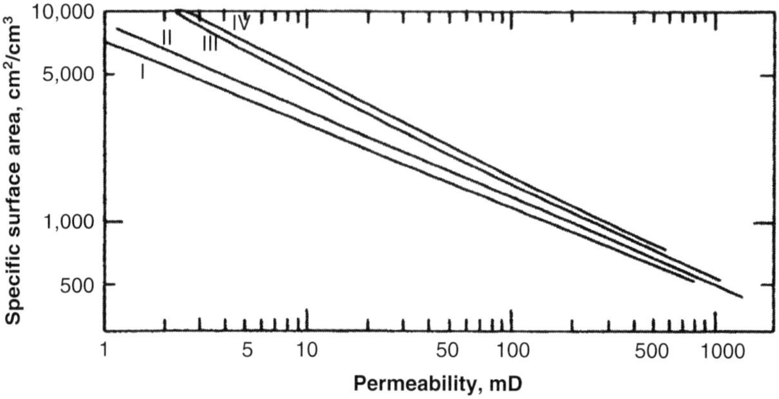

Figure 2.20 Relationship between the specific surface area and permeability of carbonate rocks from various areas in Russia. I – Vuktyl'skiy field; II – Orenburg field; III – Kuybyshev Along-Volga Region; and IV – Uchkyr area. (After Bagrintseva, 1977, p. 194, Fig. 85.)

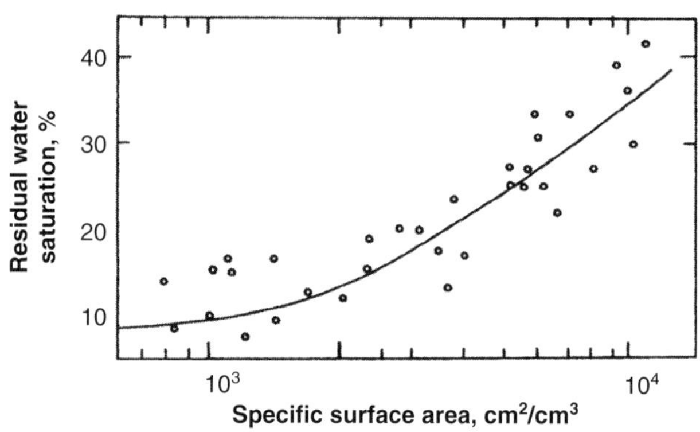

Figure 2.21 Relationship between residual water saturation and specific surface area for carbonate rocks of Orenburg Field, Russia. (After Bagrintseva, 1977, p. 107, Fig. 31.)

2.5 Interrelationship Among Prorosity, Permeability, and Specific Surface Area

Until 1990, there was no agreement among petroleum engineers and geologists whether or not there is a correlation between porosity and permeability. And how can there be a correlation between porosity and permeability in a micro-fractured dolomite reservoir with inter-rhombohedral porosity, for example. Micro-fractures could account

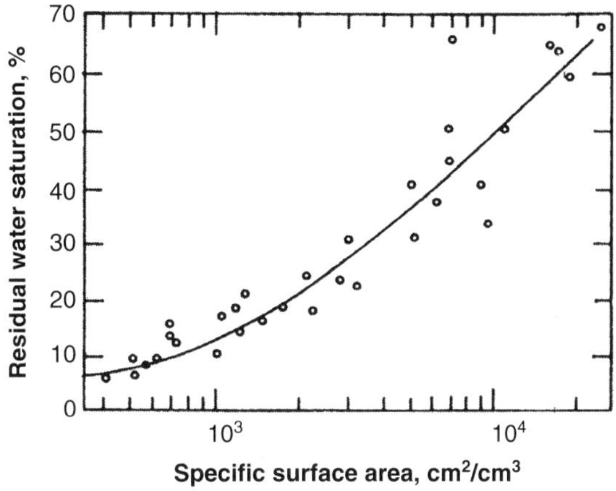

Figure 2.22 Relationship between the residual water saturation and specific surface area for carbonate rocks from Kuybyshev Along-Volga Region, Russia. (After Bagrintseva. 1977, p. 132, Fig. 44.)

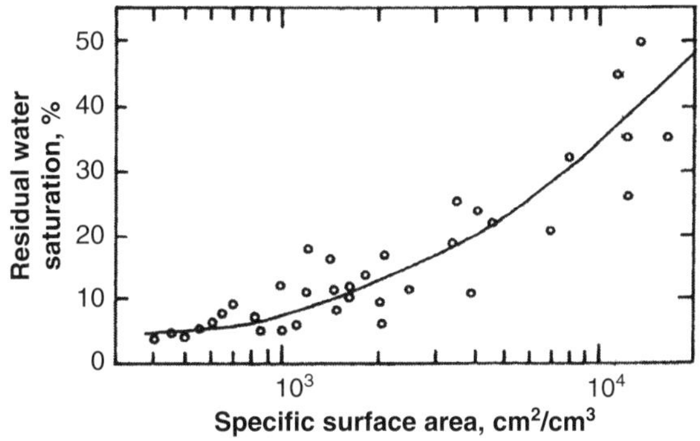

Figure 2.23 Relationship between residual water saturation and specific surface area of carbonate rocks of Vuktyl'skiy Field, Russia. (After Bagrintseva, 1977, p. 151, Fig. 58.)

for only ≈1% porosity, whereas they increase permeability considerably. In vugular carbonates, vugs and fractures could account tor approximately 3–4% ϕ.

In 1990, Chilingar *et al.* solved this problem by introducing two additional parameters, i.e., irreducible fluid saturation (which has practically no effect on fluid flow) and specific surface area (a measure of the degree of fracturing present in the rock). A few examples for carbonate rocks are presented below.

2.5.1 Vuktyl'skiy Gas-Condensate Field, Russia (Table 2.2)

Lithology varies from dolomites (porous, porous-cavernous, finely-cavernous, crystalline, slightly calcareous, crystalline dense, etc,) to true limestones (dolomitized, with very five caverns, and micro-grained). The correlation equation between the open porosity (intercommunicating, or "effective" porosity as used in USA) and permeability is:

$$\log k = 0.9532 - 2.7880 \times 10^{-2} S_{wr} - 5.5597 \times 10^{-4} s_{s} + 1.3309 \times 10^{-1}\phi + 1.1707 \times 10^{-5} S_{wr} \times s_{s} \quad (2.56)$$
$$(R = 0.997)$$

2.5.2 Central Asia (Table 2.3)

The effective (intercommunicating) porosity term is used in the USA. The authors call this porosity "open" porosity. Effective porosity as

Table 2.2 Petrophysical properties of carbonate rocks of Vuktyl'skiy gas condensate field, Russia. (After Bagrintseva, 1977, table 37, p.151.)

S_{wr} (%)	s_S (cm^2cm^{-3})	ϕ (%)	$S_{wr} s_S$	$S_{wr}\phi$	$s_S \phi$	k (mD)
16	2156	7	34496	112	15092	3.6
28	7070	8	197960	224	56560	0.4
19	3878	8	73682	152	31024	1.6
25	2058	10	51450	250	20580	11.5
12	1827	12	21924	144	21924	26
12	1113	12	13356	144	13356	76
9	1421	13	12789	117	18473	58
9	1428	13	12852	117	18564	63
28	2030	13	56840	364	26390	28
5	945	14	4725	70	13230	138
9	3668	14	33012	126	51352	9.5
7	644	14	4508	98	9016	294
3	854	15	2562	45	12810	208
16	2142	15	34272	240	32130	36
8	1001	15	8008	120	15015	167
4	532	18	2128	72	9576	1011
4	441	20	1764	80	8820	1910

Table 2.3 Petrophysical properties of carbonate reservoirs (both limestones and dolomites of Central Asia, FSU. (After Bagrintseva, 1977, table 46, p. 177.)

S_{wr} (%)	s_S (cm^2cm^{-3})	ϕ (%)	$S_{wr}s_S$	$S_{wr}\phi$	$s_S\phi$	k (mD)
32	3465	13	110880	416	45045	9.6
16	658	14	10528	224	9212	302
10	231	14	2310	140	3234	2400
25	2324	15	58100	375	34860	29
30	1750	15	52500	450	26250	55
30	3311	16	99330	480	52976	18
25	1169	16	29225	400	18704	151
63	13860	17	873180	1071	235620	1.3
22	1316	17	28952	374	22372	129
58	12880	19	747040	1102	244720	2.1
43	6650	19	285950	817	126350	7.4
37	8358	20	309246	740	167160	5.6
46	6300	22	289800	1012	138600	14
27	3542	22	95634	594	77924	44
23	1848	22	42504	506	40656	150
22	840	23	1840	506	19320	883

defined by Chilingarian (1964) is the porosity of a core containing irreducible fluid. The latter is present in minute pores (some dead-end) and fractures has very little influence on the flow. Therefore, the correlation between the open porosity, ϕ_0, and permeability, k, is not as good as the correlation between our "effective" porosity, ϕ_e, and k.

The correlation equation between the open porosity and permeability for the Central Asia, Former Soviet Union is as follows:

$$\log k = 3.8690 - 1.0536 \times 10^{-1}S_{wr} - 4.1979 \times 10^{-4}s_s + 6.5363 \times 10^{-6}S_{wr}s_s + 2.8324S_{wr}\phi \quad (2.57)$$
$$(R = 0.985)$$

2.5.3 Kuybyshev, Along-Volga Region, Russia (Table 2.4)

The correlation between the open porosity and permeability for Kuybyshev, Along Volga Region carbonate reservoirs is:

$$\log k = 2.1085 - 5.0777 \times 10^{-2}S_{wr} - 4.3785 \times 10^{-4}s_s + 7.9959 \times 10^{-2}\phi + 7.6326 \times 10^{-6}S_{wr}s_s \quad (2.58)$$
$$(R = 0.988)$$

Table 2.4 Petrophysical properties of carbonate reservoirs of the Kuybyshev, Along-Volga Region, Russia. (After Bagrintseva, 1977, table 32, p.127.) Lithology is mainly limestone.

S_{wr} (%)	s_S (cm^2cm^{-3})	ϕ (%)	$S_{wr}s_S$	$S_{wr}\phi$	$s_S\phi$	k (mD)
58	7875	10.1	456750	585.8	79537.5	0.8
7	427	13	2989	91	5551	577
22	2016	14.8	44352	325.6	29836.8	39
21	1890	17.5	39690	367.5	33075	74
16	1330	17.6	21280	281.6	23408	151
41	5285	18.1	216685	742.1	95658.5	10.4
52	10850	19.4	564200	1008.8	210490	3
15	1309	19.4	19635	291	25394.6	209
16	1253	19.7	20048	315.2	24684.1	239
17	700	20.5	11900	348.5	14350	860
42	9520	21	399840	882	199920	5
10	518	21	5180	210	10878	1671
52	7210	23.4	374920	1216.8	168714	11.9
32	5432	23.7	173824	758.4	128738.4	22
18	2310	23.7	41580	426.6	54747	123
14	749	24	10486	336	17976	1207

2.5.4 Orenburg Field, Russia (Table 2.5)

The correlation between the open porosity and permeability of the Orenburg Field, Russia, is as follows:

$$\log k = 3.4351 - 2.0442 \times 10^{-1} S_{wr} + 9.5086 \times 10^{-6} S_{wr} \times s_s + 8.0217 \times 10^{-3} S_{wr}\phi - 2.3892 \times 10^{-5} s_s\phi \quad (2.59)$$

$$(R = 9810)$$

A good example of the influence of lithology on the relationship between porosity and permeability is presented in Figure 2.24 for the Cretaceous Edwards Limestone, which exhibits intercrystalline porosity. Craze (1950) noted that as the texture changes from micro-granular to coarse-grained, the permeability increases for a given porosity. Another example of porosity – permeability relationship for carbonate rocks is presented in Figure 2.25.

Figures 2.26 and 2.27 show the relationship between porosity and permeability of the Bausteinschickten Sandstone. If this relationship

Table 2.5 Petrophysical properties of carbonate reservoirs of Orenburg Field, Russia. (After Bagrintseva, 1977, table 24, p. 107.)

S_{wr} (%)	s_S (cm^{2}cm^{-3})	ϕ (%)	$S_{wr} s_S$	$S_{wr} \phi$	$s_S \phi$	k (mD)
20	2632	9.2	52640	184	24214.4	5.5
42	10164	10.3	426888	432.6	104689.2	0.52
31	5810	11	180110	341	63910	1.93
19	2471	12	46949	228	29652	14
33	7000	12.4	231000	409.2	86800	1.89
39	9296	13	362544	507	120848	1.24
15	2212	13	33180	195	28756	22
27	5551	13.2	149877	356.4	73273.2	3.65
9	770	13.4	6930	120.6	10318	198
25	2866	13.5	71650	337.5	38691	3.4
16	1183	13.8	18928	220.8	16325.4	92
25	2285	14.5	57125	362.5	33132.5	5.34
17	3451	15.7	58667	266.9	54180.7	16
20	2261	17	45220	340	38437	39
10	616	17	6160	170	10472	590
18	1274	18.2	22932	327.6	23186.8	181
12	1050	20	12600	240	21000	258

were due to an increased cementation, the points should lie on the smooth curve of equal specific surface area or even beyond the lower left part of both figures. The downward increasing deviation of the data points from the curve indicates that the specific surface area increass with decreasing porosity as a result of the sediments becoming finer grained and more argillaceous. There is an increase in the calcareous matrix content with decreasing grain size, so that the net result is a lowering of porosity.

2.5.4.1 Classification of Fragmental (Or Calcarenites and Calcisiltites) Reservoir Rocks

The relationship among rock granulometric composition, porosity, and permeability of sandstones and siltstones is presented in Table 2.6.

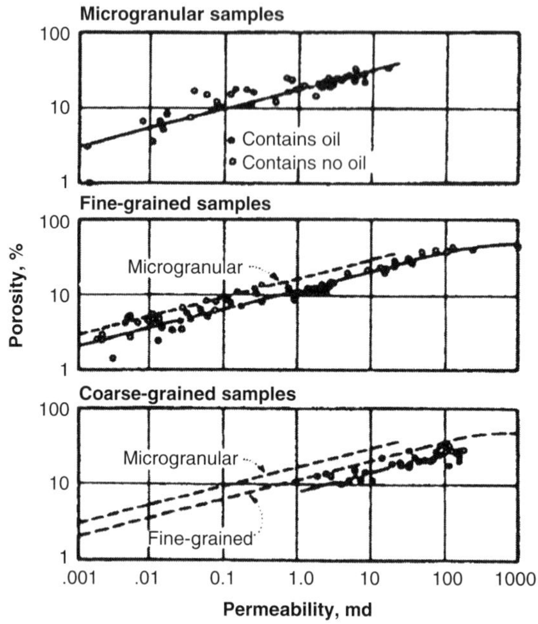

Figure 2.24 Relationship between porosity and permeability for various textural types of Cretaceous Edwards Limestone, USA. (Craze, 1950. courtesy of AIME.)

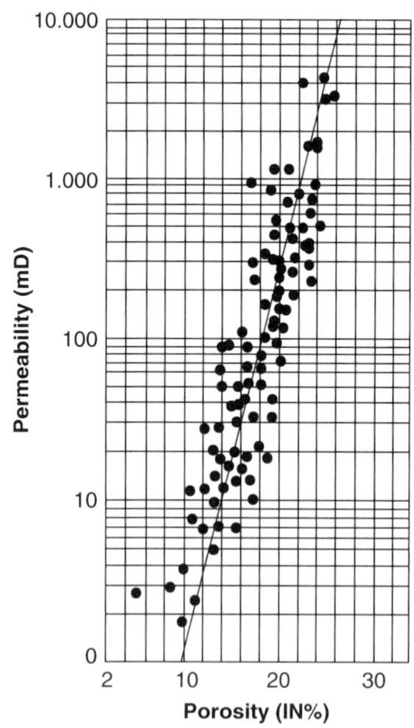

Figure 2.25 Porosity – permeability correlation for Cazaux, Albian Field. (From J.P. Dupuis, G. Oswald, and J. Sens. In: Monicard, 1980, Fig. 2.10.51, p. 88.)

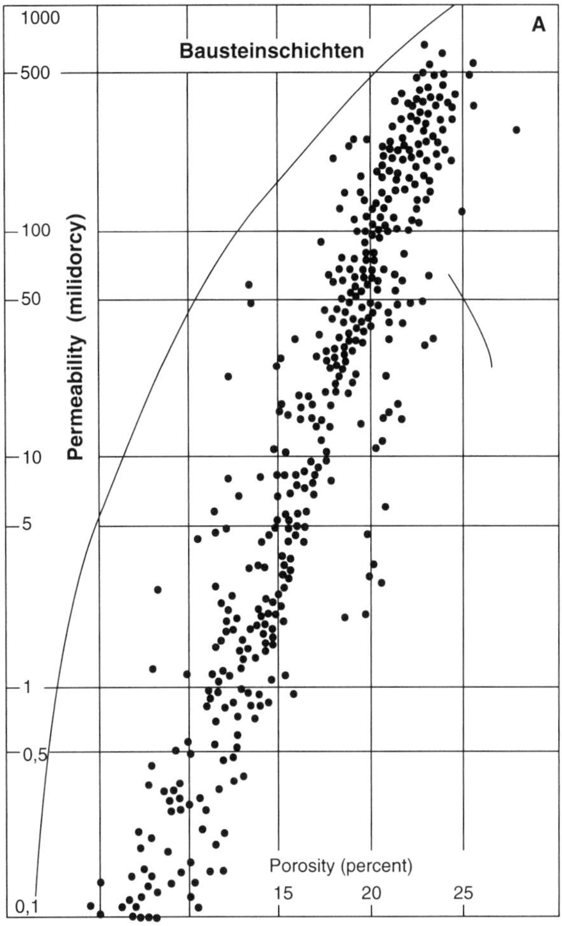

Figure 2.26 Relationship between porosity and permeability of the Bausteinschichten Sandstone [calcareous (calcite + dolomite) clayey arenites of the Tertiary Molasse]. Solid-line curve is for constant specific surface area. (After Füchtbauer, 1967a, Fig. 9, p. 360; courtesy 7th World Pet. Congr.)

2.5.4.2 *Correlation of Porosity or Permeability With Clay and Carbonate Cement Contents*

In reservoir simulations, of great importance is the correlation of porosity or permeability with clay (argillaceous) cement content, C_{clay}, and calcareous cement content, C_{carb}. According to Buryakovsky (1985b) and Buryakovsky *et al.* (1990a; 2001) the relation between

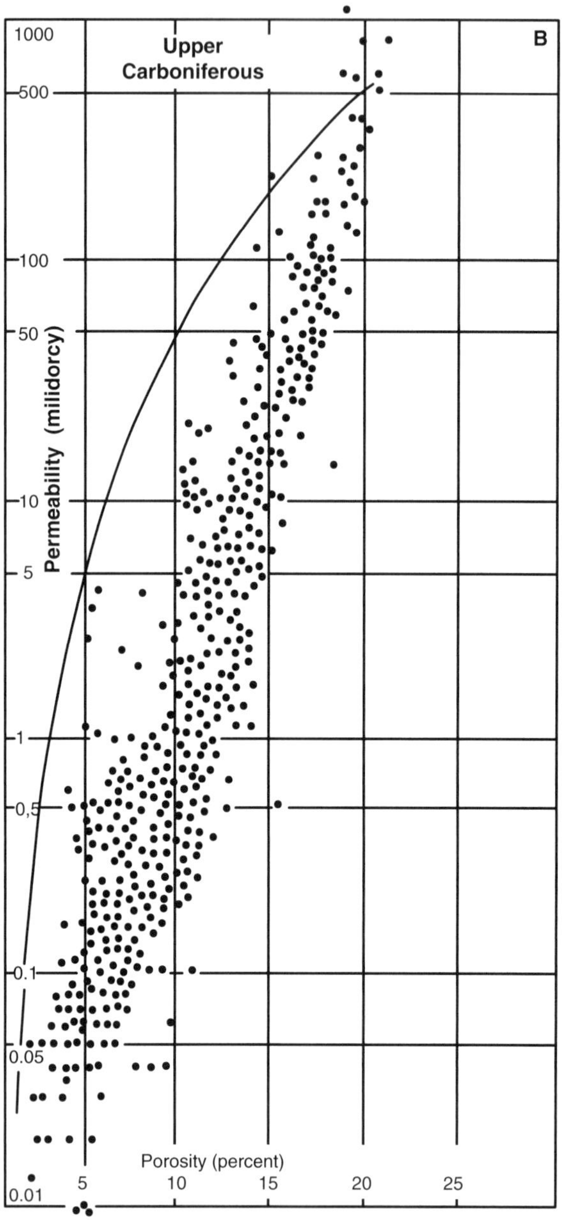

Figure 2.27 Relationship between porosity and permeability of the Upper Carboniferous sandstones from the Emsland region, Germany. Solid line is for a constant specific surface area. (After Füchtbauer, 1967a, Fig. 10, p. 361; courtesy 7th World Pet. Congr.)

Table 2.6 Classification of fragmental reservoir rocks. (Modified after Khanin, 1969, p. 234.)

Class	Name of rock and granulometric composition	Effective porosity,%	Absolute permeability to gas md	Charactersitic of rock on the basis of porosity and permeability
I	Calcarenite, medium-grinned(0.25–0.50 mm) Calcarenite, fine-grained (0.1–0.25 mm) Calcisiltite, coarse-grained (0.05–0.1 mm)	≥17 ≥20 ≥24 ≥29	≥1000	Very good
II	Calcarenite, medium-grained Calcarenite, find-grained Calcisiltite, coarse-grinned Calcisiltite, fine-grnined	15–17 18–20 22–24 27–29	500–1000	Good
III	Calcarenite, medium-grainerd Calcarenite, fine-grained Calcisiltite, coarse-grnined Calcisiltite, fine-grained	11–15 14–18 17–22 21–27	100–500	Medium
IV	Calcarenite, mediun-grinned Calcarenite, fine-grained Calcisiltite, coarse-grained Calcisiltite, fine-grained	6–11 8–14 10–17 12–21	10–100	Medium low

Table 2.6 Classification of fragmental reservoir rocks. (Modified after Khanin, 1969, p. 234.) *(continued)*

Class	Name of rock and granulometric composition	Effective porosity,%	Absolute permeability to gas md	Charactersitic of rock on the basis of porosity and permeability
V	Calcarenite, medium-grained Calcarenite, filne-grained Calcisilitite, coarse-grained Calcisiltite, fine-grained	0.5–6 2–8 3–10 4–12	1–10	Poor
VI	Calcarenite, medium-grained Calcarenite, fine-grained Calcisiltite, coarse-grained Calcisiltite, fine-grained	<0.5 <2 <3.3 <3.6	<1	Very poor

the porosity (or permeability) and the clay content (or carbonate content) can be described by the following exponential equation:

$$Y = ae^{-bX} \tag{2.60}$$

where Y is the dependent variable that maybe substituted by porosity ϕ or permeability k, $\times$ is the independent variable that can be substituted by clay content (C_{clay}) or carbonate content (C_{carb}), and a and b are empirical coefficients that are derived from a statistical analysis of available core data. For example, they were calculated for different fields of the Absheron Archipelago in Azerbaijan (Buryakovsky, 1985b).

More sophisticated statistical relationships were obtained for combined influence of clay and carbonate contents on the porosity and permeability:

$$\phi = (59.7 - 0.54C_{clay} + 0.0075 (C_{clay^2})\exp[-(0.00047C_{clay} + 0.055C_{carb}], \tag{2.61}$$

and

$$k = (6700 - 130C_{clay})\exp[-(0.000172C_{clay^2} - 0.001172C_{clay} - 0.1)C_{carb}]. \tag{2.62}$$

If C_{clay} and C_{carb} are expressed in %, then ϕ is expressed in % and k is in mD.

Buryakovsky (1985b) developed relations between permeability and porosity taking into account:

1. the clay content:

$$k = 9C_{clay}{}^{-2.35}\phi^3, \tag{2.63}$$

2. the carbonate content:

$$k = 0.0128C_{arb}\exp(0.283\phi); \tag{2.64}$$

3. or both:

$$\log k = 3.935 + 0.020\phi - 0.075C_{arb} - 0.050C_{clay} \tag{2.65}$$

If ϕ, $C_{clay,}$ and C_{carb} values are expressed in %, then k is in mD.

In the oil and gas reserve estimation and field development planning it is important to know the irreducible water saturation of the hydrocarbon-producing formation. In order to calculate the water saturation, one has to estimate the thickness of the water film covering grains (pore walls). The average thickness of the water film can be determined from the following equation (Buryakovsky, 1985b):

$$\tau_w = S_w\phi/s_g \tag{2.66}$$

where τ_w is the thickness of water film on pore and channel walls, microns; S_w is the water saturation, in percent; ϕ is the porosity, in percent; and s_g is the specific surface area per unit of grain volume, in cm^2/cm^3.

Thus, Buryakovsky (1985b) obtained the relation between the water saturation and permeability:

$$S_w = [\tau_w/(1-\phi)(\phi/K_{cf}k)^{0.5} \tag{2.67}$$

This relationship can be presented in a simpler format:

$$S_w = \tau_w (a/k)^{0.5} \tag{2.68}$$

Buryakovsky presented additional petrophysical relationships (Buryakovsky 1997, 1985b, and Buryakovsky *et al.*, 1982, 1990a).

2.6 Wettability – Capillarity

The wetability of rocks and its measurement were discussed in detail in the classical books of Tiab and Donaldson (2004) and Donaldson and Alam (2008). Wettability may be defined as the ability of a liquid to wet, or spread over, a solid surface, Figure. 2.28A shows a liquid wetting a solid. On the other hand, Figure 2.28B shows the relationship between the liquid and solid when the liquid has little affinity for the solid. In Figure 2.28C the liquid drop occupies an intermediate position. The fluid, which wets the surface more strongly, occupies the smaller pores and minute interstices.

2.6.1 Interfacial Tension and Contact Angle

The angle, which the interface makes with the solid, is called the contact angle, θ. Usually, θ is measured from the solid through the liquid phase (if the other phase is a gas) and through the water phase if oil and water are both present. In a capillary tube, shown in Figure 2.29A, the angle between the side of the tube and the tangent to the curved surface (where it intersects the side of the tube) is less than 90°. For a capillary depression, shown in Figure 2.29B, the contact angle is greater than 90°. In the case of no rise or depression, the angle is 90° (Figure 2.29C).

Interfacial tension, σ, is caused by the molecular property (intermolecular cohesive forces) of liquids. It has the dimensions of force

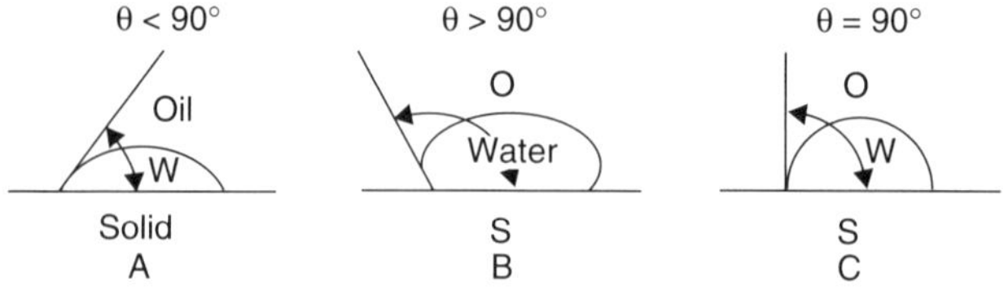

Figure 2.28 Different degree of wetting of solid by a liquid.

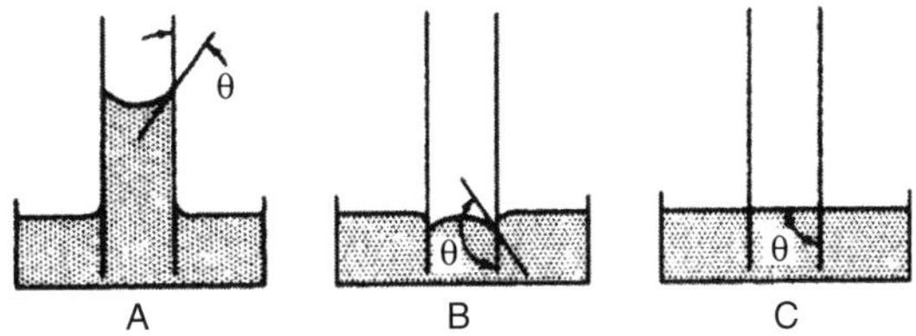

Figure 2.29 Behavior of various fluids in glass capillary tubes. A – water; B – mercury; and C – tretrahydronaphthalene (when glass is perfectly clean and liquid is pure).

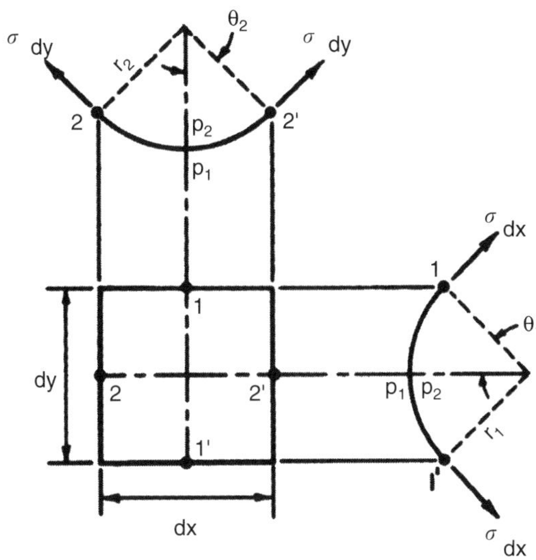

Figure 2.30 Surface tension forces acting on a small element on the surface having double curvature ($p_2 = p_1 + \gamma h$). (Binder, 1962; Vennard, 1961.)

per unit length (lb/ft or dynes/cm), or energy per unit area (ergs/cm²). On considering an element of surface having a double curvature (r_1 and r_2), the sum of the force components normal to the element is equal to zero (Figure 2.30). The interfacial tension forces balance the pressure difference, $p_2 - p_1$:

$$(p_2 - p_1)\mathrm{d}y\mathrm{d}x = 2\sigma \mathrm{d}y \sin\theta_2 + 2\sigma \mathrm{d}x \sin\theta_1 \tag{2.69}$$

If the contact angles θ_1 and θ_2 are small, the following simplifications maybe made:

$$\sin\theta_1 = \mathrm{d}y/2r_1 \tag{2.70}$$

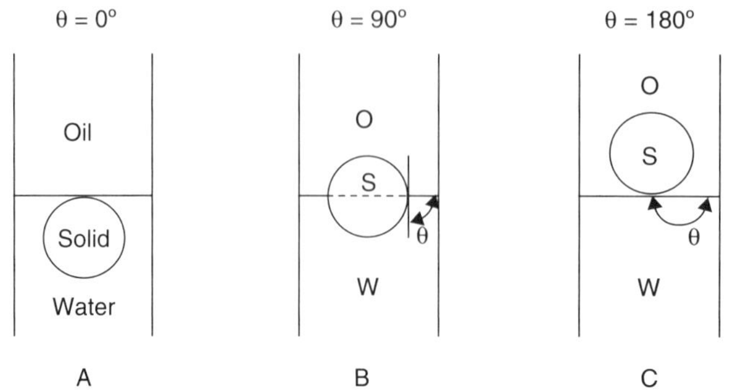

Figure 2.32 Illustrations of 0°, 90°, and 180° contact angles.

If a rock is completely water wet ($\theta = 0$), water will try to envelope all the mineral grains and force all of the oil out into the middle of the pore channel. Even though some of the oil may still be trapped in this case, the recovery would be high. However, if all the pore surfaces were completely oil wet ($\theta = 180°$), oil will try to envelop all the grains and force all the pore water out into the center of the pore channel. In this extreme case, oil recovery would be very low by water drive. Some oil-wet reservoirs are known to exist (Oklahoma City Field, OK, is a good example).

In the usual case ($0° < \theta < 180°$), to improve waterflooding recovery of oil the contact angle θ should be changed from >90° to <90°, through the use of surfactants. A surfactant would move the oil from the surface of the grains out into the center of the pore channels.

Contaminants or impurities may exist in either fluid phase and/or may be adsorbed on the solid surface. Even if present in minute quantities, they can and do change the contact angle from the value measured for the pure system (Marsden, 1968).

2.6.1.1 *Contact Angle and Interfacial Tension Effects on The Movement of Oil*

In an ideal system composed of pure liquids, the advancing contact angle should equal the receding angle. Because of the presence of impurities within the liquids, the advancing contact angle is greater in most systems. The advancing contact angle is the angle formed at the phase boundary when oil is displaced by water. This can be measured as follows: A crystal plate is covered by oil and then the water drop is advanced on it. The contact angle is the limiting angle with time after equilibrium has been established (Figure 2.33). The contact angle formed when the oil displaces the water is the receding angle

(Figure 2.34). The contact angles during the movement of water–oil interface in a cylindrical capillary having a hydrophilic surface is shown in Figure 2.35.

Inasmuch as a hydrocarbon reservoir is basically a complex system of interconnected capillaries of various sizes and shapes, knowledge on how the fluid flows through the capillaries is very important. In Figure 2.36, a conceptual model of a two-branched capillary system is presented. If a pressure drop is applied, the water will flow more readily through the large-diameter capillary than it will through the small-diameter capillary. Thus, a certain volume of oil may be trapped in the small capillary, when the water reaches the upstream fork. Poiseuille's law states that

$$q = (\pi d^4 \, \Delta p_t)/128\mu L \tag{2.84}$$

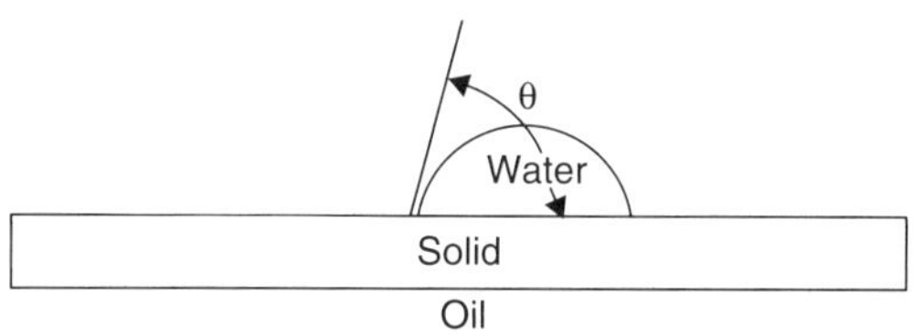

Figure 2.33 Contact angle: plate first immersed in oil followed by the placement of water drop on top.

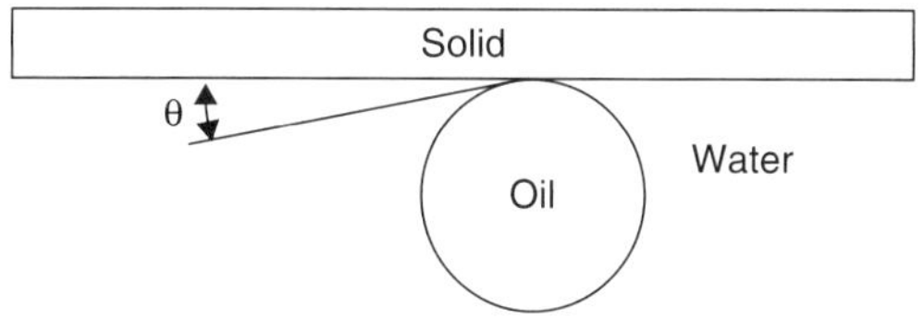

Figure 2.34 Contact angle: plate first immersed in water followed by placing a drop of oil underneath.

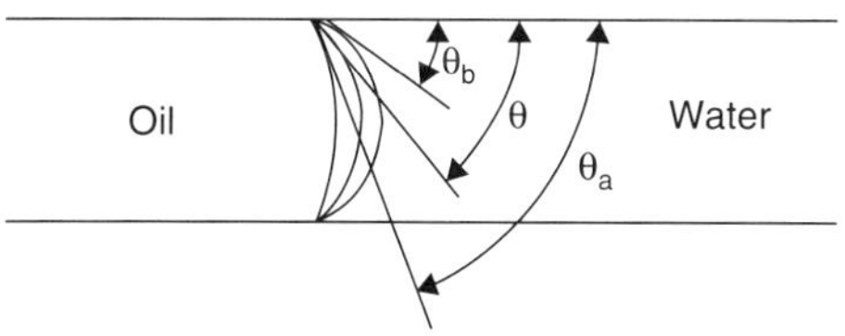

Figure 2.35 Changes in contact angle as a result of movement of water–oil interface: θ = contact angle at static position; θ_a = contact angle when oil is displaced by water (advancing angle); θ_b = contact angle when water is displaced by oil (receding angle).

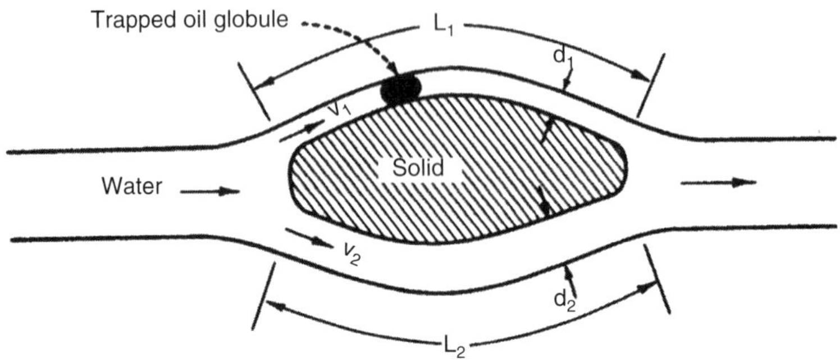

Figure 2.36 Schematic of flow through a two-branch capillary and trapping of oil in a small-diameter capillary.

and

$$v = q/A = (d^2 \, \Delta p_t)/32\mu L \tag{2.85}$$

where v = velocity, cm/s; q = volumetric flow rate, cm^3/s; A = cross-sectional area, cm^2; d = capillary diameter, cm; Δp_t = the total pressure drop in the system, dynes/cm^2; μ = fluid viscosity, cP; and L = flow path length, cm.

The capillary pressure, P_c is equal to:

$$P_c = (4\sigma \cos \theta)/d \tag{2.86}$$

where σ = interfacial tension between oil and water, dynes/cm; d = diameter of the capillary, cm; and θ = contact angle.

The total pressure drop, Δp_t is equal to:

$$\Delta p_t = \Delta p_i + P_c \tag{2.87}$$

where Δp_i is the applied pressure, dynes/cm^2. Solving for v in each capillary by combining Eqs. 2.85, 2.86, and 2.87 results in

$$v_1 = d_1^2/32\mu_1 L_1[\Delta p_i + \{4\sigma \cos \theta\}/d_1] \tag{2.88}$$

and

$$v_2 = d_2^2/32\mu_2 L_2 \, [\Delta p_i + \{4\sigma \cos \theta\}/d_2] \tag{2.89}$$

Setting $L_1 = L_2$ and $\mu_1 = \mu_2$ and dividing Eq. 2.88 by Eq. 2.88 results in the following relationship:

$$v_1/v_2 = (d_1^2 \Delta p_i + 4\sigma \cos \theta \, d_1)/(d_2^2 \Delta p_i + 4\sigma \cos \theta \, d_2) \tag{2.90}$$

Therefore, when $\Delta p_i >> P_c$,

$$v_1/v_2 \approx d_1^2/d_2^2 \tag{2.91}$$

and when $\Delta p_i << P_c$,

$$v_1/v_2 \approx d_1/d_2 \tag{2.92}$$

As shown in Figure 2.37, the sum of forces acting on the trapped oil globule maybe expressed as:

$$\Sigma F = F_1 + F_2 - F_3 \tag{2.93}$$

where:

$\Sigma F = 0.25\pi d^2 \Delta p_t$
$F_1 = 0.25\pi d^2 \Delta p_i$
$F_2 = \pi d_{(}(\sigma_a \cos \theta_a)$
$F_3 = \pi d_{(}(\sigma_b \cos \theta_b)$
and

$$\Delta p_t = \Delta p_i + (4\sigma_a \cos \theta_a/d) - (4\sigma_b \cos \theta_b/d) \tag{2.94}$$

Because the receding angle is usually less than the advancing angle, the capillary pressure not only does not help but instead hinders the flow. The term $(4\sigma_b \cos \theta_b/d)$ is usually greater than $(4\sigma_a \cos \theta_a/d)$ because $\theta_a < \theta_b$. If a surfactant were added at the left to reduce σ_a, then Δp_t would become lower and the oil globule may eventually move to the left in Figure 2.37 when Δp_t becomes negative. The quantity of the trapped oil is dependent upon the value of $\sigma \cos \theta$ at each end of the globule as well as upon Δp_i (imposed pressure drop).

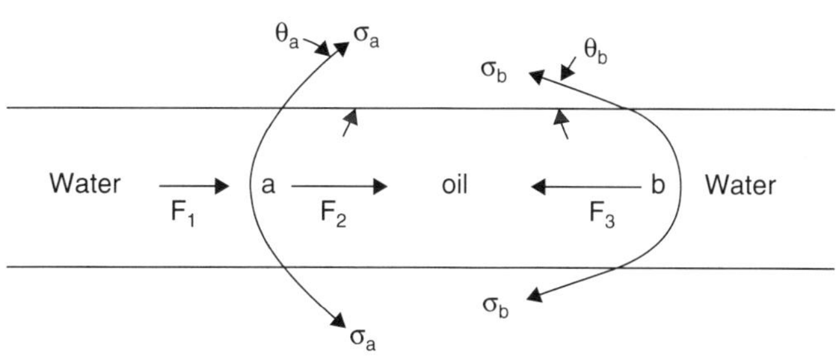

Figure 2.37 Forces acting on a trapped oil globule in a capillary.

Inasmuch as the contact angle depends upon the interfacial tensions, which in turn maybe influenced by surfactants, these chemicals may alter recovery by altering both the contact angle and interfacial tension. As the oil is displaced by water, which wets the rock surfaces, capillary pressure is the driving force. If on the other hand, water does not wet the rock surface, capillary pressure is a retarding force, which must be overcome.

The magnitude of capillary pressure in the pores having a radius of around 15 microns is not large and, therefore, capillary pressure is not an important force during the movement of the oil-water contact, providing there is no mixing. The movement of oil and water in a reservoir, however, results in the formation of water–oil and gas–water-oil mixtures (Muravyov *et al.*, 1958). The amount of gas coming out of solution during migration is greater with increasing amount of dissolved active substances, with increasing surface area of porous medium (i.e., with decreasing permeability), and with decreasing temperature. As the oil-water-gas mixtures move through the pores, the gas bubbles and water droplets are deformed on passing through constrictions (Figure 2.38) (Muravyov *et al.*, 1958).

In order to move, the gas bubble shown in Figure 2.38 must overcome the capillary pressure equal to:

$$\Delta p = p_1 - p_2 = 2\sigma/r_1 - 2\sigma/r_2 = 2\sigma(1/r_1 - 2/r_2) \tag{2.95}$$

Although the Δp may be very small for a single bubble, the cumulative resistance of many bubbles may be large (**Jamin effect**). Additional resistance to flow is created by the polymolecular layers

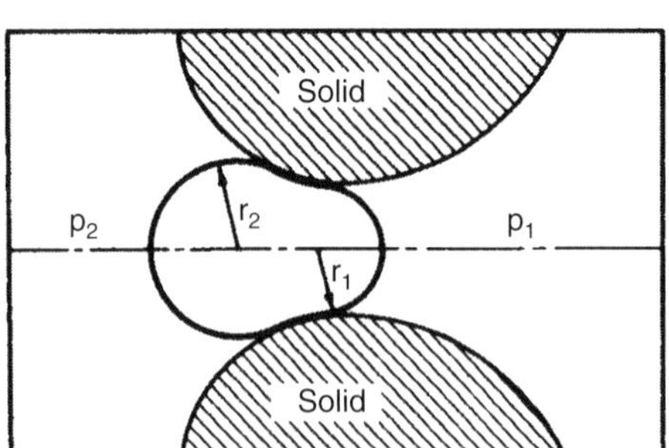

Figure 2.38 Movement of gas bubble through a constriction. (After Muravyov *et al.*, 1958.)

of oriented molecules of surface-active components in the oil, which are adsorbed on the rock surface and may be quite thick (10^{-3} to 10^{-4} cm). At a constant pressure differential, the rate of oil filtration through porous media diminishes with time and is more pronounced in the case of higher content of polar components in the crude.

2.6.2 Capillary Pressure Curves

In addition to porosity and permeability determinations, capillary pressure curves (e.g., Figure 2.39) may aid in analyzing the degree and nature of compaction, for example. The laboratory measurements of interstitial fluid content of cleaned rock samples depend

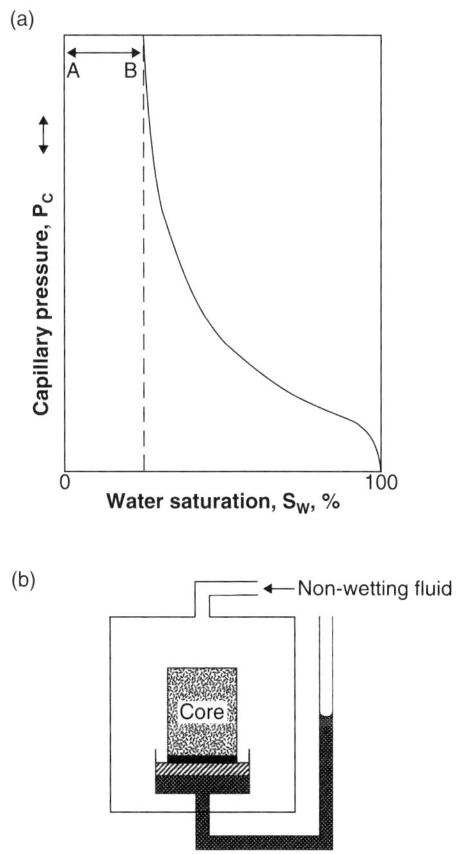

Figure 2.39 A – Capillary pressure curve. Irreducible interstitial fluid (wetting phase) = distance *AB*. B – Schematic diagram of apparatus for determining fluid (wetting phase) content at various pressures.

upon the principles of capillarity. The effect of capillary pressure and relative permeability on reservoir production is presented in Figure 2.40.

2.6.3 Compressibility

Compressibility, c, can be defined as the rate of change of volume, δV, with respect to the applied stress, σ, per unit of volume, V:

$$c = 1/V(\delta V/\delta\sigma) \tag{2.96}$$

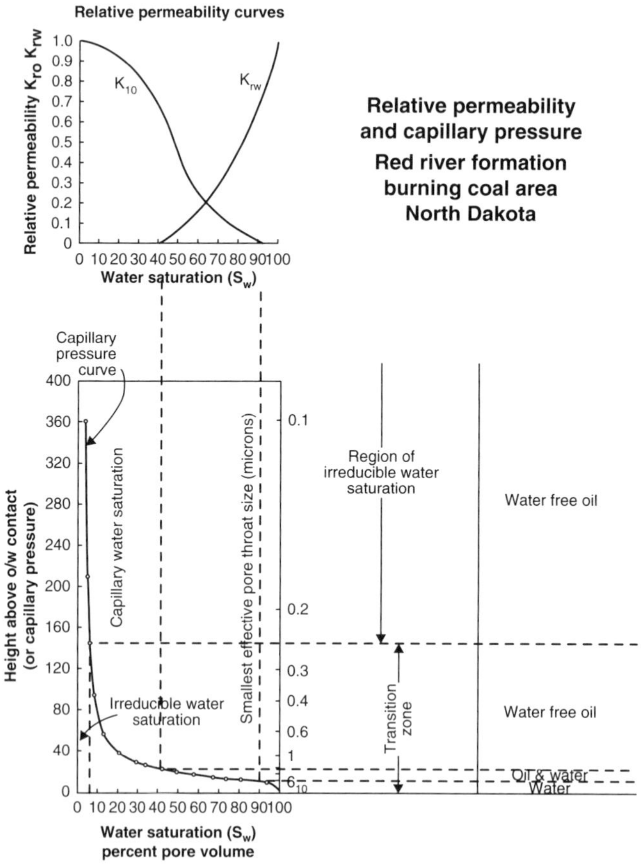

Figure 2.40 Rock productbility as determined from capillary pressure and relative permeability curves. Smallest pore throats effective at any pressure level shown on scale to the right of capillary pressure curve. (Modified after Arps, 1964; by Jodry, 1972.)

There are several different definitions of the term compressibility, which appear in the literature depending on the method of determination: (1) bulk compressibility; (2) pore compressibility; (3) formation compressibility; (4) rock solids compressibility; and (5) pseudo-bulk compressibility (Table 2.7).

Although many consolidation (compressibility) studies on clays and shales have been performed in soil-mechanics laboratories for more than 60 years, these tests have been limited largely to a low-pressure range [<1000 psi (70.3 kg/cm²)]. During the same period,

Table 2.7 Various compressibility formulas used in the literature.

Formula	
Bulk compressibility:	
$c_b = -\frac{1}{V_b}\left(\frac{\partial V_b}{\partial \bar{\sigma}}\right)_{p_p,T}$	Change in the bulk volume (∂V_b) per unit of bulk volume (V_b) per unit change in total external stress ($\bar{\sigma}$), while keeping the pore pressure (p_p) and temperature (T) constant.
$c_b = -\frac{1}{V_b}\left(\frac{\partial V_b}{\partial p_e}\right)_{\bar{\sigma},T}$	Change in the bulk volume (∂V_b) per unit of bulk volume (V_b), per unit change in effective pressure ($p_e = \bar{\sigma} - p_p$), while keeping the total external stress ($\bar{\sigma}$) and temperature (T) constant.
$c_b = -\frac{1}{e+1}\left(\frac{\partial e}{\partial p_e}\right)$ or $c_b = -\frac{1}{h}\left(\frac{\partial h}{\partial p_e}\right)$	Determined in the uniaxial compaction apparatus, if pore pressure $p_p = 0$, i.e. $\bar{\sigma} = p_e$. Void ratio, e, is equal to the volume of voids (V_p) divided by the volume of solids (V_s): $e = V_p / V_s = \phi/(1-\phi)$, where ϕ is fractional porosity. h is a sample thickness in a uniaxial compaction apparatus, i.e. thick-walled cylinder.
Pore compressibility:	
$c_p = -\frac{1}{V_p}\left(\frac{\partial V_p}{\partial \bar{\sigma}}\right)_{Pp,T}$ or $c_p = -\frac{1}{V_p}\left(\frac{\partial V_p}{\partial p_e}\right)_{\bar{\sigma},T}$	Change in the pore volume (∂V_p) per unit of pore volume (V_p) per unit change of external stress ($\bar{\sigma}$), keeping the pore pressure (p_p) and temperature (T) constant. Change in the pore volume (∂V_p) per unit of pore volume (V_p), per unit change in effective pressures (p_e), while keeping the total external stress ($\bar{\sigma}$) and temperature (T) constant.

Table 2.7 Various compressibility formulas used in the literature. *(continued)*

Formula	
Formation compressibility (pore compressibility of some authors):	
$c_f = -\dfrac{1}{V_p}\left(\dfrac{\partial V_p}{\partial p_p}\right)_{\bar{\sigma},T}$	Change in the pore volume (∂V_p) per unit of pore volume (V_p), per unit change of pore pressure (p_p), while keeping the total external stress ($\bar{\sigma}$) and temperature (T) constant.
Rock solids compressibility:	
$c_r = -\dfrac{1}{V_s}\left(\dfrac{\partial V_s}{\partial p_p}\right)_{\bar{\sigma}=Pp,T}$ $c_r = -\dfrac{1}{V_s}\left(\dfrac{\partial V_s}{\partial \bar{\sigma}}\right)_{\bar{\sigma}=Pp,T}$	Change in the rock solids volume (∂V_s) per unit of rock solids volume (V_s), per unit of external stress ($\bar{\sigma}$), at constant temperature. If a rock sample is tested without a jacket, external stress will be equal to the pore pressure ($\bar{\sigma} = p_p$).
Pseudo-bulk compressibility:	
$c_{bt} = -\dfrac{1}{V_b}\left(\dfrac{\partial V_b}{\partial p_p}\right)_{\bar{\sigma},T}$	Change in the bulk volume (∂V_b), per unit of bulk volume (V_b), per unit change of pore pressure (p_p), at constant external stress ($\bar{\sigma}$) and temperature (T).
Coefficient of compressibility:	
$a_v = -\dfrac{\partial e}{\partial p}$	Change in void ratio (∂e) per unit change of net confining pressure (∂p).

Note: In calculating bulk and pore compressibility, one can use either the *initial* bulk (V_b) or pore (V_p) volume in all cases, or measure volumes at each particular pressure for which compressibility is being calculated. The results appear to plot better in the former case.

high-pressure confining tests on consolidated sedimentary rocks have exceeded 15,000 psi (1055 kg/cm^2). Most investigators used mainly well-indurated sandstones or limestones in their laboratory experiments. Knutson and Bohor (1963) tested the oil reservoir rocks typical of the Texas-Louisiana Gulf Coast region (orthoquartzites to calcareous subgraywackes). Van der Knaap and Van der Vlis (1967) determined the compressibilities of unconsolidated clays and sands from the Bolivar Coast of Venezuela.

Carpenter and Spencer (1940) measured the "pseudo-bulk" compressibility of various consolidated sandstones in an attempt to investigate whether or not fluid withdrawal from U.S. Gulf Coast oil reservoirs and the resulting volume reduction could account for ground subsidence. They defined "pseudo-bulk" compressibility as:

$$\beta = 1/V_b(\delta V_p/\delta p_t) \tag{2.97}$$

where β is the pseudo-bulk compressibility in psi^{-1}; V_b is the original bulk volume in cm^3; δV_p is the change in void-volume in cm^3; and δp_t is the change in the applied pressure in psi. Their experiments showed that sediments compact owing to fluid withdrawal from the pore space.

Fatt (1958b) studied the relationship between compressibility and rock composition. He reported that unconsolidated sediments, which are poorly-sorted and contain clay, have higher compressibilities than do consolidated and well-sorted sands. Fatt (1958b) found that the bulk compressibilities of sandstones are a function of rock composition for a given grain shape and sorting. If sandstone are divided into two groups (one with well-sorted, well-rounded grains and the other with poorly sorted, angular grains), then for each group the compressibility is a linear function of the amount of intergranular material.

The procedure used in the laboratory by Fatt (1958b) was similar to that of Carpenter and Spencer (1940), but in the former case the fluid was expelled under constant external pressure with a reduction in pore pressure rather than an increase in the external stress. This is believed to closely duplicate petroleum-reservoir producing conditions. Fatt's (1958b) procedure was to apply a constant external stress to the core and decrease or increase the pore pressure. This apparatus simultaneously measured both the bulk- and pore-volume changes at room temperature. Volume changes of the core in the pressure cell were measured through the use of a liner potentiometer that could resolve a movement of 1×10^{-3} in. (2.54×10^{-3} cm).

Van der Knaap (1959) noted that pore compressibility increases with decreasing porosity. It has been suggested by some investigators that between certain minimum and maximum pressures, a straight line can approximate the relationship between pore compressibility and the logarithm of pressure. A straight-line relationship has been found to exist between the log of the bulk compressibility and the log of the "effective" pressure, which in this case was equal to the direct applied axial load, because pore pressure was atmospheric (Van der Knaap and Van der Vlis, 1967). Bulk compressibilities of unconsolidated clays and sands decreased with increasing overburden pressure. From their studies, Van der Knaap and Van der Vlis concluded that clay and sand layers compact almost to the same extent, the main difference being that the low permeability to water of the clay prevents instantaneous compaction and time effects become important.

In his classical paper on the compaction of freshwater-bearing alluvial clays, silts, ands silty sands in California, Meade (1968) showed that the loss in pore volume that resulted from compaction by effective overburden pressures in the range between 3 and 70 kg.cm^2 (1 kg/cm^2 = 14.223 psi and 1 psi^{-1} = 14.223 cm^2/kg) averages about 0.3 void ratio ($e = V_p/V_s$) units or about 15% of the bulk volume of the fine-grained sediments. Meade stated that when one allows for the lesser compaction of the interbedded coarser sands and gravels, the reduction of the total volume of the alluvial sediments amounts to about 12% in the pressure range of 10–70 kg/cm^2 on the east side of the San Joaquin Valley and about 10% in the 3–33 kg/cm^2 pressure range in the Santa Clara Valley, California. The factors that directly influence the compressibility of shallow marine and alluvial sediments are the average particle size, particle sorting, the amount of smectite (montmorillonite) clay, proportion of exchangeable sodium cations relative to the calcium and magnesium cations in the phyllosilicates (clay minerals), presence of diatom skeletons, and probably the mica content. Figure 2.41 presents Meade's 1968 published results.

Relationship between compressibility and applied pressure for unconsolidated sands, sandstones, shales, limestones and illite clay is presented in Figure 2.42.The compressibilities of unconsolidated sands appear to be very close to those of clays. Unconsolidated sands are as compressible as clays, or even more so. Compressibility values of sands obtained in a hydrostatic compaction apparatus are commonly about twice as high as those determined using uniaxial compaction equipment. Sawabini *et al.* (1974) found that compressibility increases with increasing feldspar content. The correlation between the compressibility and porosity is shown in Figure 2.43 (Hall, 1953). Hall's graph is widely used by petroleum engineers

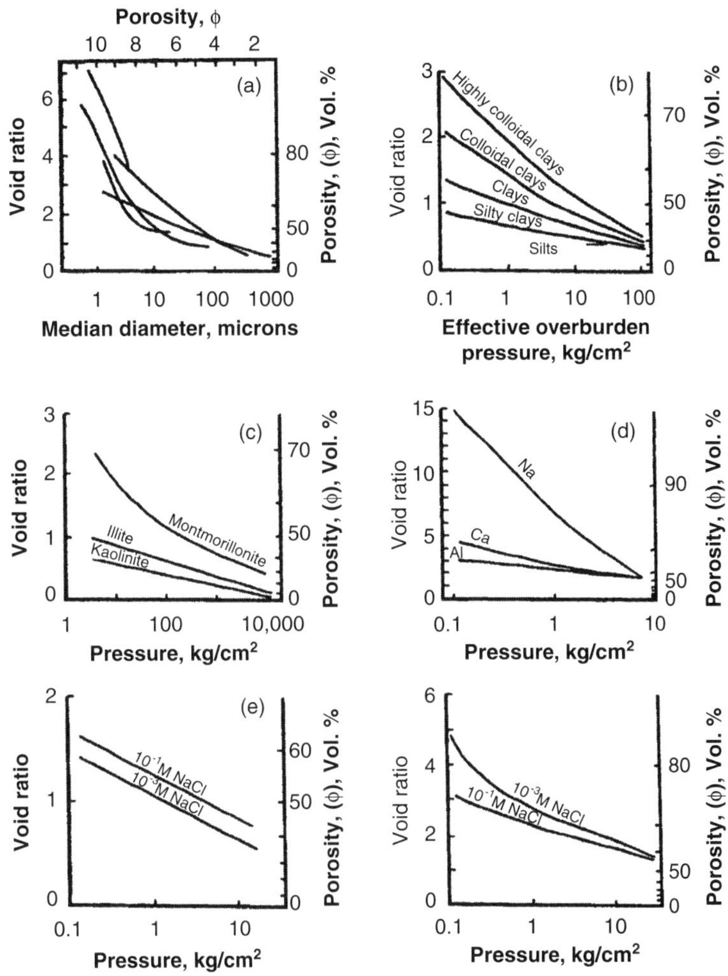

Figure 2.41 Influence of different factors on the relationship between void ratio and pressure in clayey materials. (a) Relationship between void ratio and median particle diameter at overburden pressures less than 1 kg/cm^2 (after Meade, 1964, p. B6). (b) Generalized influence of particle size (modified from Skempton, 1953, p. 55). (c) Influence of clay mineral species (modified from Chilingar and Knight, 1960, p. 104). (d) Influence of cations adsorbed by smectite (modified from Samuels, 1950). (e) Influence of NaCl concentrations in unfractionated illite, about 60% of which was coarser than 2 μm in size (modified from Mitchell, 1960, Fig. M3). (f) Influence of NaCl concentration in illite finer than 0.2 μm (modified from Bolt, 1956, p. 92). (After Meade, 1968, p. D4, Fig. 1.)

for sandstones. It is not used for unconsolidated sands and fractured formations. Instead of using such a generalized correlation, however, one should develop laboratory data for each formation being investigated.

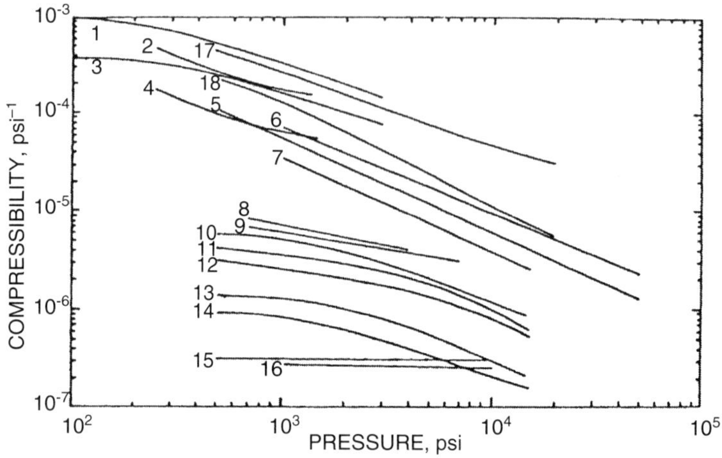

No.	Investigator	Rock type	Type of applied pressure	Compressibility	
1	Sawabini et al. (1974)	California unconsolidated arkosic sands[a]	Hydrostatic	Pore	$[-(1/V_p)(\partial V_p/\partial P_e)_\sigma]$
2	Kohlhaas and Miller (1969)	California unconsolidated sands	Uniaxial	Pore	
3	Sawabini et al. (1974)	California unconsolidated arkosic sands[a]	Hydrostatic	Bulk	$[-(1/V_b)(\partial V_b/\partial P_e)_\sigma]$
4	Kohlhaas and Miller (1969)	California unconsolidated sands	Uniaxial	Bulk	
5	Chilingarian and Rieke (1968)	Illite clay (API No. 35) (wet)°	Uniaxial	Bulk	$[-(1/e+1)(de/dP_e)]$
6	Chilingarian and Rieke (1968)	Illite clay (API No. 35) (dry)	Uniaxial	Bulk	$[-(1/h)(dh/dP_e)]$
7	Knutson and Bohor (1963)	Repetto Fm. (Grubb Zone) (wet)[a]	Net confining	Pore	$[-(1/V_p)(\partial V_b/\partial \sigma)_{pP}]$
8	Knutson and Bohor (1963)	Lansing – Kansas City Limestone (wet)[a]	Net confining	Pore	
9	Carpenter and Spencer (1940)	Woodbine Sandstone (wet)	Net confining	Pseudo bulk	$[-(1/V_b)(\partial V_p/\partial P)]$
10	Fatt (1958b)	Feldspathic graywacke (No. 10) (wet)[c]	Net confining[d]	Bulk	
11	Fatt (1958b)	Graywacke (No. 7) (wet)[c]	Net confining	Bulk	
12	Fatt (1958b)	Feldspathic graywacke (No. 11) (wet)[c]	Net confining	Bulk	$[-(1/V_p)(\partial V_b/\partial p_1)_{pP}]$
13	Fatt (1958b)	Lithic graywacke (No. 12) (wet)[c]	Net confining	Bulk	
14	Fatt (1958b)	Feldspathic quartzite (No. 20) (wet)[c]	Net confining	Bulk	$[-(1/V_p)(\partial V_b/\partial p_1)_{pP}]$
15	Podio et al. (1968)	Green River shale (dry)[b]	Net confining	Bulk	
16	Podio et al. (1968)	Green River shale (wet)[b]	Net confining	Bulk	
17	Chilingarian et al. (1973)	Montmorillonite clay saturated in seawater	Hydrostatic	Bulk	$C_b = -\frac{1}{V_b}\left(\frac{\partial V_b}{\partial p_e}\right)_\sigma$, T
18	Chilingarian et al. (1973)	Montmorillonite clay saturated in seawater	Uniaxial	Bulk	$C_b = -\frac{1}{h}\left(\frac{\partial h}{\partial p}\right)$

[a] Saturated with formation water.
[b] Saturated with distilled water.
[c] Saturated with Kerosene.
[d] Net confining pressure = external hydrostatic pressure on a jacketed specimen = $p_e = (\sigma - 0.85_{Pp})$, where σ is the total overburden stress and p_p is the pore pressure. Stresses in the triaxial apparatus of Sawabini et al. (1974) approached hydrostatic; i.e., three principal stresses in x-, y- and z-directions are equal; senior author believes that 0.85 should be equal to 1.

Figure 2.42 Relationship between compressibility (psi^{-1}) and applied pressure (psi) for unconsolidated sands, illite clay, limestones, sandstones, and shale.

2.6.3.1 *Degree of Porosity Reduction Upon Compaction*

A few examples of reduction in porosity (and thickness) of sediments upon compaction are presented here. For example, Figure 2.44 shows the different theoretical stages of compaction of argillaceous

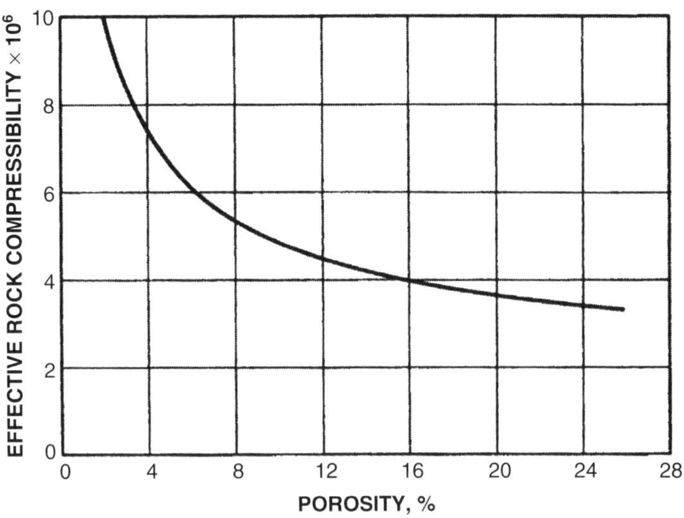

Figure 2.43 Relationship between porosity in percent and effective rock compressibility in psi^{-1} (change in pore volume/unit pore volume/psi). (After Hall, 1953, courtesy of AIME.)

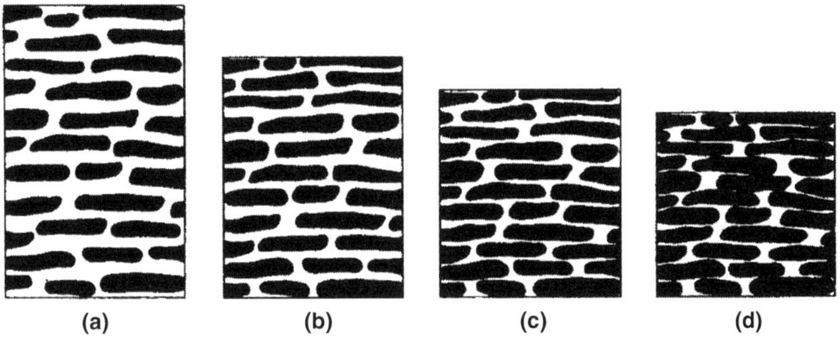

Figure 2.44 Schematic diagrams of theoretical compaction stages of an argillaceous sediment, having initial and final porosity of 50% (a) and 20% (d), respectively. Intermediate porosities are 40% (b) and 30% (c). All porosities were measured exactly.

sediment, whereas Figure 2.45 depicts compaction stages of a sand, as the interstitial fluids move out. In these figures, the final thickness of consolidated sediments (rocks) is 62.5% of the original thickness of unconsolidated sediment. The effects of cementation and other diagenetic changes were not considered in preparing Figures 2.44 and 2.45. In addition, the effect of crushing of sand grains under

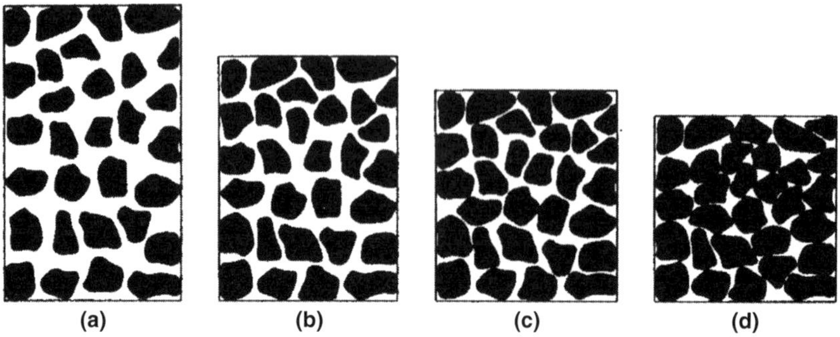

Figure 2.45 Schematic diagrams of theoretical compaction stages of sand, having initial and final porosities of 50% (a) and 20% (d), respectively. Intermediate porosities are 40% (b) and 30% (c). All porosities were measured exactly in preparing these diagrams.

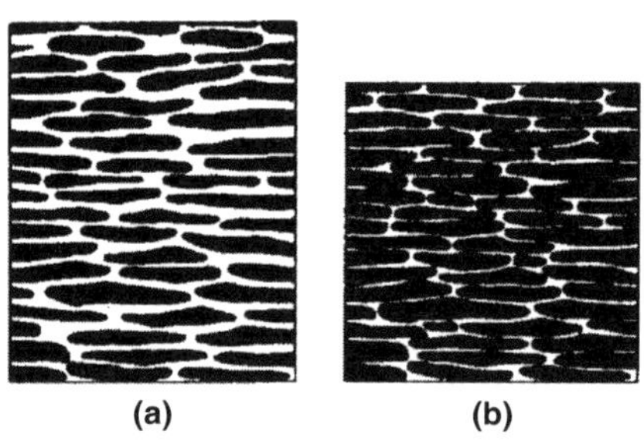

Figure 2.46 Comparison between clayey sediment having a porosity of 33.3% (a) and a similar sediment with a porosity of 20% (b). Porosities were measured exactly in preparing these theoretical diagrams.

high overburden stresses is also not included. Figure 2.46 is a schematic diagram showing the difference between a sediment consisting of flat particles (e.g., clay particles) and having a porosity of 33.3% (Figure 2.46a) and a similar sediment with a porosity of 20% (Figure 2.46b). In Figure 2.47, a sand having a porosity of 33.3% (Figure 2.47a) is compared with a better consolidated, similar sand having a porosity of 20% (Figure 2.47b)

2.6.3.2 *Compressibilities of Fractured–Cavernous Carbonates*

Tkhostov *et al.* (1970) presented an excellent discussion on compressibility of cavernous and fractured rocks. According to them, at low effective pressures (up to 200–300 kg/cm^2) the compressibility

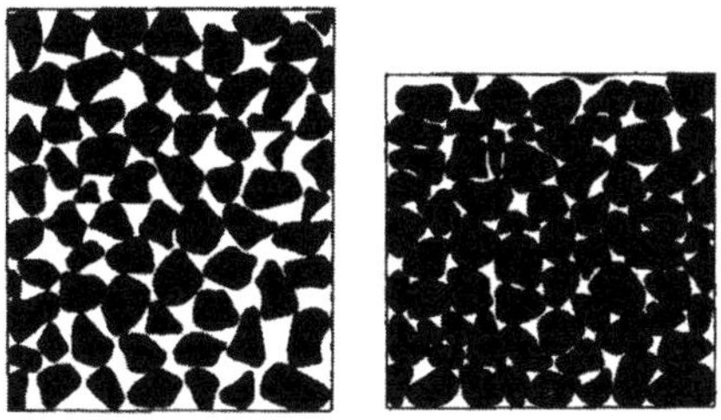

Figure 2.47 Comparison between degree of compaction af a sand having a porosity of 33.3% (a), and that of a similar sand with 20% porosity (b). Porosities were measured exactly in preparing these theoretical diagrams.

of fractures is of the order of 10^{-4} cm^2/kg, whereas that of the caverns and vugs is of the order of 10^{-5} cm^2/kg (cm^2/kg = 7.031×10^{-2} psi^{-1}).

The compressibility of carbonate rocks, c_{rc}, is equal to:

$$c_{rc} = (\phi_{fr} \times c_{fr}) + (\phi_{cv} \times c_{cv}) + c_m \tag{2.98}$$

where ϕ_{fr} = fractional porosity of fractures; c_{fr} = compressibility of fractures; ϕ_{cv} = fractional porosity of caverns and vugs; c_{cv} = compressibility of caverns and vugs; and c_m = compressibility of the matrix.

For Solnhofen Limestone, $c_m = 0.03 \times 10^{-4}$ cm^2/kg; whereas the compressibility of caverns and vugs, c_{cv}, was estimated by Tkhostov *et al.* (1970) to be equal to:

$$c_{cv} \approx 3 \times c_m \approx 3(0.03 \times 10^{-4}) = 0.09 \times 10^{-4} \text{cm}^2/\text{kg} \tag{2.99}$$

The following simplified equation for determining the secondary pore compressibility was derived by Tkhostov *et al.* (1970):

$$c_{ps} \approx [(\phi_{fr}/\phi_{ts} \times 1350/(p_t - p_p) - 0.09] \times 10^{-4} \tag{2.100}$$

where c_{ps} = compressibility of secondary pores (fractures + vugs and caverns, cm^2/kg; ϕ_{ts} = total fractional porosity of secondary pores; p_t = total overburden pressure, kg/cm^2; and p_p = pore pressure, kg/cm^2.

Relationship between the pore compressibility of carbonate rocks and effective overburden pressure is presented in Figure 2.48.

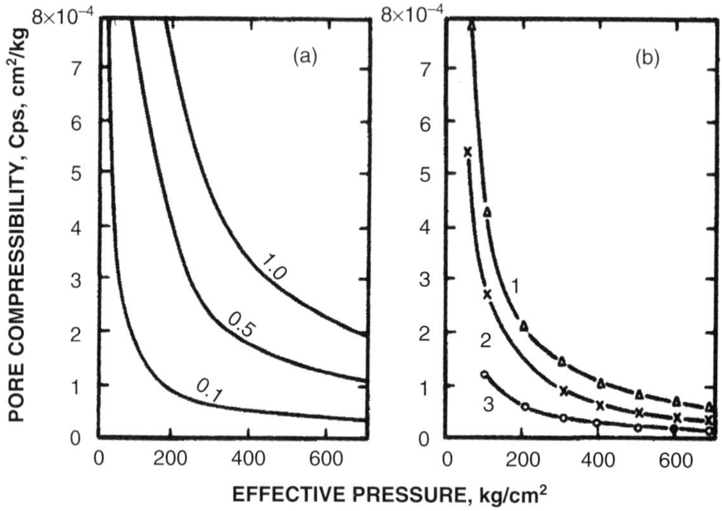

Figure 2.48 Relationship between the pore compressibility of carbonates and effective pressure ($p_e = p_i - p_p$). A–theoretical curves for fractured-cavernous reservoirs calculated on using Eq. 2.100; numbers on curves designate the ϕ_{fr}/ϕ_{ts} ratios. B–experimental curves ($p_p = 0$) for (1) limestone with $\phi = 2.01\%$; (2) marl with $\phi = 2.63\%$; and (3) limestone with $\phi = 11.31\%$. (After Tkhostov *et al.*, 1970.)

2.7 Elastic Properties

As pointed out by Donaldson *et al.* (1995), the atoms and molecules of all substances above the absolute zero of temperature exhibit random motion about a mean location. Whereas these motions tend to disintegrate the molecule, they are opposed by greater atomic binding forces that hold the particles together. The resultants of these forces acting within the cohesive structure of the substance oppose deformation of the material when it is subjected to external forces. The distance between atoms, or molecules, decreases from gases to liquids, to solids; therefore, gases exhibit a larger amount of compressibility relative to liquids and solids. The molecules of gases and liquids, however, do not have sufficient cohesive force to support an expansive force, whereas solids may be compressed or extended within the limits of their elastic properties without causing permanent deformation (Guyod and Shane, 1969; Dresser-Atlas, Inc., 1982; Ellis, 1987).

The subsurface geological formations have been subjected to some stresses during geological history. Natural forces affecting geologic formation or geologic body tend to change its dimensions and shape. Stress in the geologic formations is the ratio of applied force to the area of formation affected by this force. Each stress causes a strain (or deformation) that is the ratio of the change in dimension, volume, or shape to the original dimension, volume or shape. Stress–strain properties of rocks and their role in engineering practice were described in detail by Terzaghi and Peck (1948), Ornatskiy (1950), and Terzaghi (1961).

2.7.1 Classification of Stresses

Based on the manner in which stresses are applied or transmitted to the formations, they can be classified as *static*, *repeated*, or *impact*.

1. Static stresses are forces per unit area that are applied slowly and remain nearly constant after being applied to the geologic formation.
2. Repeated or periodical stresses are forces that are applied a large number of times, causing a stress in the substance that is continually changing, usually through some definite range.
3. Impact stresses are forces that are applied to the affected body in a relatively short period of time. The stress and strain produced by an impact may be calculated from the energy, e.g., expressed in foot-pounds, delivered to the affected body.

Stresses may be classified also as being *orthogonal* (vertical) and *lateral* (horizontal). Vertical and lateral stresses, in turn, may be classified as distributed stresses and concentrated stresses. A distributed stress may be uniformly or non-uniformly distributed. For example, if sediments are spread on the seafloor so that their thickness is constant, the seafloor is subjected to a uniformly distributed stress (load), whereas if the sediments are distributed so that their thickness is not constant, then the seafloor carries a non-uniformly distributed stress (load). A conceptual stress occurs when the area of contact with the affected body is negligible in comparison wit the whole area of this body.

Stress produced by a load can be resolved into two components: (1) compressive (or tensile) stress, which acts normal to the surface and changes the volume of the geologic body, and (2) shear stress, which acts parallel to the surface and changes the shape of the geologic body.

2.7.1.1 Stress-Strain Curve

Experiments have shown that for nearly all materials the unit stress in a material is approximately proportional to the accompanying unit strain provided that the unit stress does not exceed a value called the proportional limit.

The relation between the unit stress σ and the unit strain ε found experimentally is represented (within approximate limits) by the stress– strain curve shown in Figure 2.49a, if the material is ductile, such as plastic or sticky clay. The stress–strain curve shown in Figure 2.49b is for relatively brittle material, such as fractured carbonates.

2.7.1.2 Proportional Limit

A shown in Figure 2.49, for most formations the stress–strain curve is a close approximation to a straight line until the stress reaches a value called the proportional limit (or limit of proportionality). This unit stress is represented on the stress–strain curve by the point PL (Figure 2.49).

2.7.1.3 Yield Point

As the load on the material is increased further, causing a stress greater than the proportional limit, the stress is reached at which the material continues to deform without an increase in the load.

This stress is called the yield point and is represented on the stress – strain curve by the *CD* portion of the curve Figure 2.49.

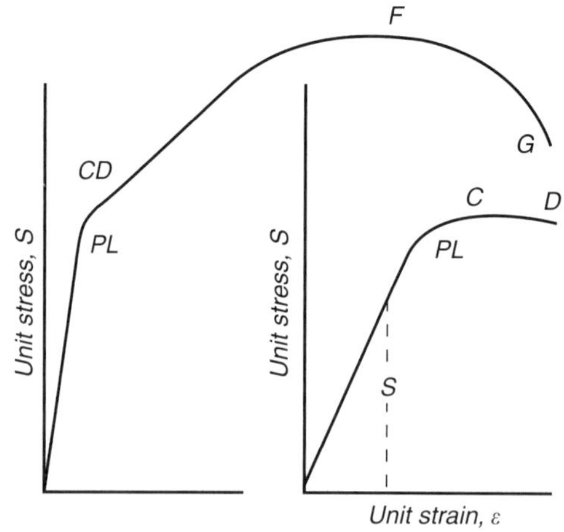

Figure 2.49 Stress versus strain graph. a – ductile material; b – brittle material.

2.7.1.4 Hooke's Law

Robert Hooke originally formulated the basic concept of linear elasticity in 1676. Over 100 years later, Thomas Young stated that linear proportionality exists between the stress and strain. Hooke's law can be expressed mathematically as:

$$\sigma/\varepsilon = a \tag{2.101}$$

where a is a constant.

Thus, the resistance to deformation, or stiffness, of the material is its elasticity, which is expressed quantitatively by Hook's law. Hooke's law states that the strain (deformation of the substance) is proportional to the applied stress (per unit area). The ratio of the applied stress (F_0/A) is the Hookean elasticity (or modulus of elasticity of the substance). Three elastic moduli have been defined, which quantitatively describe the fractional strains of length, volume, and shape within the elastic limits of substances, respectively:

$$\text{Young's modulus } (E) = (F_0/A)/(\mathrm{d}L/L) \tag{2.102}$$

where: E is in psi or Pa; F_0 is in lb or N; L is in in. or m,

$$\text{Bulk modulus } (K) = (F_0/A)/(\mathrm{d}V/V) \tag{2.103}$$

where: K is in psi or Pa; V is in in^3 or m^3

$$\text{Shear modulus } (G) = (F_0/A)/\tan\Theta \tag{2.104}$$

where G is in psi or Pa; and Θ = angle of deformation.

A fourth elastic property, known to the Poisson's ratio, is the ratio of strain perpendicular to the extensional force to that perpendicular to the compressional force:

$$\text{Poisson's ratio } (\nu) = (\mathrm{d}x/\mathrm{d}y)/(\mathrm{d}y/y) \tag{2.105}$$

The elastic properties of subsurface geological formations (Table 2.8) are affected by anisotropy, diastrophism, lithology, and overburden pressure. Crystalline rocks exhibit larger values of elastic moduli than rocks having a greater degree of diastrophism and fragmentation, or those containing colloidal materials. Poisson's ratio of sedimentary rocks ranges from 0.15 to 0.4.

Table 2.8 Compressional and shear wave velocities in various materials. (After Dresser-Atlas, Ellis, 1987.)

	Sp. Gr.	Compressive velocity (ft/s)	Shear velocity (ft/s)
Nonporous solids			
Anhydrite	2.98	20,000	11,400
Calcite	2.71	20,100	–
Dolomite	2.88	23,000	12,700
Granite	2.65	19,700	11,200
Gypsum	2.35	19,000	–
Limestone	2.71	21,000	11,100
Quartz	2.64	18,900	12,000
Salt	2.16	15,000	8000
Steel	8.93	20,000	9500
Water-saturated rocks			
Dolomite	2.88	15,000–20,000	8000–11,000
Limestone	2.71	13,000–18,500	7000–9000
Sandstones	2.65	12,000–16,500	6000–9500
Sands	2.65	9000–17,000	
Shales	2.45	7000–17,000	
Liquids			
Water, pure		4800	
Water, 100 K ppm NaCl		5200	
Water, 200 K ppm NaCl		5500	
Drilling mud		6000	
Crude oil		4200	
Gases			
Air		1100	
Hydrogen		4250	
Methane		1500	

Poisson's ratio of some weak, unconsolidated, porous rocks, however, may approach zero.

2.7.1.5 Practical Use of Stress–Strain Properties of Rocks

Stress–strain properties of rocks are used in petrophysical studies. It is known that porosity and permeability of reservoir rocks are adversely affected by compaction. Compaction due to weight of the overburden squeezes the sand grains closer together and, at greater depths, may crush and fracture them. This will result in smaller pores and, therefore, lower porosity. More importantly, there is a drastic decrease in permeability. To determine the porosity and permeability one has to conduct the experimental study of cores under high pressure and temperature. Knowledge of the stress–strain properties of rocks helps to analyze and interpret well-logging data.

2.8 Acoustic Properties

The mechanical vibrations or stresses can produce elastic (or acoustic) waves of different frequencies. For example, the frequencies' of waves from earthquakes are principally in the range of 10^{-3} to 1 Hz.

There are three types of acoustic waves: compression–dilation (P-waves, primary or longitudinal waves), shear waves (S-waves, secondary or transverse waves), and surface waves (L-waves). These different waves travel with different velocities depending on the density, porosity, elastic properties, and compressibility of the rock through which they pass (Table 2.8).

P-waves oscillate in the direction of propagation by compression and dilation. They travel through any material (gas, liquid or solid) and are faster than the other types of waves. The velocity of the compression–dilation waves is given by $(G/\rho)^{0.5}$, where G is the Young's modulus of elasticity and ρ is the mass density for the solid. S-waves oscillate normal to the direction of propagation. They pass through solids.

The shear wave velocity is given by $(G/\rho)^{0.5}$, where G is the shear modulus. L-waves travel in the surface layers of the Earth and have very long periods; they travel more slowly than the P or S waves.

Acoustic waves are reflected by solid surfaces. This is due to their traveling at different velocities in different media. They are, therefore, refracted on the boundaries between solid surfaces. Reflection and refraction of acoustic waves obey the optic laws. The amplitude of the wave is not maintained, mainly because energy is dispersing over the expanding surface of a sphere (nearly), which increases the area as the square of the distance from the source as it propagates from the source. Thus, the wave is attenuated and becomes weaker with its distance from the source.

Acoustic waves are widely used in both surface and subsurface operations that include, respectively, seismic exploration and borehole seismic and logging methods. Seismic exploration is used to map subsurface geologic structure and stratigraphic–lithologic features with the aim of discovering reservoirs and traps with potential oil and gas reserves.

Seismic method in an attempt to image the subsurface, during the earliest phases of implementation used a two-dimensional (2-D) approach. However, the oil and gas reserves we find or evaluate are accumulated in three-dimensional (3-D) traps. It was only in 1970 when Walton (1972) presented the concept of 3-D surveys. In 1975, 3-D surveys were first performed on a normal contractual basis. The following year, Bone (1976) presented new technology to the seismic acquisition and processing specialists.

Analysis of 3-D seismic data with several types of seismic attributes, such as envelope amplitude, instantaneous phase, instantaneous frequency, polarity, velocity of acoustical waves, etc., can reveal geologic features that control the location of productive reservoirs. Although each attribute when used alone has some level of ambiguity, it is important to note that when a number of attributes with different mathematical algorithms yield similar results, the accuracy and reliability of the geologic interpretation are enhanced.

There are several hydrocarbon indicators, which indicate the presence or absence of hydrocarbon accumulations. These include:

1. Local acoustic wave amplitude increase ("bright spot") or decrease ("dim spot").
2. Acoustic wave phase change.
3. Acoustic wave frequency change.
4. Horizontal event because of gas-water, gas-oil or oil-water contact (flat spot").
5. Lower velocity than in laterally-equivalent sediments.
6. Decrease in acoustic wave amplitude below the hydrocarbon accumulation ("shadow zone").
7. Apparent sag below because of increased time in transiting the hydrocarbon accumulation ("velocity sag").

The essence of the 3-D method is 3-D data acquisition followed by the processing and interpretation of a closely spaced data volume. The fundamental objective of the 3-D seismic method is increased resolution, which has both vertical and horizontal aspects. Sheriff (1985) discussed the subject qualitatively, whereas Embree (1985) presented a quantitative approach. Figure 2.49 summarizes two key issues related to resolution: migration and deconvolution that are the principal techniques for improving, respectively, both horizontal and vertical resolution.

2.8.1 Borehole Seismic and Well Logging Methods

Check shot surveys were used to obtain travel times and interval velocities. The 2-D VSP's (vertical seismic profiles) and 2-D high-resolution cross-well data also have been recorded.

In the past, borehole methods have been relegated to a secondary role in seismology because they generate only 1-D or 2-D images. Another limiting feature in borehole methods is that the fundamental designs of borehole measuring systems only allowed a small number of geophones to be deployed in the borehole.

At present time, a new type of borehole seismic (acoustic) receiver array has been introduced that currently has the much more three-component (3-C) geophone levels in a single borehole (Paulsson *et al.*, 2001). The design can be modified to allow as many as 400 to 1000 three-component geophone levels when fully deployed. The fundamental difference between the new and the old borehole

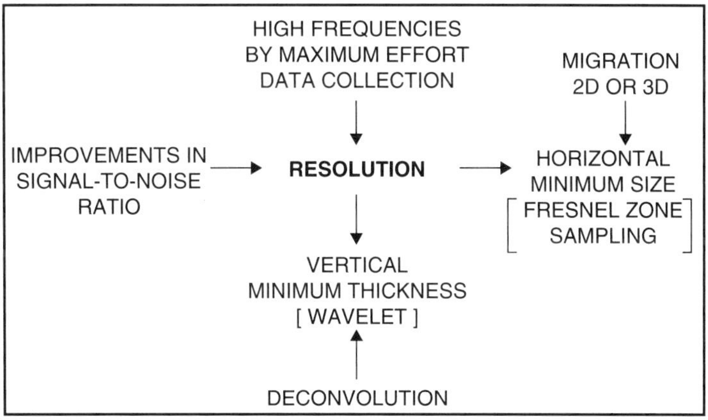

Figure 2.50 Factors affecting horizontal and vertical resolution. (Modified after Brown, 1986.)

arrays is that the new array is deployed on production tubing, whereas the old style arrays were deployed using wireline technology. Geophones for the new array can easily be deployed in horizontal wells because they are conveyed on standard production tubing using the same method used to deploy electric submersible pumps. The advantage of deploying a large number of borehole seismic (acoustic) receivers is that large amounts of reflection coverage can be obtained per one shot, thus making borehole method commercially feasible.

2.8.2 Practical Use of Acoustic Properties of Rocks

Acoustic properties of rocks are widely used in discovering and assessing the potential of reservoirs. After the initial reservoir discovery, acoustic properties of the rocks are used for formation evaluation.

The key seismic survey attribute that is used for formation evaluation is seismic reflection amplitude. The principal reservoir properties, which affect the seismic amplitude, can be divided into two groups:

- **Group A**: nature of fluid, lithology, and pressure.
- **Group B**: hydrocarbon saturation, porosity, and net pay thickness.

The properties in Group A are those that affect the reservoir as a whole. The properties in Group B are the ones, which can vary laterally and/or vertically over short distances and, therefore, significantly affect the formation evaluation, reserve estimation, and further field development and production

Any one or more of properties of Group B can cause some changes in amplitudes of reservoir reflections. The amplitude of a seismic "bright spot" is higher where (1) hydrocarbon saturation is higher, (2) porosity is higher, and (3) the net pay thickness is greater. Other independent seismic attributes such as instantaneous frequency and phase, interval velocity, etc. offer little help for reservoir properties evaluation.

Acoustic properties of rocks are used in acoustic logging (sonic log that provide a recording, versus depth, of the time, Δt, required for compressional acoustic wave to traverse one unit measure (foot or meter) of subsurface formation. Known as the *interval transit time*, Δt, is the reciprocal of the velocity of the compressional acoustic wave. The interval transit time for a given rock depends upon its lithology and porosity. Its dependence upon porosity, when the lithology is known, makes the acoustic log very useful as a porosity log. Integrated acoustic transit times are helpful in interpreting seismic records. Average values of acoustic velocity v_{ma} and transit time Δt_{ma} for common rock matrix and well casing steel are presented below:

Material	v_{ma}	Δt_{ma}
	(ft/sec)(m/sec)	(μsec/ft) (μsec/m)
Unconsolidated sand	18,000/5,486	55.5/182
Consolidated sand	19,500/5,944	51.0/167
Limestone	22,000/6,706	47.5/156
Dolomite	23,000/7,010	43.5/143
Anhydrite	20,000/6,096	50.0/164
Salt	15,000/4,572	67.0/220
Shale	6,000–16,000 ft/sec	167–62.5 μsec/ft
Casing (iron)	17,500/5,334	57.0/187

With increasing porosity of rocks, the velocity of acoustic waves travel through a given rock decreases with a corresponding increase in the Δt_{ma} value.

After numerous laboratory experiments, Wyllie (1956) and Gregory and Gardner (1958) showed that in clean and consolidated formations with uniformly distributed small pores there is a linear relationship between porosity and transit time:

$$\Delta t_{log} = \phi \Delta t_{fl} + (1 - \phi)\Delta t_{ma} \tag{2.106}$$

or

$$\phi = (\Delta t_{log} - \Delta t_{ma})/(\Delta t_{fl} - \Delta t_{ma}) \tag{2.107}$$

where ϕ is the fractional porosity; Δt_{log} is the reading on the sonic well log; Δt_{ma} is the transit time of the rock matrix; and Δt_{fl} is the transit time of the rock-saturating fluid (about 189 μsect/ft or 620 μsec/m), which corresponds to the velocity through fluid of about 5300 ft/sec or 1615 m/sec.

The following equation enables one to determine the matrix travel time:

$$\Delta t_{ma} = 10^6 / v_{ma} \tag{2.108}$$

where velocity of sound (P waves), v_{ma}, can be determined from the following equation.

$$v_{ma} = [(K + 0.75G)) / \rho_{ma}]^{0.5} \tag{2.109}$$

where K = bulk modulus; G = shear modulus; and ρ_{ma} = matrix density.

As pointed out by Tiab and Donaldson (2004), the presence of fractures, vugs, gas, and shale complicates the sonic porosity measurements. For detailed analysis of this problem, one should consult the classical book by Tiab and Donaldson (2004, pp. 289–292).

Because of more detailed understanding of the subsurface formations, 3-D seismic surveys and borehole seismic and well logging methods have been able to contribute significantly to solving the problems of field exploration, appraisal, development, and production (Tegland, 1977; Paulsson *et al.*, 2001). Sheriff and Geldart (1983) wrote that "...there seems to be unanimous agreement that 3-D surveys result in clearer and more accurate pictures of geologic detail and that their costs are more than repaid by the elimination of unnecessary development holes and by the increase in recoverable reserves through the discovery of isolated reservoir pools which otherwise might be missed."

2.9 Electrical Resistivity

The major electrical properties of rocks are their *resistivity* and *spontaneous potential*. The resistivity of subsurface formations is their ability to impede the flow of electric current through the liquid-saturated porous media. The resistivity unit is the Ohm-meter^2/meter, usually expressed as Ohm-m. The resistivity in Ohm-m is the resistance in Ohms of a one-meter cube when the current flows between opposite faces of the cube. Subsurface formation resistivities usually fall in the range of 0.2 to 1000 Ohm-m. The reciprocal of resistivity is the electrical conductivity, which is expressed in mhOs per meter or Simens per meter.

Most subsurface formations are made up of rocks that will not conduct electrical current when dry. Current flows through the interstitial water made conductive by salts in solution. These salts dissociate into positively-charged cations (mainly Na^+, K^+ and Ca^{++}) and negatively charged anions (mainly, Cl^-, SO_4^{--}, HCO_3^-). Under the influence of an electrical field these ions move, carrying an electrical current through the solution. Other factors being equal, the greater the salt concentration, the lower the resistivity of formation water and hence the formation. Besides the influence of salt concentration in formation water, the strong influence is produced by formation temperature. The higher the formation temperature, the lower is the resistivity or greater is the conductivity of formation water.

Clay mineral content also contributes to formation conductivity. Clay conduction differs from the saline water conduction in that the current is not carried by ions moving a solution. Rather, conduction is an ion-exchange process whereby (usually the positively charged) ions move under the influence of the impressed electric field between exchange sites on the surface of the clay particles. The more the clay content in a clastic rock (i.e., the amount of clay particles), the higher the ion-exchange capacity and the higher the clay conductivity (or lower the resistivity) of clayey formations.

Surface conductance at the clay–water interfaces is an important factor in the effect of clayeness on the electrical conductivity. The net effect of clay conductivity depends on the amount, type, and distribution of clay particles, and on the nature and relative amount of the formation water.

Very important is the relationship between formation resistivity factor, $F_{R,}$ and porosity in saturated rocks. Empirical formula for clean (not argillaceous) granular rock was proposed by Archie (1942) as follows:

$$F_R = a\phi^{-m} \tag{2.110}$$

where F_R is the formation resistivity factor ($=R_0/R_w$), where R_0 is the resistivity of a formation 100% saturated with formation water and R_w is the formation water resistivity; ϕ is the porosity in decimal fraction; m is the cementation factor (or lithology exponent), which is a function of the shape and distribution of pores; and a is the proportionality constant (Table 2.9).

Table 2.9 Equations for determining formation resistivity factor (F_R) based on lithology. (Modified after Tiab and Donaldson, 2004, p. 836.)

Formula	Lithology
$F_R = 0.81/\phi^{2}$	Consolidated sandstones
$F_R = 0.62/\phi^{2.15}$	Consolidated sandstones (Humble formula)
$F_R = 1.65/\phi^{1.33}$	Shaly sands
$F_R = 1.0/\phi^{1.5}$	Unconsolidated sands
$F_R = 1.97/\phi^{1.29}$	Unconsolidated Miocene sands, US Gulf Coast
$F_R = 1.451/\phi^{1.70}$	Calcareous sands
$F_R = 1.01/\phi^{2}$	Limestones and dolomites

Also, very important is the relationship between formation resistivity index (I_R) and water saturation:

$$I_R = (1 - S_{0g})^{-n} \tag{2.111}$$

and

$$\log I_R = -n \log S_w \tag{2.112}$$

where I_R is the formation resistivity index ($=R_t/R_0$), where R_t is the true formation resistivity of oil- or gas-saturated rock; R_0 is the resistivity of porous rock 100% saturated with formation water; S_w is the water saturation; S_{0g} is the oil or gas saturation and n is the saturation exponent.

As an example, for the Pliocene Productive Series of offshore Azerbaijan fields, $a = 1$, $m = 1.58$ and $n = 1.61$ under surface conditions, and $m_{p,t} = 1.79$ and $n_{p,t} = 1.74$ under subsurface conditions. Figure 2.51 illustrates the relationships between formation resistivity factor and porosity; Figure 2.52, between formation resistivity index and oil/water saturation; and Figure 2.53; between formation resistivity index and permeability. For a detailed discussion of resistivity, the reader is referred to the classical book of Tiab and Donaldson (2004).

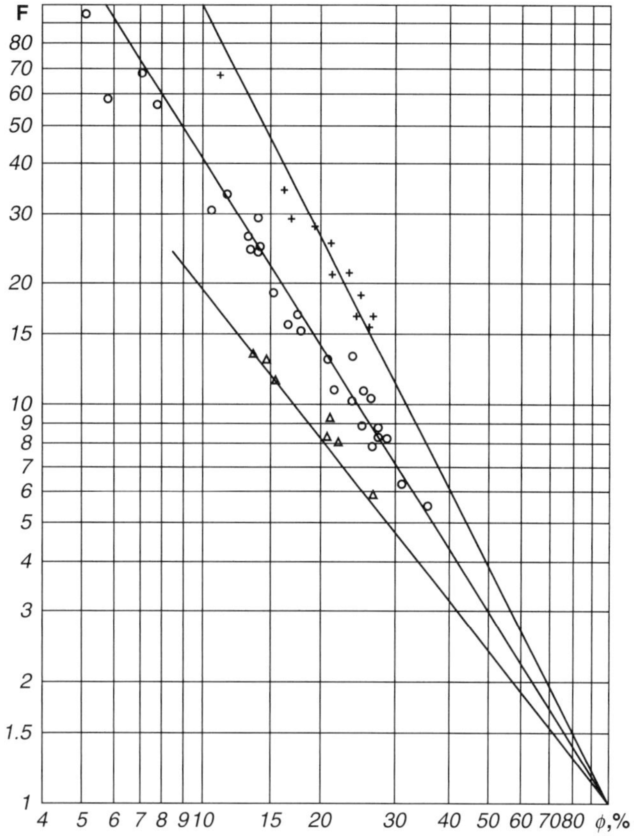

Figure 2.51 Relationship between the formation resistivity factor and porosity for Pliocene Productive Series of Azerbaijan. (After Buryakovsky, 1977.) 1 – Unconsolidated sand; 2 – poorly-cemented sandstone; and (3) – well-cemented sandstone.

2.9.1 Spontaneous Potential

Spontaneous potential (SP) is caused by electromotive forces (emf) of *electrochemical and electrokinetic origins* in the subsurface formations.

2.9.1.1 Electrochemical Component

The electrochemical component of the SP occurs at the interface between permeable (e.g., porous sand or sandstone) and impermeable (e.g., clay or shale) layers both saturated with the NaCl at different concentrations. Due to the layered structure of the formation and the charges on the clay surfaces, clay layers are permeable to the

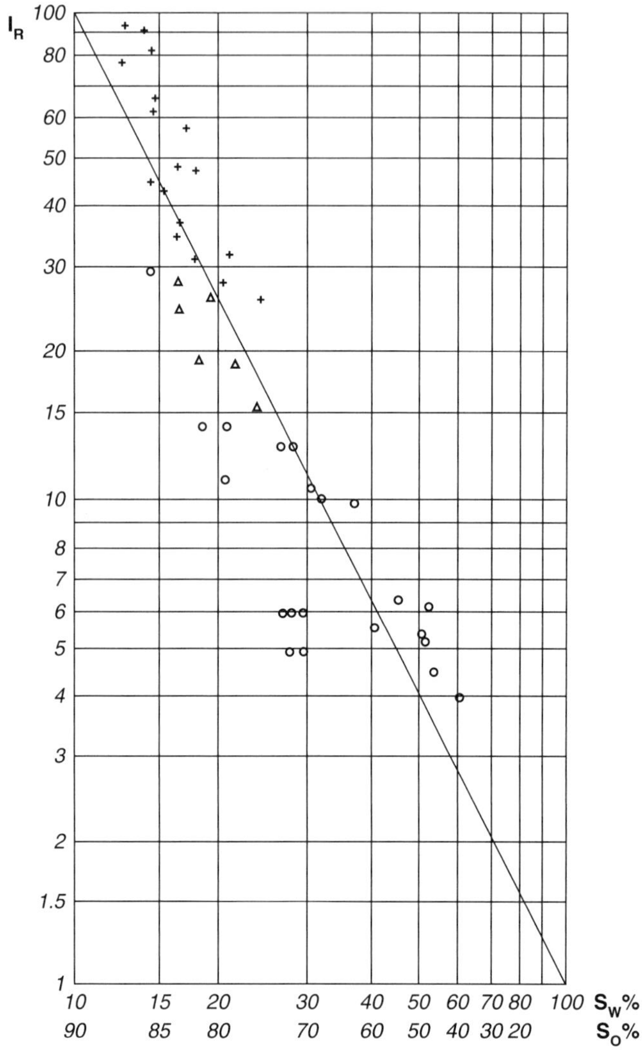

Figure 2.52 Relationship between the formation resistivity index and oil/water saturation for Azerbaijan's Pliocene Productive Series. (After Buryakovsky, 1977.) KS – Kirmaku Formation; PK – Podkirmaku Formation; and KaS – Kala Formation.

Na^+ cations but impermeable to the Cl^- anions. When a clay layer separates NaCl solutions of different salinity, the Na^+ cations move through the clay from more concentrated to less concentrated solution. The movement of charged ions is an electric current, and the force causing them to move constitutes an electric potential across

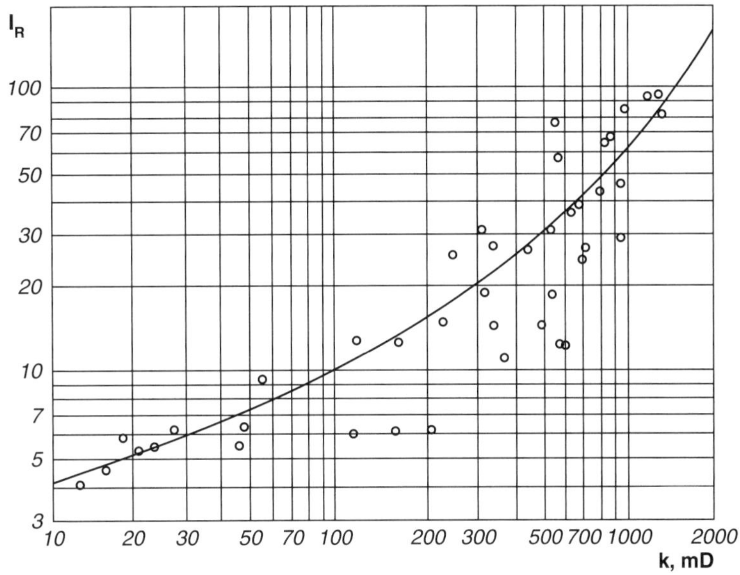

Figure 2.53 Relationship between the formation resistivity index and permeability for Pliocene Production Series of Azerbaijan. (After Buryakovsky, 1977.)

the clay. Inasmuch as the clay passes only the cations, it resembles ion-selective membrane, and the potential across the clay layer is called the *membrane potential.*

Another component of the electrochemical potential is produced at the edge of the invaded zone, where the mud filtrate and the formation water are in direct contact. Here the sodium and chlorine ions can transfer from either solution to the other. Because Cl^- ions have a greater mobility than Na^+ ions, the net result is a flow of negative charges (Cl^-) from the more concentrated solution to the less concentrated solution. This result in a current flow in the direction opposite to the current flow due to membrane potential.

The resulting potential is only about one-fifth of the membrane potential.

If the permeable formation is not clayey, the total electrochemical emf, corresponding to these two phenomena, is equal to:

$$E_{ch} = -K\log(a_w / a_{mf}) \tag{2.113}$$

where E_{ch} is the total electrochemical emf; a_w and a_{mf} are the chemical activities of the two solutions (formation water and mud filtrate) at formation temperature; and K is the coefficient proportional to the absolute temperature. For saline (NaCl) formation water and mud filtrate, K is equal to 71 at 25°C (77°F).

The chemical activity of a solution is roughly proportional to its salt content (or to its conductivity) or inversely proportional to its resistivity that is expressed as follows:

$$V_{ch} = -K\log(R_{mf}/R_{w}) \tag{2.114}$$

where V_{ch} is the electrochemical component of the spontaneous potential in clean formations, and R_{mf} *and* R_{w} are the resistivities of the mud filtrate and formation water, respectively.

2.9.1.2 *Electrokinetic Component*

The electrokinetic component of the SP is produced by the flow of an electrolyte through a porous media. It depends on several factors, among which are the differential pressure producing the flow and the resistivity of the electrolyte.

Movement of filtrate through the mud cake or a permeable formation produces an electrokinetic emf. Usually, little or no electrokinetic emf is generated across the permeable formation itself, because practically all the differential pressure (between the borehole and formation) occurs across the much less permeable mud cake. The remaining differential pressure across the formation is normally not large enough to produce an appreciable electrokinetic emf.

Appreciable electrokinetic effect may be observed in very low-permeability formations and shales, in which a rather great portion of the pressure differential is applied across these formations. If the formation permeability is so low that no mud cake is formed, the total pressure differential is applied to the formation itself. If the formation water is brackish, then the mud is resistive, and the formation is clean with some porosity, the electrokinetic effect may be quite large.

As an example, in order to determine relationships between reservoir-rock properties and spontaneous potential data, more than 860 cores, recovered from the Productive Series of Azerbaijan and adjacent offshore areas, were analyzed. Almost 4000 analyses of sands, silts, and shales were made (Buryakovsky *et al.*,

1986). The cores were recovered from two regions of the South Caspian Basin:

1. Offshore areas of the Baku Archipelago and South Absheron Zone.
2. Onshore areas of the Lower Kura Depression.

The important parameters luencing the downhole spontaneous potential (so-called surface activity parameters) are the cation exchange capacity Q_{100} (in mg-equivalents of exchangeable ions per 100 g of rock) and diffusion/adsorption factor A_{da} (in mV). The reservoir rock properties studied were as follows: ϕ,- porosity in percent; k – permeability in mD; $C_{carb'}$ - carbonate cement content in weight %; K_{sh} – shale/clay content in volume %, and η.- relative shale-clay content, dimensionless:

$$K_{sh} = C_{sh}(1 - \phi/1 - \phi_{sh}) \tag{2.115}$$

$$\eta = K_{sh}/(K_{sh} + \phi) = C_{sh}(1 - \phi) \div [C_{sh}(1 - \phi) + \phi] \tag{2.116}$$

where ϕ_{sh} is the shale/clay porosity; when ϕ_{sh}=0, $K_{sh} = C_{sh}(1 - \phi)$.

Figure 2.54 shows correlations between reservoir rock properties and surface activity parameters of rocks of the Productive Series of Azerbaijan (Buryakovsky *et al.*, 2001), whereas the interrelationship among the reservoir rock properties, surface activity parameters, and relative shale/clay content is presented in Figure 2.55.

2.9.1.3 Practical Use of Electrical Properties of Rocks

The electrical resistivity and spontaneous potential (SP) are important petrophysical characteristics used in geophysical investigation of subsurface formations, especially for qualitative and quantitative estimation of reservoir-rock properties. For quantitative evaluation, petrophysical parameters should be calibrated using core-analysis data.

Resistivity is used to:

1. Detect reservoir and non-reservoir beds and permit their correlation.
2. Locate boundaries and determine thicknesses of such beds.
3. Determine quantitative values of reservoir-rock porosity and permeability.
4. Determine the nature and amount of reservoir-rock saturation with formation fluids.

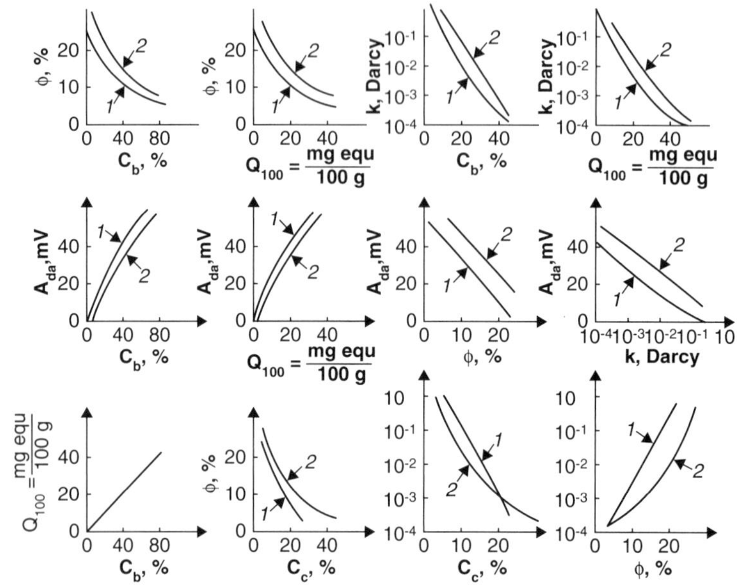

Figure 2.54 Correlations between reservoir-rock properties and surface activity parameters of rocks of the Productive Series. (Modified after Buryakovsky *et al.*, 1986a.) 1 – Baku Archipelago and South Apsheron Offshore Zone; 2 – Lower Kura Depression. C_s = clay/shale content; C_c = carbonate content.

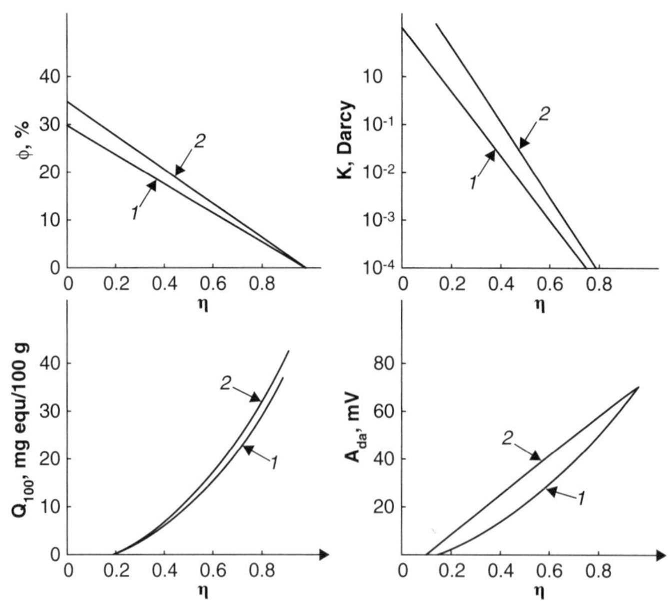

Figure 2.55 Interrelationships among the reservoir-rock properties, surface activity parameters, and relative clay content of the Productive Series rocks. (Modified after Buryakovsky *et al.*, 1986a.) 1 – Baku Archipelago and South Apsheron Offshore Zone; 2 – Lower Kura Depression. η – see Eq. 2.116.

Spontaneous potential is used to:

1. Detect the permeable beds and permit their correlation.
2. Locate boundaries and determine thicknesses of such beds.
3. Determine values of formation-water resistivity, R_w.
4. Give qualitative indications of bed clayeness and porosity.

The combination of resistivity and SP data plus results of some other well-logging methods permits to produce the detailed stratigraphic framework and give description of penetrated section.

2.9.1.4 *Sonic Velocity Estimation in Shales in Abnormally-Pressured Formations From Resistivity Data*

Kerimov *et al.* (1996) proposed a method for estimating (1) the sonic velocity v_a in abnormally high-pressured shales using resistivity logs and, thus, (2) shale bulk density. Analysis of the well-log data for productive strata in Azerbaijan showed that there is a poor correlation between the sonic velocity and other well-log data, such as resistivity, SP, neutron, and gamma-ray data. Introducing the normal trend of sonic velocity v_n and resistivity ρ_n, allowed Kerimov *et al.* (1996) to express sonic velocity (and, therefore, the bulk density of shales) as a non-linear function of resistivity with a good correlation between the normalized velocity and normalized resistivity.

The best-fit regression equation is of the following form:

$$v_a / v_n = 0.63 + 0.37(\rho_a / \rho_n)^{3.26} \tag{2.117}$$

where ρ_a and ρ_n are the resistivities of abnormally-pressured and normally-compacted shales, respectively. The coefficient of correlation between parameters v [$= v_a / v_n$] and ρ [$=(\rho_a / \rho_n)^{3.26}$] is 0.87 and the mean squared error for v_a is 190 m/s. Thus, the v_a can be estimated from the (ρ_a / ρ_n) ratio and the v_n obtained from the normal compaction trend in the area studied.

2.10 Radioactivity

In 1896, Antoine Becquerel found that compounds of uranium emitted rays that gave an impression on a photographic plate covered with black paper. The emitted rays were able to pass through thin sheets of metal and other substances that are opaque to light. These rays also possess the important property of discharging a body that is electrified, either positively or negatively. It was soon found that the property of emitting penetrating radiation is inherent also in other chemical elements such as, for example, thorium.

By studying radioactive minerals, which contained uranium and thorium, Pierre and Marie Curie discovered in 1989 that these minerals often contain a substance, which is much more radioactive than either uranium or thorium. This element was finally isolated and called radium.

Such property of emitting penetrating radiation is called radioactivity. The physical notion of radioactivity is confined in the atomic structure of chemical elements.

2.10.1 Atomic Structure

All elementary forms of matter are composed of very small unit quantities called atoms. The atoms of a given element all have the same size and weight. The atoms of different elements have different sizes and weights. Atoms of the same or different elements unite with each other to form very small unit quantities called molecules. Molecules form chemical compounds.

Atoms are composed of a positively charged central core or nucleus, surrounded by a cloud of negatively-charged electrons. The charges are equal, so that the atom is electrically neutral.

Nucleus is the central part of the atom, which makes up most of the atom's weight and is positively charged. Atomic nuclei are made up of two kinds of fundamental particles: *protons* and *neutrons*.

Protons are small particles with a positive charge equal numerically to the negative charge of an electron, 1.602×10^{-19} Coulomb A single proton makes up the entire nucleus of the ordinary hydrogen atom. The nuclei of atoms of other elements contain a number of protons.

Neutrons are particles with no electric charge, but with a mass approximately the same as that of the proton, about 1.67×10^{-24} gram. In nature, neutrons are locked up in the nucleus of an atom, but they can be discharged out from the atom's nucleus during various kinds of atom-smashing events.

Electron is the smallest known particle having a negative charge. The cloud part of an atom outside the nucleus is made up of electrons. The number of which, being equal to the protons in the nucleus, is the same as the atomic number for the atom. However, the electron weight is only about one two-thousandth part of the atom's total. A few electrons may readily be removed from, or added to, an atom leaving it with surplus of positive or negative charge. An electric current in any conductor consist of the motion of electrons through the material of the conductor.

Atomic weight is used to denote the weight of any atom as measured on an arbitrary scale based on the weight of an oxygen atom. On this scale, the figure 16 is chosen as the approximate weight of an oxygen atom. Adopting this convention makes it possible to express the weights of atoms of the other elements very nearly as whole numbers. These whole numbers are referred to as the rough atomic weight. The figure giving the rough atomic weight of an atom on this scale is also the total number of protons and neutrons contained in the atom's nucleus.

Atomic number refers to the number of protons in the nucleus and the number of electrons outside the nucleus. The number of neutrons can thus be found by subtracting the atomic number from the atomic weight. For example, in uranium-238, the total number of protons and neutrons is 238. The number of protons alone, given by the atomic number of uranium, is 92. Hence, by subtraction, the number of neutrons is 146 (238-92).

The most interesting atoms are the lighter and the heavier ones. The lightest atoms are hydrogen, helium, and lithium, whereas the heaviest atoms of naturally-occurring elements are radium, actinium, protactinium, and uranium.

The hydrogen atom has an atomic weight, an atomic number, and an atomic charge of unity. At a normal condition, the hydrogen atom consists of a positive nucleus and one electron in orbital motion around this nucleus. When the electron is by any means torn away from the hydrogen atom, the nucleus alone remains with a single positive charge on it and is called the positive hydrogen ion, which is the same as a proton.

Helium has an atomic weight of 4 and an atomic number and nuclear charge of 2. The helium nucleus, constitutes most of the helium atom's mass. The helium atom consists of 2 neutrons and 2 protons. Outside of the nucleus are 2 electrons. These electrons are not considered as point charges revolving as planets around the nucleus, but as diffuse rings of electricity occupying the entire orbit.

Isotopes involve an element's different atomic weight. The number of protons in the nucleus defines what chemical element the nucleus represents and is called the atomic number of the element; the number of protons plus neutrons defines the atomic weight of the element. Atoms, however, may have the same number of protons but a different number of neutrons, in which case they are atoms of the same element but of different weights. Such different atoms are called isotopes. The isotopes of one element differ from one another in their physical, but not their chemical properties. Some isotopes are stable, whereas other isotopes are unstable (*radioactive isotopes*).

Radiation is a process that combines the emission transmission, and absorption of radiant energy. Neutrons with their lack of charge readily penetrate other nuclei. The resulting nuclei formed are sometimes stable, but often unstable. These unstable nuclei decompose spontaneously and may emit protons or neutrons, or clusters of two protons and two neutrons, called *alpha particles*. They may emit *beta particles* (electrons), a process called beta-decay; or they may radiate *gamma rays* that are electromagnetic waves.

Alpha particles are very easily absorbed by thin metal foil or by few inches of air. They affect a photographic plate, cause many bodies to fluoresce brilliantly, and ionize the air though which they pass. They consist of positively-charged particles projected from the parent atom with velocity about one-tenth the velocity of light, and are deflected by the electromagnetic field.

Beta particles are negatively-charged and are projected from the atom of the radioactive substance with a velocity that is nearly, but not quite, as great as the velocity of light. In the electromagnetic field they are deflected just as cathode rays. Because of their larger velocities and smaller mass, they are much more penetrating than the alpha particles. Beta particles produce much less ionization in the gas through which they pass than do the alpha particles and are less active photographically than alpha particles

Gamma rays are extremely penetrating, and are not deflected by the electromagnetic field. Their nature is entirely different from that of alpha and beta particles. These rays are electromagnetic pulses like very penetrating X-rays.

The process of spontaneous disintegration or radioactive decay leads to transformation of one nucleus to another. Due to this disintegration, the original elements are gradually transmuted into other elements having different chemical properties. The ultimate product of radioactive disintegration is lead. An expression describing the radioactive decay phenomenon is called the *decay law*. If the time of decrease of a quantity is proportional to the quantity at that time, then the decay is exponential, i.e.,

$$N(t) = N_0 e^{-\lambda t} \tag{2.118}$$

where $N(t)$ is the quantity at time t, λ is the decay constant, and N_0 is the value of $N(t)$ at time $t = 0$.

The length of time it takes a sample of radioactive element or radioactive isotope to decrease to half of its original amount by radioactive decay is called the *half-life* The half-life number is a physical constant characteristic of the isotope. It is independent of the particular amount originally present and of external conditions such as temperature and pressure. The half-life, $T_{1/2}$, is related to the decay constant λ by the following expression:

$$T_{1/2} = 0.693/\lambda \tag{2.119}$$

2.10.1.1 *Practical Use of Radioactive Properties of Rocks*

Gamma rays and neutrons have found a wide use in geophysical investigation of the subsurface formations.

Some radioactive elements or radioactive isotopes emit gamma rays (burst of high-energy electromagnetic waves) spontaneously. The radioactive potassium isotope having an atomic weight of 40, and the other radioactive elements of the uranium and thorium series, emits nearly all of the gamma radiation encountered in the Earth.

Each of these elements or their isotopes emit gamma rays, the number and energies of which are distinctive of each element or isotope. As an example, potassium (K-40) emits gamma rays of a single energy at 1.46 MeV, whereas the uranium and thorium series emit gamma rays having various energies. The energy of uranium series varies in the limits of 0.1 to 1.75 MeV, with modal value of about 0.6 MeV. The energy of thorium series changes from 0.15 to 1.0 MeV with most frequent value of 0.25 MeV.

In passing through subsurface formations, gamma rays experience successive *Compton-scattering* collisions with the atoms of the formation, losing energy with each collision. Finally, after the gamma ray has lost enough energy, it is absorbed via the photoelectric effect. The amount t of absorption varies with formation density. Two formations have the same amount of radioactive material per unit volume, but having different densities, will show different radioactivity levels. The less dense formation will appear to be more radioactive.

In subsurface formations, the gamma rays normally reflect the clay content of the formations, because the radioactive element tend to concentrate in clays owing to high specific surface area of the very fine particles of clay minerals. Conversely, clean formations (sand, sandstone or carbonate rocks) usually have a very low level of radioactivity.

Neutron activity is a result of the electrically neutral particles having a mass almost identical to the mass of a hydrogen atom. High-energy (fast) neutrons emitted from a radioactive source collide with nuclei of the formation constituents. With each collision a neutron loses some of its energy.

The amount of energy lost per collision depends on the relative mass of the nucleus with which the neutron collides. The greater energy loss occurs when the neutron strikes a nucleus of practically equal mass, i.e., a hydrogen nucleus. Collision with heavy nuclei does not slow the neutron down very much. Thus, the slowing-down of neutrons depends largely on the amount of hydrogen in the formation. Within a few microseconds, the fast neutrons are slowed down by successive collisions to thermal velocities, corresponding to energies of around 0.025 eV. The neutrons are diffused randomly, without losing any more energy until being captured by the nuclei of atoms such as chlorine, hydrogen, silicon, etc. The capturing nucleus becomes intensely excited and emits a high-energy gamma ray of capture.

When the hydrogen concentration of the material surrounding the neutron source is large, then most of the neutrons are slowed down and captured within a short distance of the source. Conversely, if the hydrogen concentration is small, the neutrons travel father from the source before being captured.

In subsurface formations, the neutron activity normally reflects the amount of hydrogen present in the formation. In clean formations the pores of which are filled with water or oil, the neutron activity reflects the amount of liquid-filled porosity.

Radioactivity relationships are used to determine clay content, porosity, permeability, and connate (interstitial) water content of the rocks. All relationships are based on core and log analysis and are represented as petrophysical curves.

Table 2.10 presents the results of an experimental study of cores from Azerbaijan to determine the total radioactivity and concentration of main radioactive isotopes in different types of siliciclastics rocks. The presence of radioactive isotopes is associated with clay minerals: total and partial radioactivity increases with increasing clay mineral content. Total radioactivity values ranged from 0.6 to 4.1 pulse/minute. *g*; content of potassium, C_K, from 0.1 to 2.6%; content of thorium, C_{Th}, from 1.1×10^{-4} to 12.4×10^{-4} %; and content of uranium, C_U, from 0.2×10^{-4} to 2.8×10^{-4}%. These results are in agreement with radioactivity of other types of siliciclastic rocks, and are significantly lower than the Clarke values (i.e., the average percent of an element in the Earth's crust). Interpretation of this data indicated that the total radioactivity of studied rocks was mainly due to their potassium and thorium contents.

Determination of the clay content in the reservoir rocks from the gamma-ray log is based on the relationship between the natural radioactivity of rocks and their clay content. The precondition is the absence of dispersed radioactive isotopes or local accumulations of radioactive minerals.

2.10.1.2 *Increase in The Clay Content Causes an Increase in the Radioactivity*

To obtain the relationship between natural radioactivity and clay content, the relative gamma-ray factor ΔI_y, is used according to the following equation:

$$\Delta I_\gamma = K(I_\gamma - I_{min})/(I_{max} - I_{min}) \qquad (2.120)$$

where: I_y, I_{max}, and I_{min} are the GR log readings for (a) the interval under study, (b) the reference shale interval, and (c) the reference non-shaly clean sand; *K* is the coefficient reflecting the content of fine-particle fraction in the reference shale layer. Through the use of the relative gamma-ray factor, it is possible to avoid many errors caused by the measuring equipment, accuracy of measurement, well design, and lateral shale radioactivity fluctuation.

If lithologies of reference shale and sand layers are constant for a given stratigraphic section, then the coefficient *K* will also be constant and can be disregarded. If one considers the minimum reading as zero and then expresses all values in relative units, the previous equation will become:

$$\Delta I_\gamma = I_\zeta / I_{max} \qquad (2.121)$$

Table 2.10 Total radioactivity and content of main radioactive isotopes in sedimentary rocks of Absheron Archipelago, Azerbaijan.

Lithology	Total radioactivity pulse/min*g		C_K, %		$C_{Th} \times 10^4$%		$C_U \times 10^4$%	
	Range	Mean	Range	Mean	Range	Mean	Range	Mean
Siltstone	1.26–2.98	1.83	0.2–1.8	0.87	2.5–6.3	4.4	0.1–2.5	1.4
Sandy-clayey siltstone	1.23–2.99	2.11	0.1–1.95	1.2	1.8–7.4	4.4	2.4–3.0	2.6
Clayey-sandy siltstone	0.60–1.10	0.92	0.1–0.6	0.21	1.1–4.5	2.9	0.3–0.9	0.3
Clayey siltstone	2.3–3.85	3.0	0.8–2.2	1.4	3.4–10.4	6.1	1.1–2.8	1.7
Silty clay	3.35–4.1	3.82	1.9–2.6	2.2	7.1–12.4	9.7	0.7–2.2	1.7

When calculating ΔI_{ς}, corrections are introduced in the I_y and I_{max} values. These corrections take into account the borehole conditions: deviation of the actual borehole diameter over the studied and reference intervals from nominal one, and the effect of the equipment inertia depending on the interval thickness. For thin and medium-thick layers, the lowering of the GR readings is observed with the increase in the speed v of the measuring equipment as it is logging up the borehole, and of the constant τ characterizing the integrating circuit. The true value of I_γ is calculated by using the following equation:

$$(\Delta I_\gamma)' = \Delta I_\gamma / v_\gamma \tag{2.122}$$

where v_γ is the GR reading decrease ratio determined from the product $v\tau$ and interval thickness with the use of the graph proposed by Vendelshtein *et al.* (1984).

Determination of porosity from the GR logs in based on two relationships: (1) between the core porosity ϕ and clay content C_{cl} (or $C_{cl\,v}$), and (2) between the clay content C_{cl} (or $C_{cl\,v}$) and natural radioactivity from core data q_γ. Determination of the connate water content from GR logs is based also on two relationships: (1) between the content of connate water from core data, $S_{w,res}$, and clay content C_{cl} (or $C_{cl\,v}$), and (2) between the clay content C_{cl} (or $C_{cl\,v}$) and natural radioactivity from core data q_γ. To transfer from the radioactivity derived from the data (q_γ) to the relative radioactivity from logs $(\Delta I_\gamma)'$, the correlation between these two parameters can be established.

2.10.2 Radioactivity Logging Applications

Radioactive logging is conducted by *gamma-ray, gamma-gamma-ray, neutron-gamma-ray, neutron-neutron, density, and pulsed neutron-capture* sondes for measuring the natural or induced radiation.

Gamma-ray log signature is based on the capture of bursts by high-energy electromagnetic waves that are emitted spontaneously by some radioactive elements. Radioactive potassium isotope K^{40} and radioactive isotopes of the uranium and thorium series emit almost all amount of gamma radiation encountered in the Earth. The gamma-ray log is particularly useful for delineating the alteration of clay-rich (highly radioactive) and clay-poor (low radioactivity) lithologies.

The gamma-ray (GR) sonde is the well-logging tool, comprising a scintillometer, used to measure the natural radioactivity of the rocks exposed in a wellbore. The gamma-ray log can be used in cased wells, which makes it very useful in completion and workover operations. It is frequently used in combination or as a substitute for the SP log in cased holes where the SP is unavailable, in open holes where the SP is unsatisfactory, or to check for radioactive sands that appear as shales on the SP signature.

Neutron-gamma log signature is produced by neutrons that are electrically neutral particles, each having a mass almost identical to the mass of a hydrogen atom. High-energy ("fast") neutrons are continuously emitted from a radioactive source mounted in the sonde. As the neutrons collide with nuclei of materials in a formation, each neutron loses some of its energy with each collision. The greatest energy loss occurs when the neutron strikes a hydrogen nucleus. Thus, the slowing-down of neutrons depends largely on the amount of hydrogen in the formation.

Within a few microseconds the neutrons are slowed down by successive collision to "thermal" velocities, corresponding to energies of around 0.025 electron volts. The neutrons then diffuse randomly, without losing any more energy, until they are captured by the nuclei of atoms such as chlorine, hydrogen, silicon, etc. The capturing nucleus becomes intensely excited and emits the high-energy rays. Depending on the type of neutron log tool, either gamma rays (gamma-ray logging – NGR) or neutrons themselves (neutron-neutron logging – NNL) are counted by a detector in the sonde.

The NGR sonde is an instrumental package for measuring the amount of hydrogen present within the rocks surrounding a borehole. The sonde contains a neutron source and a gamma-ray detector. Typical neutron source used in the NGR package is (1) polonium–beryllium (Po+Be), (2) plutonium–beryllium (Pu+Be), or (3) americium–beryllium (Am+Be), with a neutron spectrum of continuous energy up to 11 MeV and having three maximums at 0.4, 3, and 4.5 MeV. Source capacity is 3.7×10^{16} neutrons per second for Pu+Be source. The neutron source is located 600 mm from the detector. Spacing of the combined GR and NGR device is 60 cm. Generally, GR and NGR logging are conducted in cased holes.

Neutron-neutron log (NNL) sonde comprises a neutron source and detector. The number of neutrons back scattered to the detector is proportional to the number of hydrogen atoms within the rocks surrounding a borehole. The NGR and NNL logs are used to determine the porosity of rocks because pore space is filled wither with water containing salts derived from hydrochloric acid or with hydrocarbons.

Gamma-ray (GR) and neutron–gamma (NGR) logs are recorded for correlation purposes at an up the borehole speed of 400 to 800 m/hr. Detailed logging is performed at a speed of 200 to 400 m/hr. Both GR and NGR tools are calibrated. Typical recording scales are 0.5 or 1 μ-roentgen/hr for the GR and 0.05 or 0.1 arbitrary units per 1 cm for the NGR.

Formation density log is used as a porosity tool. Other applications include detection of gas, determination of hydrocarbon density, evaluation of argillaceous sands and complex lithologies, etc. The formation density logging is conducted using medium-energy gamma rays, which are emitted by a radioactive source applied to the borehole wall in a shielded sidewall skid. These gamma rays are the high-velocity particles, which collide with the electrons in the formation. At each collision, a gamma ray loses some, but not all, of its energy to the electrons, and then continues to radiate with diminished energy. This type of interaction is known as Compton scattering. The scattered gamma rays reaching the detector, at a fixed distance from the source, are counted as an indication of formation density.

The number of Compton-scattering collisions is related directly to the number of electrons in the formation. So, the response of the formation density tool is determined as the *electron density* (the number of electrons per cubic centimeter), and electron density, in turn, is related to the true bulk density of the formation.

In order to minimize the influence of the drilling mud column, the source and detector (mounted on a skid) are shielded. The openings of the shields are applied against the well's wall by means of an eccentric arm. If the contact between the skid and the formation is not perfect, a special correction should be applied to the actual readings.

Pulsed neutron-capture log (thermal decay time log) is used to determine the residual oil saturation in cased holes where reservoir characteristics are known from previous logs and where drilling mud filtration has dissipated. A neutron generator is used to produce fast, high-energy (about 14 MeV) neutrons, which are slowed to thermal energy level and captured by nuclei resulting in gamma-ray emission. Spacing of thermal neutron detector is around 43 cm. Pulsed neutron log is recorded at a speed up the well at 100 to 200 m/hr.

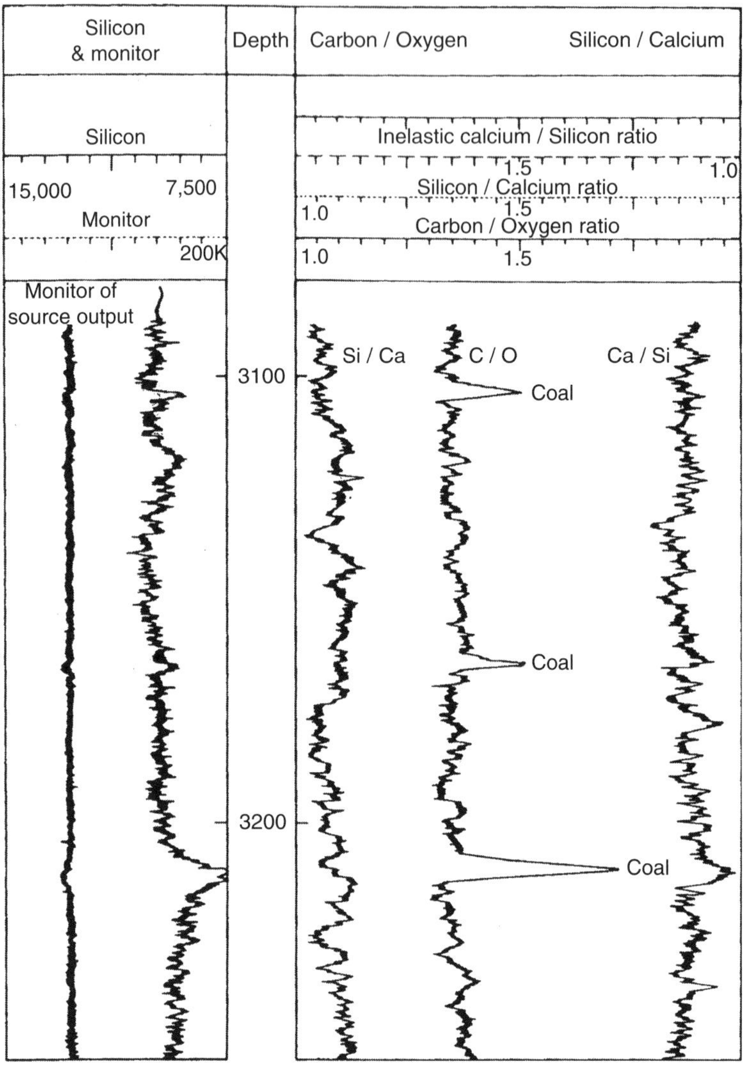

Figure 2.56 Continuous carbon/oxygen log response over a depth range of 3090 to 3250 ft in the McAlester Formation, Oklahoma. Note that the carbon/oxygen log does not respond to gas zones in clastic rocks. The silicon curve is used to identify natural gas in clastic rocks having a signature that is deflected to the left on the well log when gas is present. The well encountered high-volatile A bituminous coals (the coal rank designation hvAb) in the Barringer well 1-11 all of which contained gas as determined by desorption in the amounts of about 309 ft^3/ton to 570 ft^3/ton.

Carbon–oxygen log utilizes a high-energy pulse neutron source, which produces 14 MeV neutrons by the deuterium–tritium reaction. Under high-energy neutron irradiation, elements emit gamma rays with specific energy levels (Culver *et al.*, 1974). The carbon–oxygen logging tool allows a measurement of the relative elemental amounts of carbon and oxygen present in the formation. This log is valuable in determining coal beds with possible associated methane gas and is used to identify existing coal beds from other carbon deposits such as petroleum based shales and low-density lignites. Inasmuch as the carbon–oxygen log measures the elemental carbon and oxygen directly, coal detection behind casing for sizes of 4-in or greater, became feasible (Rieke *et al.*, 1980). An example of the carbon-oxygen log was identification of three coalbeds in the McAlester Formation, penetrated by the Barringer 1-11 well, Pittsburg County, Oklahoma, USA, between the depths of 3213 and 3217 feet (Figure 2.56). The well was air drilled and cored the 18 coal zones in order to obtain the amount of gas in each of the beds along with their rank.

2.11 Chemistry of Waters in Shales versus those in Sandstones

It is commonly assumed that the chemistry of interstitial waters in shales is the same as those in associated sandstones. Rieke and Chilingarian (1974) and Chilingar and Rieke (1975) showed that this assumption is incorrect. The intestinal water in shales is fresher than those in associated sandstones (Table 2.11). In addition, the interstitial water in undercompacted shales (associated with overpressured formations) is saltier than those in well-compacted similar shales (same depth, mineralogy, etc.).

The pore water in the sandstones has higher salinity than those found in either type of the associated shales. The importance of these findings is applicable in well logging. Erroneous electric log interpretations could result if the R_w (resistivity of the formation water) in the shales is assumed to be equal to those in the associated sandstones.

Table 2.11 Chemistry of pore water in associated undercompacted and well-compacted shales and sandstones from various parts of the world where overpressured formations are present. The results show that the overpressured (undercompacted) shales have slightly higher chloride ion concentrations than those in comparable (at about the same burial depth) well-compacted shales having similar mineralogy. (After Chilingar and Rieke, 1994, table 5-3, p. 119.) (Courtesy of Elsevier.)

Number of Samples Tested	Depth, ft.	Chlorinity, mg/1		
		Well-Compacted Shales	Undercompacted Shales	Associated Sandstones
3/3/3	2,000–3,000	3,000–4,000	8,000–20,000	70,000–80,000
4/2/2	3,000–4,000	2,000–3,000	10,000–30,000	70,000–90,000
3/3/2	4,000–5,000	1,600–3,500	10,000–40,000	75,000–90,000
2/2/3	5,000–6,000	1,500–3,500	9,000–35,000	60,000–200,000
6/2/3	6,000–7,000	3,000–6,000	8,000–10,000	70,000–130,000
3/3/4	7,000–8,000	4,000–8,000	5,000–9,000	90,000–135,000
3/4	8,000–9,000	10,000–20,000	–	90,000–100,000
4/3/4	10,000–11,000	2,000–3,000	10,000–14,000	15,000–70,000
5/3/2	11,000–12,000	2,000–3,000	8,000–14,000	13,000–17,000
7/3/4	12,000–13,000	1,500–3,000	8,000–14,000	11,000–30,000
2/2/2	13,000–14,000	2,500–4,500	10,000–14,000	11,000–50,000
2/4	14,000–15,000	10,000–14,000	–	90,000–120,000

3
Seismic Parameters

Carl Richter

3.1 Introduction

Major progress was made recently in the use of seismic data to obtain information about reservoir fluids, their movement in the reservoir over time, and the changes in rock properties within the reservoir. Three-dimensional seismic data make it possible to visualize reservoirs in 3-D space. Quantifying rock physics properties, will improve recovery of oil and gas from complex reservoirs. Today, exploration coupled with the improved oil and gas recovery technologies relies on seismic characterization of petrophysical properties, such as porosity, permeability, fracture detection and delineation, pore pressure, matrix mineralogy, and fluid saturation components.

Quantifying the relationships between petrophysical and seismic properties can add a significant amount of information to any seismic interpretation, because seismic velocities are sensitive to reservoir parameters, such as porosity, lithology, pore fluid type, saturation, and pore pressure. Rock physics uses predominantly seismic properties to characterize and understand hydrocarbon reservoirs.

The goal of this chapter is to provide an overview of rock physics as it relates to reservoir characterization and the interpretation of seismic data. The main goals are to illustrate the extraction of important data, such as the porosity, permeability, and the composition of pore fluids, to improve commercial drilling decisions.

Many reservoirs exhibit significant heterogeneity in porosity, permeability, sand and clay contents, and other properties. This can cause great complexity in reservoir recovery processes, such as migration of gas caps in reservoirs with discontinous shales, or the tracking of injected water, steam, or temperature during recovery in reservoirs having large variations of permeability. Wheras it is

Leonid Buryakovsky, George V. Chilingar, Herman H. Rieke and Sanghee Shin
Petrophysics, (151–182) © Scrivener Publishing LLC

impossible to determine the variability from wireline logs or core data, it can be extracted from geophysical measurements, especially seismic data. The relationships between velocity and porosity, porosity and permeability, P- and S-wave velocity and saturation, which were ill-defined in the past, now exhibit some degree of correlation (e.g., Mavko *et al.*, 1998; Mavko, 2000, 2005; Avseth *et al.*, 2005).

Petrophysics integrates all kinds of log and core data to obtain information about the properties of reservoir rocks. Determining the correlation between seismic attributes and rock properties involves the application of the elastic rock and pore fluid properties, and the development of models for rock–fluid interactions. Rock physics developed a precise relationship between rock properties and seismic attributes and made it possible for geoscientists to tie rock properties together with seismic data. As the information about porosity, fluids, and lithology become accessible, the seismic interpretation is augmented.

3.2 Elastic Properties

The stress state in a rock pore is presented in Figure 3.1. Understanding of seismic waves requires a comprehension of the elastic properties of matter responsive to stress. In the elastic field, where materials react to stress by temporary deformation, the linear relationship between stress and strain (Hook's Law) is specified by an elastic modulus, which is the ratio between the stress and the resultant

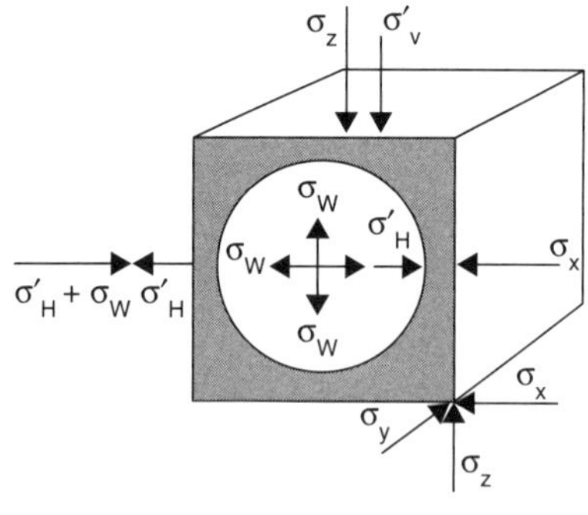

Figure 3.1 Stress state in rock. Diagrammatic sketch of the stress state in a rock body underground, where σ'_v is the effective (intergranular) stress in the vertical direction, σ'_H is the horizontal effective stress, σ_W is the pore water stress, and σ_Z is the total vertical stress component. The total horizontal stress component in the x-direction σ_X is equal to $\sigma'_H + \sigma_w$.

strain. For the discussion of seismic properties, only two of the elastic moduli are of importance: the bulk modulus or compressibility, κ, and the shear or rigidity modulus, μ. The bulk modulus is the stress/strain ratio for hydrostatic pressure applied to a cubic element:

$$\kappa = \frac{stress}{strain} = \frac{P}{\Delta v / v} \tag{3.1}$$

where $\Delta v/v$ represents the volume change of the cube as a function of applied stress P.

The shear modulus is the ratio of the shearing stress, τ, to the tangent of resultant angle of defrmation, θ (Figure 3.2):

$$\mu = \frac{\tau}{\tan \theta} \tag{3.2}$$

where θ is the shear angle (deformation angle). Liquids cannot be sheared and, therefore, the shear stress of liquids is equal to zero.

The seismic velocity of P-and S-waves is a function of density, ρ, and the shear and compressibility moduli of the material. The seismic velocity of P-waves, v_p, increases with increasing compressibility, κ, and shear, μ, moduli, and decreases with density of the material:

$$\upsilon_p = \sqrt{\frac{\kappa + 1.33\mu}{\rho}} \tag{3.3}$$

Knowledge of density and bulk and shear moduli can, therefore, be translated directly into P-wave velocity.

The velocity of S-waves, v_s, increases with increasing shear modulus and decreases with density. It is not dependent on the compressibility and, therefore, can be easily calculated from only two parameters:

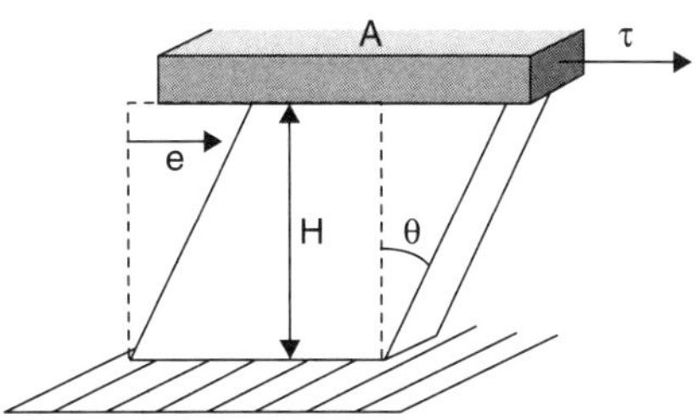

Figure 3.2 Shear or rigidity modulus, μ: ratio of shearing stress to resultant shear strain. μ is zero in liquids. H = height of element; A = surface area; e = resultant shear strain; θ = shear angle; τ = shear stress.

$$\upsilon_s = \sqrt{\frac{\mu}{\rho}} \tag{3.4}$$

Because the shear modulus, μ, of liquids is zero, S-wave velocity in liquids is zero. Inasmuch as the shear modulus influences not only the S-wave velocity, but also the P-wave velocity, materials with lower rigidity also exhibit lower P-wave velocities.

3.3 Velocity and Rock Properties

Seismic velocity is determined by the density of a rock and its elastic moduli, which are influenced by numerous parameters, such as porosity, type of material, temperature, and pressure. These parameters and their effects on the seismic velocity of rocks are presented below.

Porosity is defined as the ratio of the pore volume, V_p, to the total bulk volume, V_t:

$$\phi = \frac{V_p}{V_t} = \frac{V_t - V_s}{V_t} \tag{3.5}$$

where V_s is the volume of solid phase(s). Reservoir rocks are distinguished by high porosity.

Differences in seismic velocities of various rock types have long been used to decipher the seismic velocity layering in the body of Earth. Major discontinuities occur where rock types with different velocities occur, e.g., at the mantle/core boundary or where the material is in a different state, e.g., at the outer core/inner core boundary or at the low-velocity layer where about 1% of the upper mantle is in a liquid state. The P- and S-wave velocities of different geological materials are presented in Table 3.1. Most sedimentary rocks have P-wave velocities of less than 3,000 m/s, except for certain limestones, dolomites, salt, and igneous and metamorphic rocks, which have velocities of up to 6,000 m/s.

Laboratory experiments have shown that increasing effective pressure (the difference between the confining pressure and pore pressure) will increase the P- and S-wave velocities as presented in Figure 3.3, because voids, such as fractures which may be closed, increase the rigidity modulus (e.g., Mavko, 2000, 2005). The number of cracks is proportional to the amount of velocity change with

Table 3.1 P-wave and S-wave velocities for various materials (data compiled by Bourbié *et al.*, 1987).

Type of Formation	P-wave Velocity (m/s)	S-wave Velocity (m/s)	Density (g/cm^3)	Density of Constituent Mineral (g/cm^3)
Scree, vegetal soil	300–700	100–300	1.7–2.4	–
Dry sands	400–1200	100–500	1.5–1.7	2.65 (quartz)
Wet sands	1500–2000	400–600	1.9–2.1	2.65 (quartz)
Saturated shales and clays	1100–2500	200–800	2.0–2.4	–
Marls	2000–3000	750–1500	2.1–2.6	–
Saturated shale and sand sections	1500–2200	500–750	2.1–2.4	–
Porous and saturated sandstones	2000–3500	800–1800	2.1–2.4	2.65 (quartz)
Limestones	3500–6000	2000–3300	2.4–2.7	2.71 (calcite)
Chalk	2300–2600	1100–1300	1.8–3.1	2.71 (calcite)
Salt	4500–5500	2500–3100	2.1–2.3	2.1 (halite)
Anhydrite	4000–5500	2200–3100	2.9–3.0	–
Dolomite	3500–6500	1900–3600	2.5–2.9	2.8–2.9 (dolomite)
Granite	4500–6000	2500–3300	2.5–2.7	–
Basalt	5000–6000	2800–3400	2.7–3.1	–
Gneiss	4400–5200	2700–3200	2.5–2.7	–
Coal	2200–2700	1000–1400	1.3–1.8	–
Water	1450–1500	–	1	–
Ice	3400–3800	1700–1900	0.9	–
Oil	1200–1250	–	0.6–0.9	–

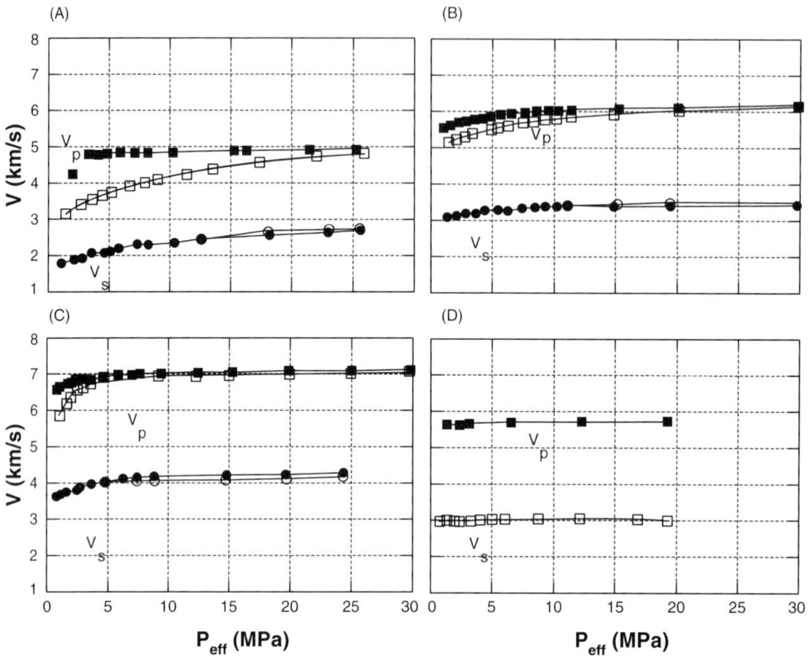

Figure 3.3 Relationships between the P- and S-wave velocity and effective pressure for different materials. Velocities are higher for saturated rocks and at higher pressure. (Based upon data by Mavko, 2005.) A = Bedford Limestone, B = Westerly Granite, C = Webatuck Dolomite, and D = Solnhofen Limestone.

pressure, whereas the shape of the cracks determines how much pressure difference is needed to reach the high-pressure asymptote. Most reservoir rocks show a characteristic velocity-versus-effective pressure curve that reaches an asymptote at high pressures.

Laboratory experiments also show that pore fluids play a significant role in determining the velocity dependency on pressure (Figure 3.3). Velocities of saturated rocks are higher than those of dry rocks, with the type of pore fluid determining the velocity. The bulk modulus stiffens with a less compressible pore fluid. Bulk density increases with increasing fluid saturation, and because velocity depends on the ratio of elastic moduli to density, the velocity can either increase or decrease. Figure 3.4 shows the dependence of bulk modulus and velocity on pressure (Mavko, 2005). The velocities of dry, oil-saturated, and water-saturated sandstone overlap and cross over and, therefore, do not discriminate well. Acoustic impedance (the product of density and P-wave velocity) does not relate to individual rock properties; however, the harder a rock, the

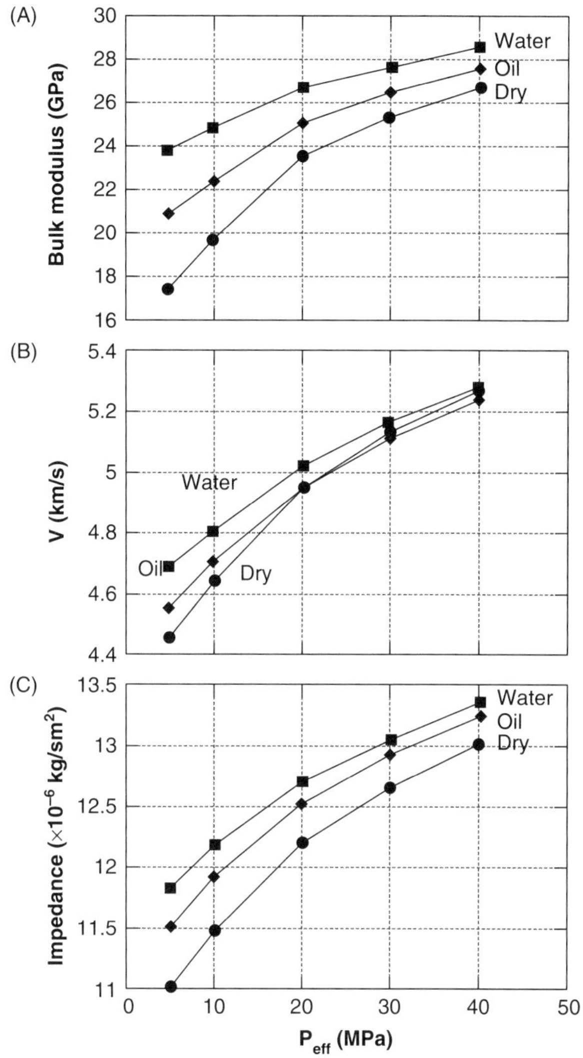

Figure 3.4 Dependence of bulk modulus, velocity, and impedance on pressure. Velocity does not discriminate the effects of different pore fluids, whereas seismic impedance (product of velocity and density) does. (Based upon data by Mavko, 2005).

higher the acoustic impedance. Figure 3.4c shows that the use of seismic impedance instead of seismic velocity removes the ambiguities and show the effect of different pore fluids.

If the P-wave velocity over S-wave velocity ratio or Poisson's Ratio is considered (Figure 3.5), then the effect of pore fluids on

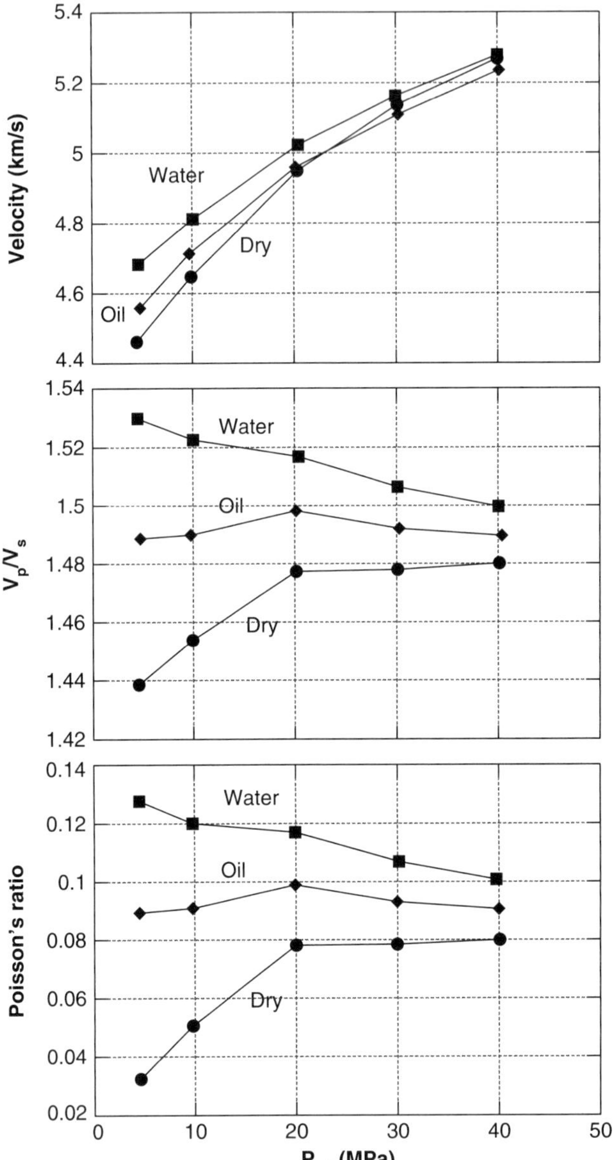

Figure 3.5 The effect of fluid composition on velocity (minimal). V_p/V_s ratio and the Poisson's ratio distinctly separate the effect of the pore fluids. (Based upon data by Mavko, 2000).

seismic velocity shows the greatest discrimination. Some sands, however, show a significant difference in P-wave velocities depending on the type of pore fluid (Figure 3.6).

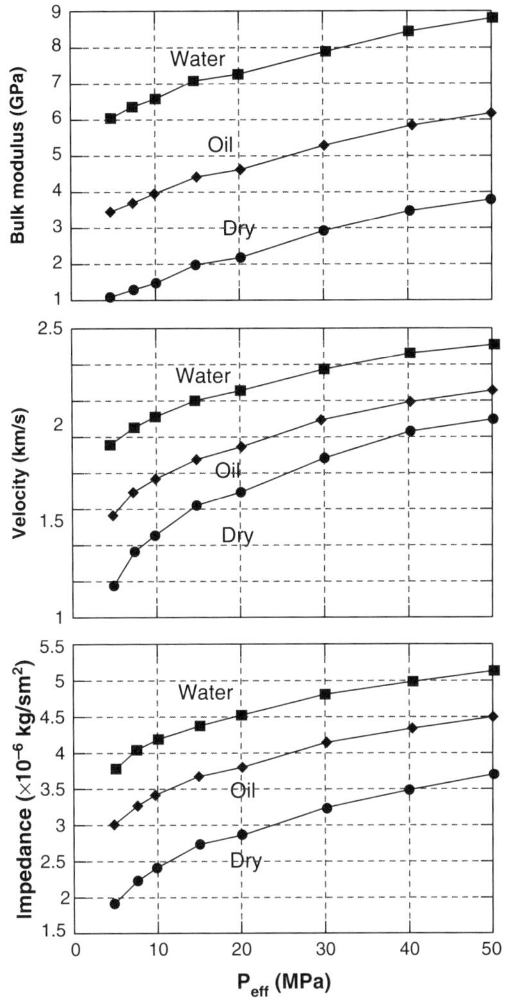

Figure 3.6 Effect of the fluid composition on properties of Ottawa Sand (Illinois, USA), bulk modulus, P-wave velocity, and impedance show significant separation. (Based upon data by Mavko, 2000).

3.4 Pore Pressure

Increasing pore pressure decreases seismic velocities by opening fractures and vugs. This can also make the pore fluids less compressible resulting in increased velocity. P-wave velocity differences between the dry and fluid-saturated rocks are determined by the effect the pore fluid has on both density, ρ, and the bulk modulus, κ.

Whereas density increases with increasing pore fluid saturation, the bulk modulus stiffens with the addition of pore fluids. Because the P-wave velocity is a function of both density and bulk modulus (Eq. 3.3), the effect of pore fluid can cause the P-wave velocity either to increase or to decrease.

Han (1986) investigated the effect of porosity and increasing pressure on the bulk and shear moduli (Figure 3.7). The moduli change significantly with no or little change of porosity, which demonstrates that closing-down of fractures with increasing pressure has a significant effect on the moduli and, hence, on velocity. Figure 3.8 is a schematic summary diagram, which shows the relation between the rock's velocity and the effective pressure. Highest velocity is attained in the minerals. However, the existence of

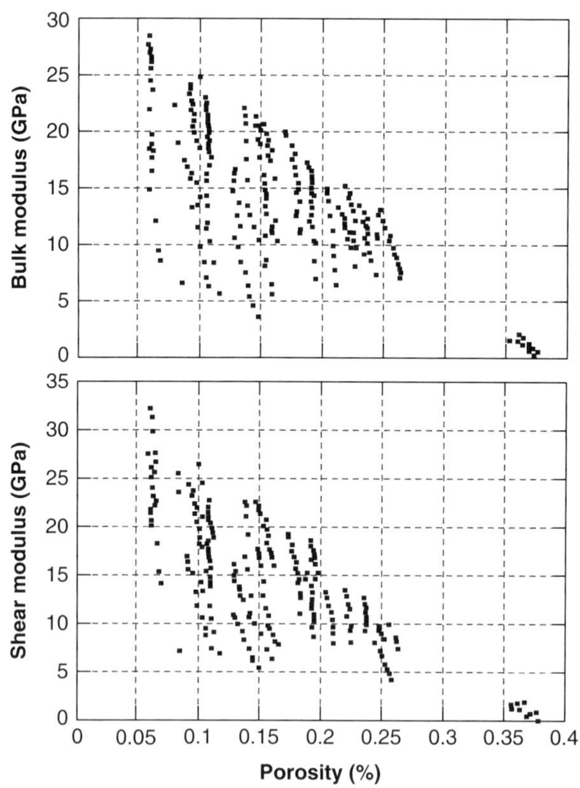

Figure 3.7 Effect of porosity and pressure on the bulk and shear moduli. (Based upon data by Han, 1986).

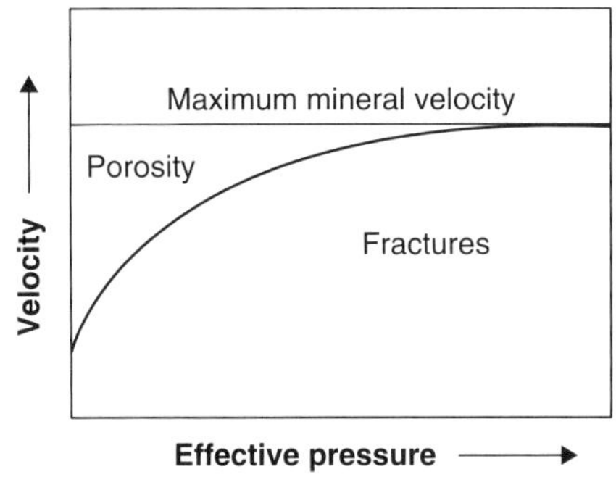

Figure 3.8 Schematic diagram showing the effect of increasing pressure on velocity. Fracture porosity is increasingly reduced until a maximum velocity (defined by the mineral composition and porosity) is reached.

porosity will lower this maximum velocity, which is asymptotically reached with increasing effective pressure. The velocity increase of both S- and P-waves as a function of increasing pressure is caused by the closure of small fractures and vugs, also known as "soft or crack" porosity. Soft porosity decreases with depth (increasing confining pressure) and increases with high pore pressure.

Figure 3.9 shows the results of an experiment illustrating the relationship between the seismic velocities and the effective pressure (both increasing and decreasing) while keeping the pore pressure constant (Jones, 1983). Minor hysteresis was observed.

Seismic velocity increases with increasing confining pressure, P_c, in sandstones and shales, and decreases with increasing temperature (Figure 3.10). Tosaya *et al.* (1985) demonstrated that the seismic velocity of the oil-saturated rocks exhibits the most significant temperature dependence, probably resulting from decreasing oil viscosity coupled with the increasing rock compressibility.

Batzle and Wang (1992) utilized a combination of thermodynamic relationships, empirical trends, and new and published data on the effects of pressure, temperature, and composition on the seismic properties of hydrocarbon gases, oils, and brines. Figure 3.11 illustrates the decrease in gas density with increasing temperature and increase in density with increasing pressure for two hydrocarbon gases having different gravities. The effect of the same parameters on the bulk modulus of the gas is similar (Figure 3.12). Heavier gases are less compressible and exhibit a higher bulk modulus.

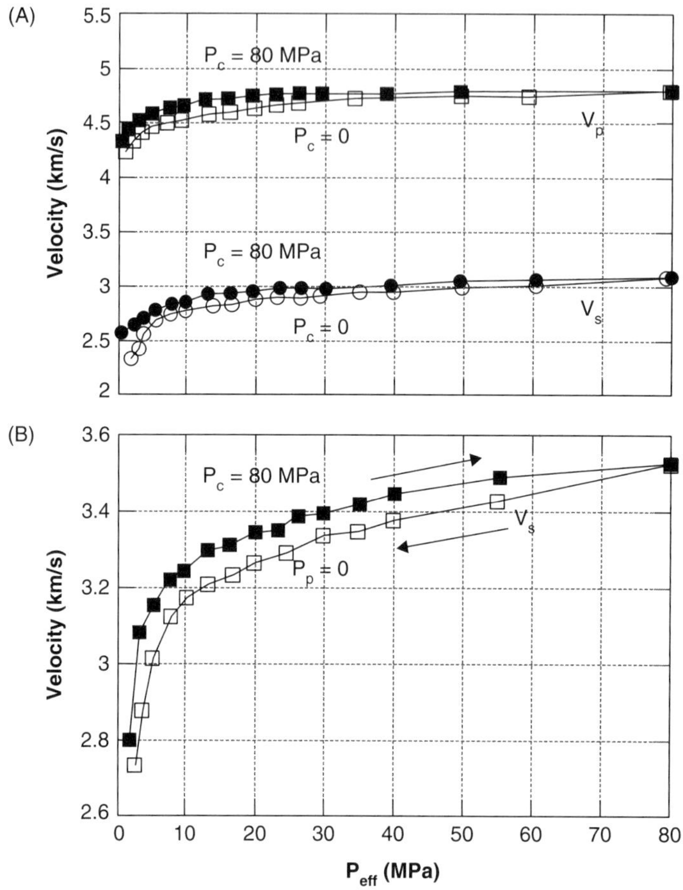

Figure 3.9 Effect of effective pressure (total overburden pressure minus pore pressure) on the velocity in sandstone and granite samples. The effective pressure was increased and then decreased while the pore pressure was kept constant. (Based upon data by Jones, 1983.) A = St. Peter Sandstone; B = Sierra White Granite; P_c = confining pressure; P_p = pore pressure.

Figure 3.13 shows the variation of calculated viscosity of the light and heavy gases with temperature. The dependence of density and bulk modulus of crude oil on the temperature, pressure, and composition shows the same trends as in the case of gases (Figures 3.14 and 3.15). Both density and the bulk modulus decrease with increasing temperature and increase with increasing pressure.

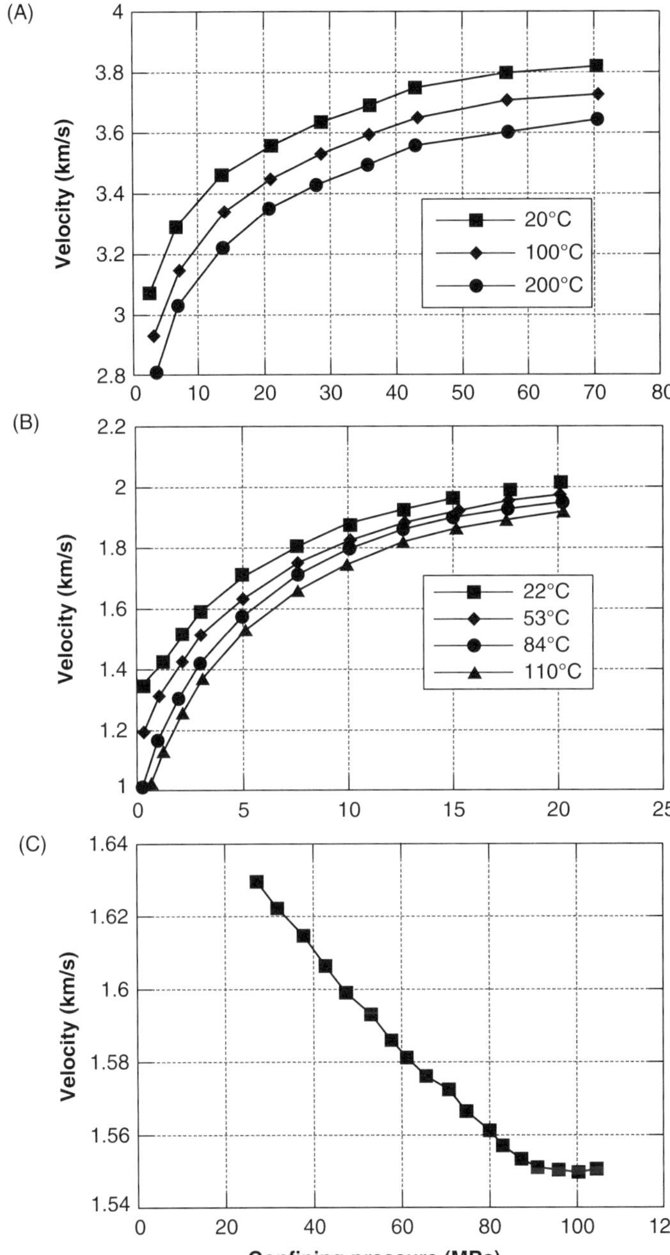

Figure 3.10 Relationship between the confining pressure and velocity for Berea Sandstone at different temperatures. Velocity increases with increasing pressure and decreases with increasing temperature. (Modified after Mavko, 2000; based upon data by Mobarek, 1971, and Jones, 1983.) A = Dry Berea Sandsone; B = Water-saturated Berea Sandstone; C = Water-saturated Berea Sandstone.

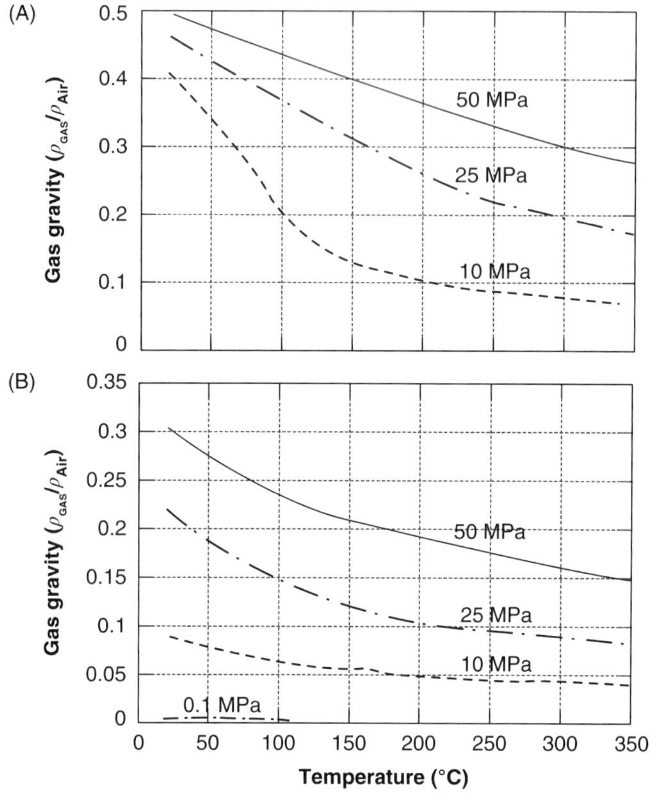

Figure 3.11 Hydrocarbon gas gravity versus temperature at different pressures and composition. (A) light gas with $\rho_{gas}/\rho_{air} = G = 0.6$ at 15.5°C and 0.1 MPa, and (B) heavy gas with $G = 1.2$. (Based upon data by Batzle and Wang, 1992).

3.5 Seismic Anisotropy

The directional dependence of elastic properties of individual mineral grains (their seismic anisotropy) was discovered at the end of the nineteenth century. The seismic properties of most rocks are not isotropic. The layering (bedding) of sedimentary rocks gives rise to anisotropy, which cannot be neglected. Alford (1986) demonstrated that the S-wave data cannot be processed without taking anisotropy into account. In the same year, Thomsen (1986) developed equations that describe the velocities of wave propagation in the transversely isotropic media. Materials develop anisotropic properties because of preferred orientation of minerals, fractures, and vugs.

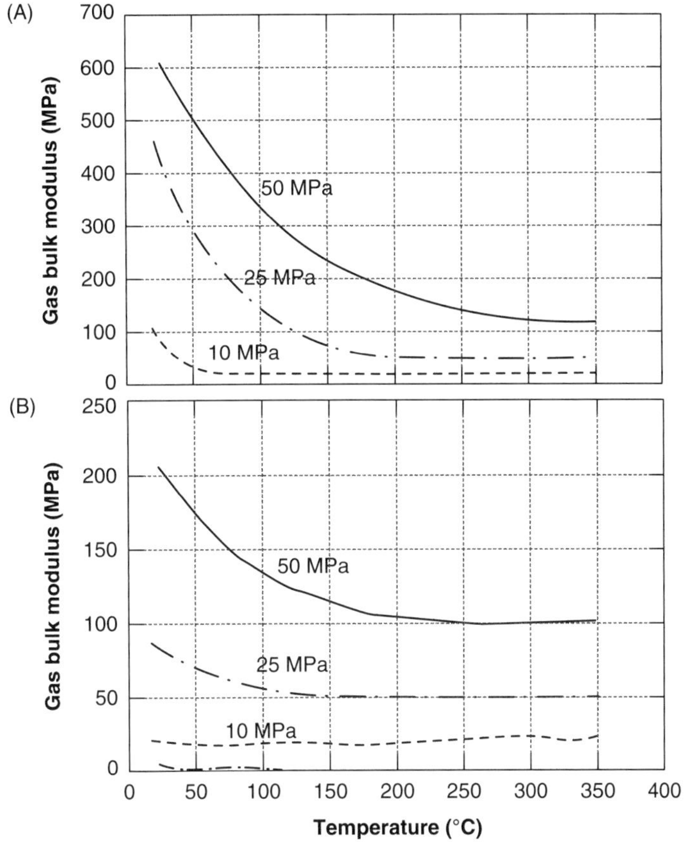

Figure 3.12 Variation of the gas bulk modulus of hydrocarbon gasses with pressure and temperature. A = Gas gravity $G = 0.6$; B = Gas gravity $G = 1.2$. (Based upon data by Batzle and Wang, 1992).

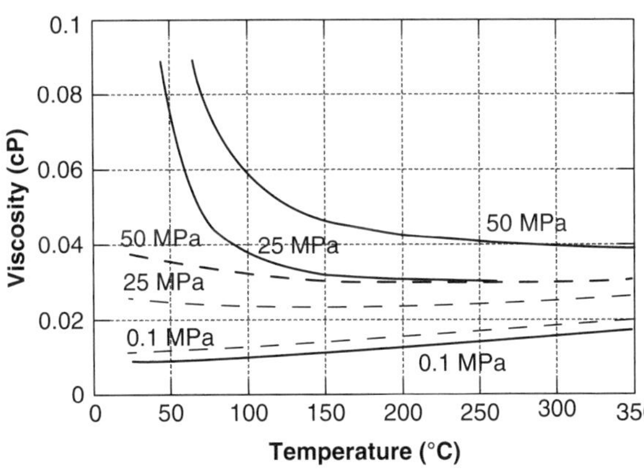

Figure 3.13 Variation of the viscosity of hydrocarbon gases with temperature. (Based upon data by Batzle and Wang, 1992).

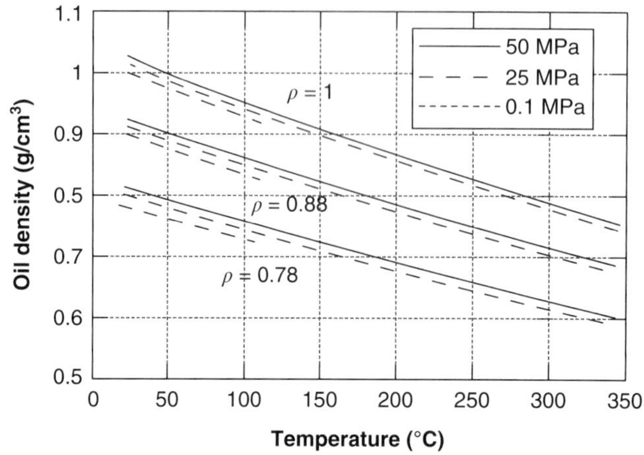

Figure 3.14 Density of oil (g/cm³) versus temperature at different pressure for three different oils having different gravity. (Based upon data by Batzle and Wang, 1992). ρ = specific gravity compared to water.

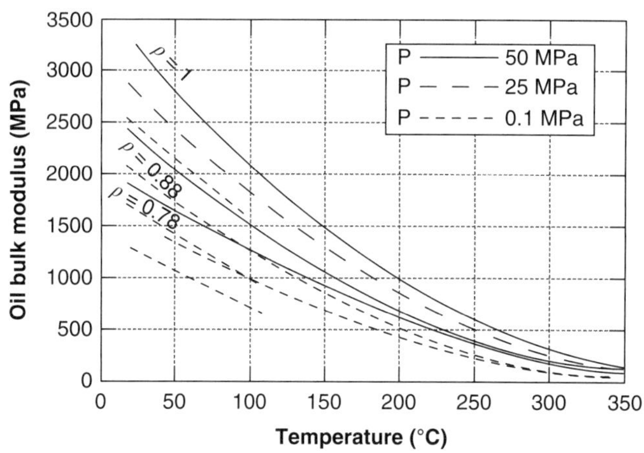

Figure 3.15 The bulk modulus of oil versus temperature at different pressure for three different oils with different gravity. (Based upon data by Batzle and Wang, 1992). ρ = specific gravity compared to water.

Nur (1969) and Nur and Simmons (1969) investigated the velocity anisotropy of granite by manipulating the fracture alignment by uniaxial stress (Figure 3.16). Without pressure, the granite samples behave isotropically, i.e., the velocity shows no directional dependence. With increasing uniaxial pressure, however, fractures normal to the stress axis close-down and cause significant anisotropy.

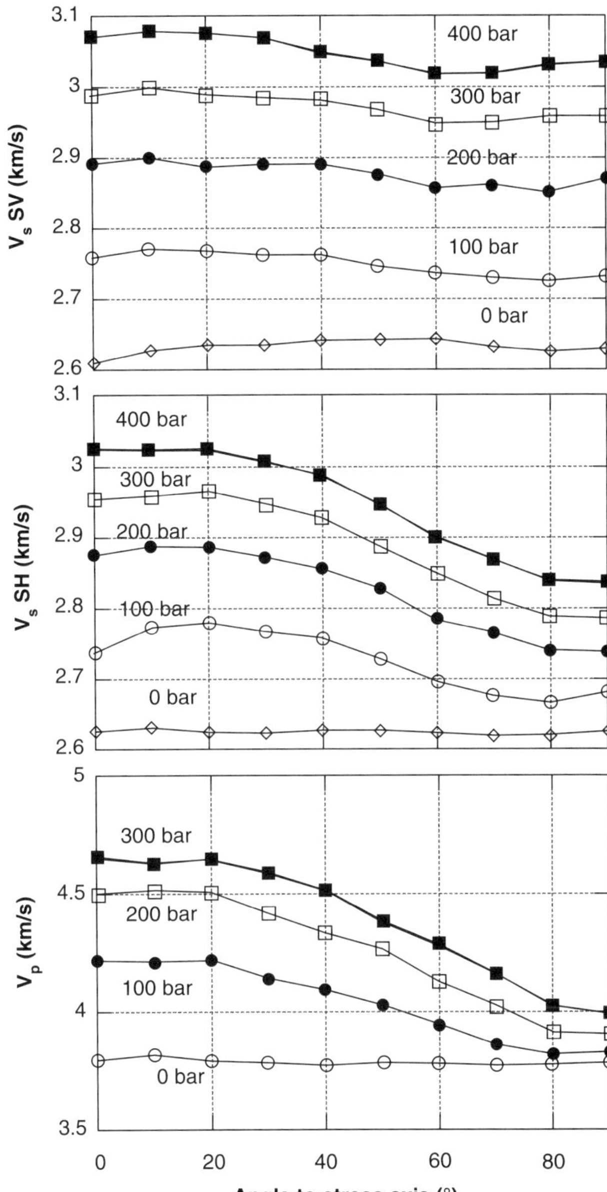

Figure 3.16 Velocity anisotropy of granite sample from Barre, Vermont. Fracture alignment was varied by uniaxial stress (bars). (Based upon data by Nur, 1969).

Any preferred orientation of mineral grains caused by compaction, bedding, or tectonic stress will cause elastic and seismic anisotropy. In general, seismic velocities are higher parallel to the bedding than

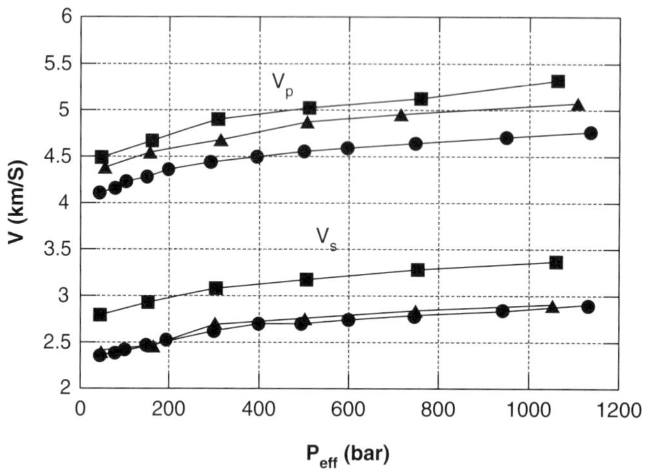

Figure 3.17 Velocity, *V*, anisotropy in Cotton Valley Shale (Texas): S- and P-wave velocities as a function of orientation and pressure. Circles = parallel to bedding; squares perpendicular to bedding; triangles = 45° to bedding. (Based upon data by Tosaya, 1982).

perpendicular to it. Figure 3.17 shows P- and S-wave velocities of a shale measured as a function of orientation and pressure. The difference between the P- and S-wave velocity parallel (V_{11} and V_{12}) and perpendicular to bedding (V_{33} and V_{13}) is quite pronounced. Measured at a 45° angle, the resulting velocity lies between the velocities of the extremes.

3.5.1 Effective Medium Theories

Effective medium theories describe the macroscopic properties of a medium based on the properties, relative fractions of constituents, and geometric distribution of its components. Modeling of the elastic moduli of rocks requires knowledge of parameters that are either difficult to obtain or unknown. In this case, if given the composition, it is possible to determine the upper and lower limits of the moduli. Published effective medium theories model the effective elastic moduli of rocks (e.g., Mavko *et al.*, 1998). Contact models use separate elastic grains that are in contact, whereas the inclusion models describe the rock as an elastic block of minerals with holes. It is impossible, however, to adequately incorporate geometric details of a rock aggregate into these theoretical models. If the volume fractions of the constituents and their elastic properties are known (for

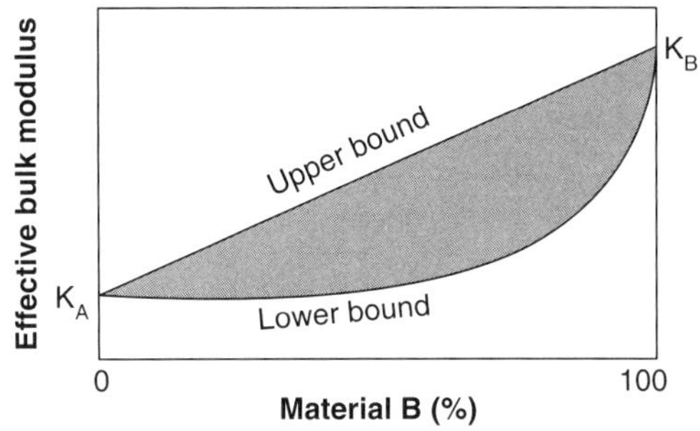

Figure 3.18 Upper bounds and lower bounds of the bulk modulus of a mixture of two different materials (A and B). The bounds provide limits, the actual value of which depend on the geometry of components. (Modified from Avseth, 2005). Shaded area = mixture range of the two materials.

example, from downhole well logs) then it is possible to obtain the upper and lower bounds of their moduli, even though the geometrical details about their mineral arrangement are unknown. It turns out that these bounds are very reliable approximations despite the lack of geometrical information. The concept of bounds is illustrated for a mixture between two components, e.g., mineral and oil, in Figure 3.18. The upper and lower bounds will bracket the effective bulk modulus for any volume fraction of the two materials. The precise values of modulus will depend on the geometry of the rock components.

3.5.1.1 Voigt and Reuss Averages

The elastic moduli of a single crystal will give an anisotropic elastic moduli tensor reflecting the symmetry of the crystallographic lattice. An isotropic average can be performed on this tensor, which will give an isotropic average elastic modulus tensor. Utilization of the Voigt (1910) and Reuss (1929) methods (Watt and Peslnick, 1980; Watt, 1980), enables one to determine the upper and lower bounds. The Voigt average (the upper bound) is defined as:

$$M_V = \sum_{i=1}^{N} f_i M_i \tag{3.6}$$

where M_v is the effective modulus of the composite; M_i and f_i are the modulus and volume fraction of the i-th constituent. The mixture is elastically softer than the arithmetic average of the constituent moduli calculated using the Voigt method; therefore, the Voigt bound represents the upper bound. The lower bound is given by the Reuss average M_R:

$$\frac{1}{M_R} = \sum_{i=1}^{N} \frac{f_i}{M_i} \tag{3.7}$$

Again, the mixture of components is elastically stiffer than the harmonic average of all moduli. The modulus, M, in the Voigt and Reuss formulas can be used to determine the shear and bulk moduli. Other moduli can then be calculated using these two moduli. The effective bulk and shear moduli for materials in which one component is not a solid is shown in Figure 3.19. The lower bound represents the effective moduli of suspended particles in a fluid. The shear modulus value is zero because liquids exhibit no resistance to shear, $\mu = 0$. The lower bound represents the effective moduli of a suspension of solids in a fluid, and can be used to describe sediments. Real systems, on the other hand, will never be as stiff as the upper bound as shown in Figure 3.19.

Because the Voigt and Reuss averages estimate the upper and lower bounds, an estimate of the actual value can be obtained

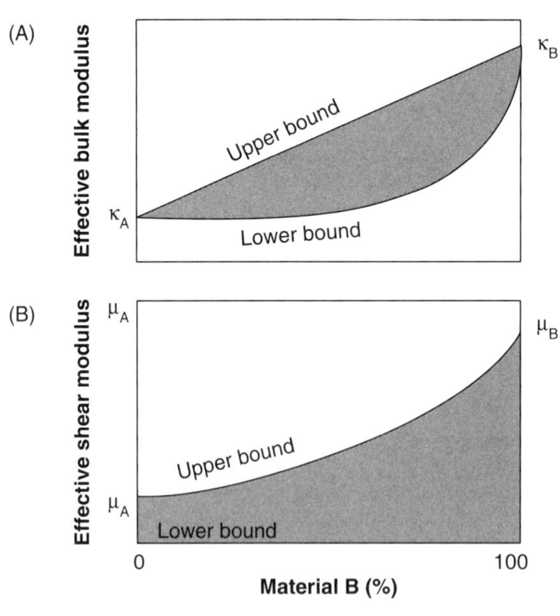

Figure 3.19 Upper bounds and lower bounds of the (A) bulk modulus, κ, and (B) shear modulus, μ, of two different materials (A and B). (Modified from Avseth, 2005).

by taking the average of the two, i.e., the Voigt-Reuss-Hill average:

$$M_{VRH} = \frac{M_V + M_R}{2} \tag{3.8}$$

3.5.1.2 Hashin–Shtrikman Bounds

The Voigt and Reuss bounds, which are relatively easy to calculate and provide a good estimate in certain cases, do not necessarily represent the best values. The Hashin–Shtrikman bounds (Hashin and Shtrikman, 1963) for isotropic elastic mixtures provide the narrowest possible bounds, without specifying the geometry of the components. For a two-component system, the Hashin–Shtrickman bounds, $HS\pm$, are calculated using the following formulas:

$$K^{HS\pm} = K_1 + \frac{f_2}{(K_2 - K_1)^{-1} + f_1(K_1 + 4\mu_1/3)^{-1}} \tag{3.9}$$

$$\mu^{HS\pm} = \mu_1 + \frac{f_2}{(\mu_2 - \mu_1)^{-1} + 2f_1(K_1 + 2\mu_1)/\left[5\mu_1(K_1 + 4\mu_1/3)\right]} \tag{3.10}$$

where K_1 and K_2 are bulk moduli of indiviual components; μ_1 and μ_2 are the shear moduli of individual components; and f_1 and f_2 are the volume fractions of the components. Subscript 1 refers to a shell, whereas subscript 2 refers to a sphere as shown in Figure 3.20.

The bounds are calculated by changing the material subscripted 1 with the material subscripted 2. To illustrate the Hashin–Shtrikman bounds, Figure 3.20 shows an assembly of spheres enclosed by a spherical shell of different material. Spheres and shells have volume fractions f_1 and f_2. The upper bound is calculated when the softer material is in the sphere, whereas the lower bound is calculated when the stiffer material is in the sphere. Most rock-forming minerals have very similar elastic properties. The upper and lower bounds of mixed solids are, therefore, close together (Figure 3.21A). Consequently, it is sufficient to use an average mineral modulus:

$$M_{AM} = \frac{M^{HS+} + M^{HS-}}{2} \tag{3.11}$$

where M_{AM} = average mineral modulus; M^{HS+} = upper Hashin–Shtrikman bound; M^{HS-} = lower Hashin–Shtrikman bound. If both

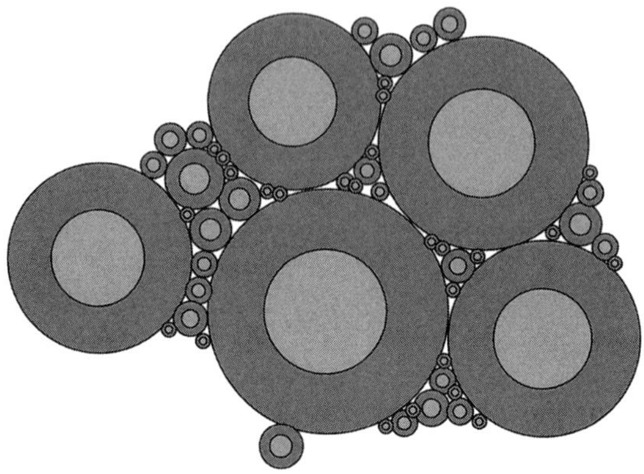

Figure 3.20 An assembly of spheres enclosed by a spherical shell of different material as illustration of the Hashin–Shtrikman bounds. (Modified from Avseth, 2005).

components have significantly different elastic properties, such as solid and liquid (Figure 3.21), the bounds are separated and predictions become more difficult.

3.5.1.3 P-wave Velocity – Porosity Relations

P-wave velocity as a function of porosity is shown in Figure 3.22 for water-saturated sediments, in addition to the Voigt and Reuss bounds for quartz–water mixtures. Velocities were calculated from the moduli of the bounds. The solid squares representing the unconsolidated deposits fall on the Reuss bound. Lithification processes, such as compaction, dewatering, and cementation, increase the sediment strength and give rise to higher velocities, moving the velocities off the Reuss bound. With increasing lithification, the velocities move towards the mineral velocity at zero porosity.

Han (1986) determined V_p and V_s velocities for water-saturated sandstones and found in general a correlation between velocity and porosity (Figure 3.23). As expected, an increasing porosity leads to decreasing velocities. The scatter of points is caused by the varying clay content. Han determined the following empirical relationships between velocity, porosity, and clay content:

$$V_p = 5.59 - 6.93\phi - 2.13C, \text{ and} \tag{3.12}$$

$$V_s = 3.52 - 4.91\phi - 1.89C, \tag{3.13}$$

where C is the clay content, fraction.

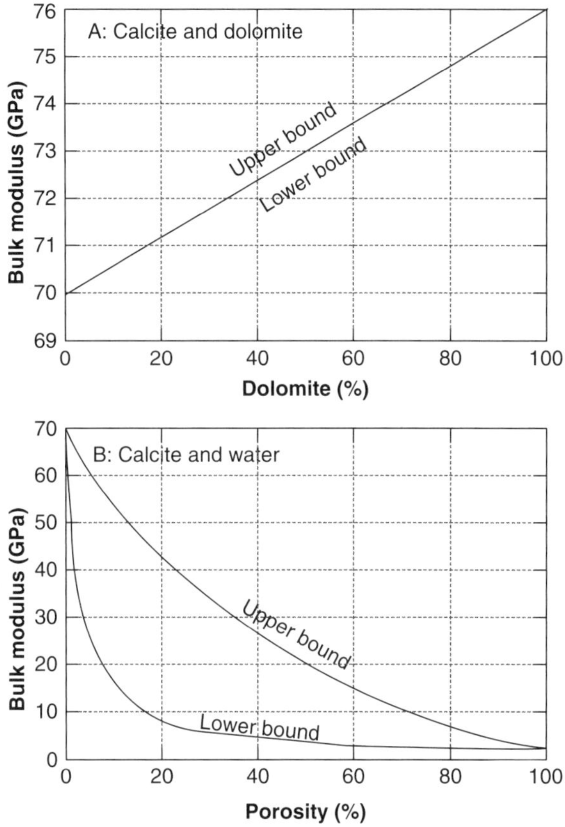

Figure 3.21 Illustration of the bounds of (A) elastically similar (calcite and dolomite) and (B) elastically different materials (calcite and water). For the elastically similar materials, the bounds are close together, whereas for the elastically different materials the bounds are spread apart. (Modified from Avseth, 2005).

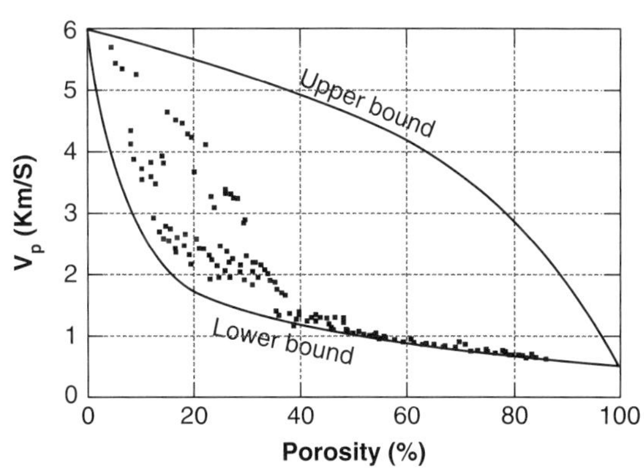

Figure 3.22 P-wave velocity versus porosity for water-saturated sediments compared with the Voigt–Reuss bounds. (Based upon data by Yin, 1992; Han 1986; and Hamilton 1956; compiled by Marion, 1990).

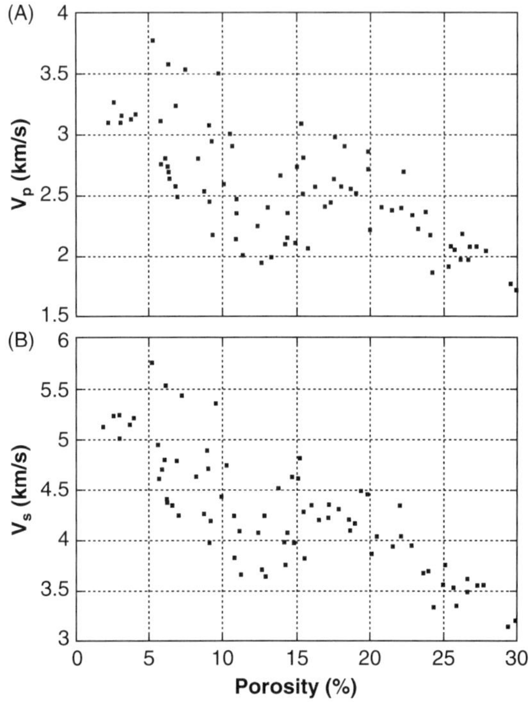

Figure 3.23 P- and S-wave velocities as a function of porosity for water-saturated sandstones at an effective pressure of 40 MPa. (Based upon data by Han, 1986).

These equations are useful in understanding the dependence of velocity on porosity and the effect that increasing clay contents have on lowering velocity.

3.5.2 The Effect of Pore Space and Pore Geometry on Moduli

Baechle *et al.* (2006) investigated velocity as a function of pore space and pore shape in carbonate rocks from different areas (Figure 3.24). They found that velocity is not only a function of total porosity, but also of the predominant type of porosity.

For a dry rock, relationship between the bulk modulus and porosity is as follows:

$$\frac{1}{\kappa_{dry}} = \frac{1}{\kappa_{mineral}} + \frac{\phi}{\kappa_{\phi}} \tag{3.14}$$

where κ_ϕ quantifies the stiffness of the pores, which is the inverse of the dry rock space compressibility at a constant pore pressure, κ_{dry}

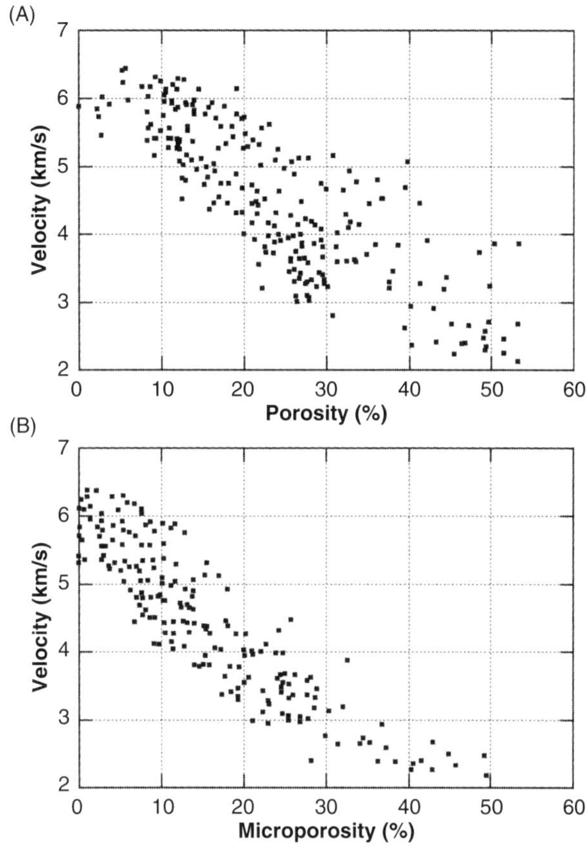

Figure 3.24 P-wave velocity as a function of (A) porosity and (B) microporosity at 10 MPa effective pressure. (Based upon data by Baechle *et al.*, 2006.) Microporosity is defined as subtracting the image macroporosity (minimum pore diameter of 30 μm detected by digital image analysis) from the total plug porosity.

is the dry rock bulk modulus, κ_{mineral} is the mineral bulk modulus, and ϕ is the porosity.

For a liquid or gas-saturated rock the relationship is similar:

$$\frac{1}{\kappa_{sat}} = \frac{1}{\kappa_{mineral}} + \frac{\phi}{\tilde{\kappa}_{\phi}} \tag{3.15}$$

where the compressibility (as presented in equation 3.1) of the pore space, $\tilde{\kappa}_{\phi}$, is approximately equal to the compressibility of dry rock plus the compressibility of the fluid. The composition of the pore fluid has, therefore, a significant effect on the sonic velocity in rocks, especially in soft rocks with relatively small $\kappa\phi$.

3.5.3 Gassmann's Equations

Gassmann's (1951) equations predict changes in velocity that result from varying pore-fluid saturations. Often, however, the input parameters are unknown and must be estimated, giving unrealistic results. The equations relate the bulk and shear moduli of a saturated porous medium to the moduli of the same medium in a dry state. The effective bulk modulus κ_{sat} of the saturated rock is given by the following equation:

$$\frac{\kappa_{sat}}{\kappa_{\min} - \kappa_{sat}} = \frac{\kappa_{dry}}{\kappa_{\min} - \kappa_{dry}} + \frac{\kappa_{fluid}}{\phi(\kappa_{\min} - \kappa_{fluid})} \quad (3.16)$$

where $\kappa_{\min}$, κ_{fluid}, and κ_{dry} are the bulk moduli of the mineral grains, fluid, and dry rock.

According to Gassmann (1951), the shear modulus is mechanically independent of the properties of fluids present in the pore space:

$$\frac{1}{\mu_{sat}} = \frac{1}{\mu_{dry}} \quad (3.17)$$

The primary assumptions behind Gassmann's relations are (1) the porous medium contains only one type of solid component with a homogeneous mineral modulus, (2) the pore space is statistically isotropic, and (3) the pore pressures are in equilibrium throughout the pore space. Arns *et al.* (2002) compared the numerically predicted moduli of the Fontainebleau Sandstone obtained from microtomographic images for dry, water-saturated and oil-saturated conditions to those obtained using Gassmann's equations (Figure 3.25). Both the predicted bulk and shear moduli appear to be in agreement with those obtained using Gassmann's equations.

Both V_p and V_s must be known in order to determine the bulk (κ) and shear moduli (μ) using Gassmann's equations:

$$\kappa = \rho(v_p^2 - \frac{4}{3} v_s^2) \quad (3.18)$$

$$\mu = \rho v_s^2 \quad (3.19)$$

Thus, the change of bulk modulus can be determined using the following equation:

$$\frac{\kappa_{sat}}{\kappa_{\min} - \kappa_{sat}} = \frac{\kappa_{dry}}{\kappa_{\min} - \kappa_{dry}} + \frac{\kappa_{fluid}}{\phi(\kappa_{\min} - \kappa_{fluid})} \quad (3.20)$$

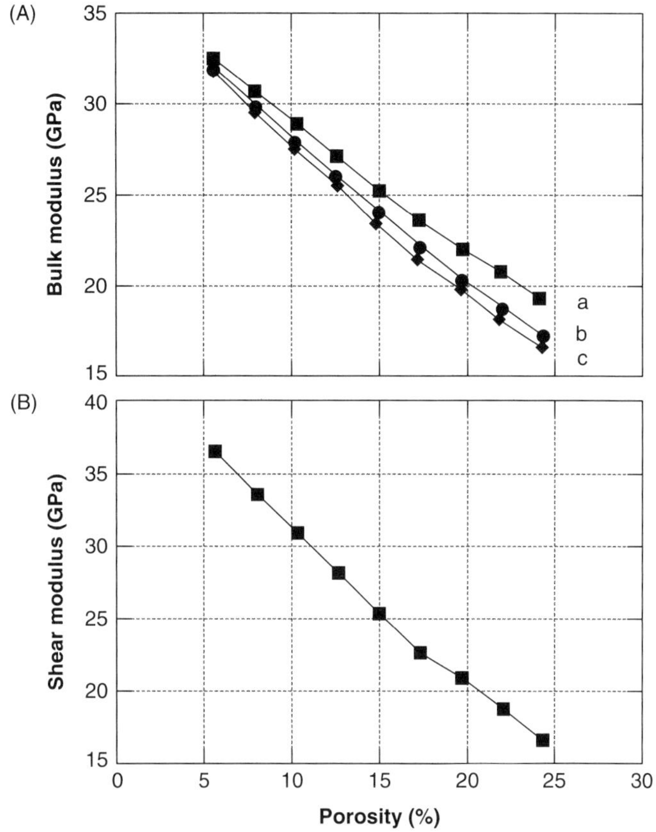

Figure 3.25 (A) Comparison between the bulk modulus data obtained by Gassmann's equations and simulations for (a) water-saturated, (b) oil-saturated, and (c) dry mixtures. The fit between the calculated and experimental data is excellent. (B) The shear modulus is independent of the pore fluid as predicted by Gassmann (1951) equations. (Based upon data by Arns *et al.*, 2002).

The shear-wave velocity is usually unknown and can, therefore, be approximated by using the following equation:

$$\frac{M_{sat}}{M_{\min} - M_{sat}} \approx \frac{M_{dry}}{M_{\min} - M_{dry}} + \frac{M_{fluid}}{\phi(M_{\min} - M_{fluid})} \tag{3.21}$$

where M is the P-wave modulus equal to ρV_P^2; M_{min} = P-wave modulus of the minerals, M_{fluid} = P-wave modulus of the fluid.

This approximation is very good and agrees closely with calculations using Gassmann's equation.

3.5.4 Bounding Average Method (BAM)

A simple and elegant method of solving the fluid substitution problem was proposed by Marion (1990). It is based on the assumption that the position of modulus between the Hashin–Shtrikman bounds is defined by the geometry of the two phases. If fluid is replaced, the geometry will not be affected and the relative position between the Hashin–Shtrikman bounds remains the same. Figure 3.26A shows the Hashin–Shtrikman bounds and the bulk modulus of a

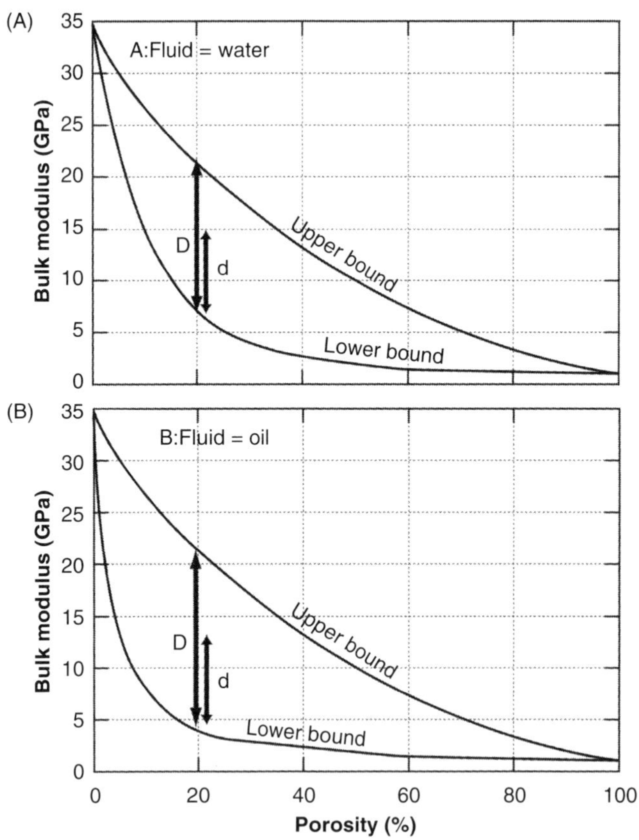

Figure 3.26 Bounding average method: Hashin–Shtrikman bounds and the bulk modulus of a sample at a certain d/D ratio are shown. When Fluid 1 (water) is replaced with Fluid 2 (oil), the geometry does not change and the relative position of the sample (d/D ratio) did not change due to fluid substitution. The new bulk modulus can thus be predicted within the bounds of the new system. (Modified from Mavko, 2001; based on data by Marion, 1990).

sample at d/D ratio (D = modulus difference between the upper and lower bounds; d = modulus difference between the sample and lower bound) for a mineral/water system. In Figure 3.26B, the water has been replaced by oil and the Hashin–Shtrikman bounds were recalculated. Because the geometry did not change during the fluid substitution, one can assume that the d/D ratio remains the same and that the new bulk modulus can be predicted within the bounds of the new fluid/mineral system. An example of the effectiveness of the Bounding Average Method (BAM) is shown for a sample of the Westerley Granite saturated with water in Figure 3.27. Measured and calculated BAM velocities plotted as a function of pressure show small differences and demonstrate the effectiveness of the method.

Figure 3.28 presents a comparison between the Gassmann and BAM modulus predictions and measured dry and saturated moduli of ten different clay-free sandstones. Although both predictions are close to the measured values, the BAM model appears to be more accurate in all but one instance.

3.5.5 Kuster and Toksöz Theory

The basic problem in understanding and modeling reservoirs is how elastic properties vary with porosity. To solve this problem, several theories have been developed. The Kuster–Toksöz (1974) formulations are based on the long-wavelength, first-order scattering

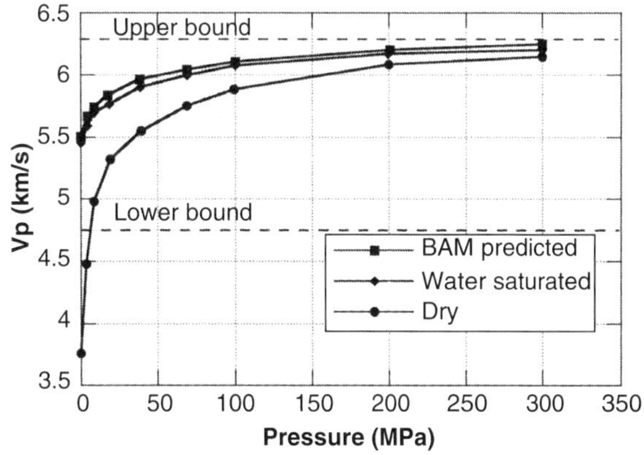

Figure 3.27 BAM-predicted and measured velocities (lower two curves) for the Westerly Granite (Rhode Island) show excellent agreement. (Modified from Mavko, 2005; based upon data by Nur and Simmons, 1969, and Marion, 1990).

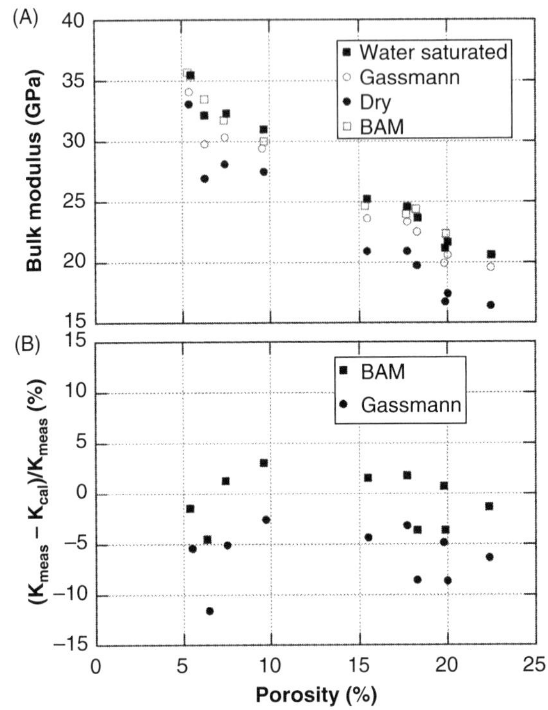

Figure 3.28 Comparison between the Gassmann and BAM modulus predictions and the real measurements demonstrate that both predictions give good results. The BAM-predicted values, however, are closer to the actual data in most instances. (Based upon data by Marion, 1990).

theory. They assume that the pores are widely dispersed and do not overlap.

$$(\kappa_{KT} - \kappa_m)\frac{\left(\kappa_m + \frac{4}{3}\mu_m\right)}{\left(\kappa_{KT} + \frac{4}{3}\mu_m\right)} = \sum_{i=1}^{N} x_i(\kappa_i - \kappa_m)P^{mi} \tag{3.22}$$

$$(\mu_{KT} - \mu_m)\frac{(\mu_m + \zeta_m)}{(\mu_{KT} + \zeta_m)} = \sum_{i=1}^{N} x_i(\mu_i - \mu_m)Q^{mi} \tag{3.23}$$

$$\zeta = \frac{\mu(9K + 8\mu)}{6(K + 2\mu)}. \tag{3.24}$$

where subscripts m and i describe the matrix and dispersed materials. The expressions P^{mi} and Q^{mi} are variables describing specific shapes, such as spheres, needles, disks, and fractures. For example, for discs:

$$P^{mi} = \frac{\kappa_m + \frac{4}{3}\mu_i}{\kappa_i + \frac{4}{3}\mu_i} \tag{3.25}$$

$$Q^{mi} = \frac{\mu_m + \zeta_i}{\mu_i + \zeta_i} \tag{3.26}$$

The differential effective medium theory can also describe the non-overlapping pores (Zimmerman, 1984) or overlapping pores (Norris, 1985). The coherent potential approximation (CPA) is one example of a self-consistent model in which the matrix and pores are treated symmetrically (Berryman, 1992).

Appendix A
Historical Review

(Milestone Developments in Petrophysics)

By Donald G. Hill

A.1 Introduction

The first wireline, or geophysical, well log was recorded September 5, 1927, at the Pechelbronn Oil Field in Alsace-Lorraine, France. The technology of well logging and the science of petrophysics have progressed, since then, via nearly equal amounts of corporate vision, industry support, scientific understanding, technological breakthroughs, and incremental improvements. Tables A.1 through A.7, highlight some of the major developments, or milestones, which have profoundly impacted petrophysics and well logging. In many cases, developments far from the petroleum industry have greatly accelerated the development of both well logging and petrophysics. In all cases, significant developments within petrophysics and well logging have met previously recognized needs.

A.2 Initial Phases of Development

The first oil well was logged, using wireline techniques, in 1927, at Pechelbronn Field, Alsace-Lorraine, France, by Henri Doll, Roger Jost and Charles Scheibli, employees of a small geophysical exploration firm, *Société de Prospection Électrique* (Bateman, 2009), or *pros*, founded only a year earlier, by two brothers: Conrad and Marcel Schlumberger, to develop electrical methods of mapping the subsurface.

The Pechelbronn well log was the first attempt at detailed borehole measurements attempted by Pros. This activity occurred because of the vision of the Pechelbronn Field management, who commissioned the work to be done. Before this could happen, however, many other events had to occur.

Table A.1 Pre-logging petrophysical milestones (from a variety of sources).

Date	Event
1830	First attempts to measure Spontaneous Polarization generated by metallic sulfide deposits.
1900	Gamma ray discovered.
1911	Conrad Schlumberger begins research in electrical prospecting techniques at Ecol des Mines, Paris.
1912	Initial equipotential mapping field work by Conrad Schlumberger, at the family's Val-Richer estate, Normandy. First observation of Induced Polarization. First Schlumberger electrical prospecting patent filed.
1913	First successful use of Mes a la Mass technique to outline a buried conductive ore body. Conrad Schlumberger observed SP signals over buried sulfide ore bodies.
1919	Paul Schlumberger underwrites Conrad Schlumberger's electrical prospecting research. Brother Marcel joins in the effort. Earl P. Halliburton founded New Method Well Cementing Co.
1923	Formation of Societe de Prospection Electrique, Procedes Schlumberger (Pros). Surface electrical surveys in Belgian Congo, Canada, Romania, Serbia, & Union of South Africa.

The English canal builder, William Smith, had established that geologic formations could be correlated over great distances by observing the character of the rocks and their included fossils as exposed in outcrops, and road cuts, canal walls, quarry faces, and building excavations. Smith called his correlation technique: *Stratigraphy* (Winchester, 2001).

Oil "Wildcaters" initially drilled their exploration (wildcat) wells on dreams and hunches. While some continued to do this, others soon realized that they could increase their odds by employing geologists to help them locate their wells, using William Smith's stratigraphic, and other geologic, techniques. Subsurface geologists were eager to apply stratigraphic correlation principles, in the subsurface. To do so, however, required that whole cores be cut, during drilling operations – a very slow and costly operation, or that the correlations be based on drill cuttings (chips) collected during the

Table A.2 Petrophysical milestones: 1927–1946 (from a variety of sources).

Date	Event
1927	First electric log (station measurements) at Pechelbronn Field, France.
1928	Introduction of 3-conductor logging cable.
1929	"Electrical Coring" paper presented at A.I.M.E meetings. Electrical logging introduced to California, USSR, and Venezuela.
1930	Introduction of SP log. Electrical logging introduced to Indonesia.
1931	Introduction of manual (hand cranked) continuous pen recording.
1932	Introduction of the Normal Electrode Array resistivity log. Introduction of directional Surveys. Introduction of temperature logs. Introduction of bullet perforation. Bill Land and Walt Wells form Lane-Wells Bullet Perforation Company. Chadwick discovers the Neutron.
1933	First neutron log evaluation.
1934	Schlumberger Well Surveying Corporation formed. Introduction or automatic (galvanometer) recording.
1935	Introduction of sidewall cores. Introduction of photographic recording.
1936	Introduction of dual galvanometer recording.
1937	Introduction of 3 galvanometer recording.
1938	Introduction of 4-conductor cable. Introduction of neutron log. Nuclear Magnetic Resonance (NMR) discovered
1939	First gamma-ray log publication.
1940	Introduction of armored cable. Introduction of gamma ray logs. Jess Hall Sr. founded Weatherford Spring Co.
1941	Introduction of three-arm (SP) Dipmeter.
1942	Publication of Archie equations.
1945	Introduction of focused electrical logs. Introduction of the R_{x0}/R_t (Rocky Mountain) Interpretation Technique.
1946	Introduction of three arm Dipmeter with short (micro) resistivity pads an caliper Logs.

Table A.3 Petrophysical milestones: 1947–1960 (from a variety of sources).

Date	Event
1947	Introduction of Induction log. Introduction of 9-pen galvanometers.
1948	Schlumberger opens Research Center (later renamed Schlumberger-Doll Research Center) in Ridgefield, Connecticut. Transistor invented.
1949	Introduction of Micro-Log, with caliper.
1950	Introduction of guard resistivity logs, or Laterolog-3 (LL3). Introduction of gamma ray logs. Introduction of gamma ray neutron log (GNT) – gamma ray log combination.
1951	Introduction of continuous recording three-arm dipmeters, with microresistivity pads. Introduction of microlaterolog.
1954	Introduction of Continuous Velocity Logs (CVL), later called Acoustic or Sonic logs. Introduction of powered down-hole caliper arms. Introduction of Nuclear Magnetic Resonance (Proton Precession) magnetometer.
1955	Introduction of wireline formation testing. Introduction of compensated (dual detector) neutron log.
1956	Introduction of five-coil Induction. Publication of Wyllie Time Average equation. Introduction of (single detector) density logs. Establishment of API calibration pits. Publication of first Nuclear Magnetic Resonance (NMR) logging paper.
1957	Introduction of FORTRAN programming language. Publication of first density logging paper
1958	Introduction of Laterolog-8. Introduction of cased-hole density logs for determining top of cement. Sputnik launched. Integrated circuits invented.
1959	Introduction of six coil Induction logs. Introduction of compensated (dual detector) density logs.
1960	Introduction of chlorine logs.

Table A.4 Petrophysical milestones: 1961–1970 (from a variety of sources).

Date	Event
1961	Introduction of cement bond logging. Introduction of production logging. Introduction of (analogue) computer processed dipmeter data.
1962	Introduction of Sidewall Neutron Porosity (SNP) logs. First Nuclear Magnetic Resonance (NMR) prototype tool. Introduction of Neutron-Density cross-plot analysis.
1963	Introduction of Dual Induction logs. Introduction of borehole compensated (BHC) sonic logs.
1964	Introduction of compensated (dual detector) density logs. Introduction of neutron lifetime logs (NLL). Introduction of IBM 360 computer series. Introduction of punched paper (TT) tape recording. Introduction of BASIC programming language.
1965	Schlumberger opens Clamart, France, Engineering Center. The California Research Co. and Shell Development Co. begin laboratory studies of Nuclear Magnetic Resonance. First computerized seismic data processing
1966	Introduction of compensated (dual detector) neutron log. Introduction of high-resolution dipmeter.
1967	Introduction of borehole televiewer (BHTV).
1968	Introduction of truck quantizer (TQ5). Introduction of truck tape recorder (TTR). Publication of Waxman and Smitts Shaly-Sand Model.
1969	Introduction of spectral (K-U-T) gamma ray logs. Moon Landing.
1970	Introduction of triangular core slicer. Introduction of first integrated digital FE sandstone analysis system (SARABAND®).

normal drilling operations – which could be very difficult to do, using only a hand lens.

The Schlumberger brothers: Conrad, a Physics Professor at l'École des Mines, and Marcel, a consulting Mining Engineer, were unlikely entrepreneurs. In the eight years, prior to logging the Pechelbronn well, they had exhausted ₣500,000, from their father and the dowries of both of their wives, trying to start up a geophysical contracting company offering to locate metallic ore bodies,

Table A.5 Petrophysical milestones: 1971–1980 (from a variety of sources).

Date	Event
1971	Introduction of 4-arm dipmeter. Introduction of Induction sonic combination stack. Introduction of first complex mineralogy digital analysis FE analysis system (CORiBAND®)
1972	Introduction of dual laterolog. Introduction of cased-hole TDT neutron logs.
1975	Introduction of Direct Digital Logging (DDL®), first in-truck digital computer system. Introduction of quicklook computer log analysis system (Cyberlook®).
1976	Introduction of Carbon/Oxygen (C/O) Logging. Introduction of 5 ¼″ Floppy Disk recording media.
1977	Introduced Cyber Service Unit (CSU®) second in-truck digital computer system. Introduction of dielectric constant logs. Introduction of vertical seismic profiling (VSP). Introduction of Dual Water interpretation model. SANDIA National Laboratory high temperature microcircuit project started.
1978	Introduction of photoelectric effect density logs (LDT® or Z-Logs®). Introduction of Motorola 68000 microprocessor.
1979	First Commercial Measurements While Drilling (MWD) service. First MWD log (gamma ray).
1980	Introduction of first reliable borehole gravity meter (BHGM). Introduction of 5 MB hard drives, for micro-computers. Introduction of over determined logic in FE.

using surface geophysical methods. They and their small, but loyal, staff had enjoyed some technical success locating commercial ore bodies. Because a good mine can last a lifetime, however, there was not a lot of repeat business.

One of the geophysical techniques utilized by the Schlumberger brothers was called surface resistivity measurements. Figure A.1 shows the surface resistivity electrode array (now called a

Table A.6 Petrophysical milestones: 1981–1990 (from a variety of sources).

Date	Event
1981	Introduction of induced gamma ray spectrometry logs. Introduction of MSDOS. Introduction of 3 ½″ floppy Drives.
1982	Introduction of array sonic logs (AST®). Introduction of Cray XMP "super computer". Introduction of Commodore 64 microcomputer.
1983	Introduction of improved dielectric constant (DPT®) logs. Introduction of MS WORD® word processor.
1984	Introduction of 6-Arm dipmeter. Introduction of Shiva dipmeter processing logic. Introduction of Array Induction (AIT®) logs, with amplitude *and* phase measurements. Introduction of Apple Macintosh® microcomputer, with "mouse" cursor controller.
1985	Introduction of Formation Micro-Imager (FMI®). Start of the Internet. Evaluation of non-chemical density source (LINAC borehole accelerator).
1986	Introduction of Phasor® (multiple frequency) Induction logs. Introduction of modern NMR logging tool. Introduction of INTEL 386 (32 Bit) microprocessor chip.
1988	Introduction of acoustic dipole source.
1989	Introduction of high-resolution (400KHz) Induction tools. Introduction of INTEL 486 (62 Bit) microprocessor chip.
1990	Introduction of MS WINDOWS microcomputer operating system. Introduction of the Circumferential Acoustic Scanning Tool (CAST®).

Schlumberger Array), utilized by the Schlumberger brothers. Electrical power was introduced, to the subsurface, via the A and B current electrode dipole and the potential gradient (voltage) was measured via the M and N potential electrode dipole centered within the current electrode dipole. The current streamlines and equal-potential surfaces, for a homogeneous, isotropic half-space, are as shown in Figure A.1. Thus, the resistivity, R, of this material

Table A.7 Petrophysical milestones: 1991–2006 (from a variety of sources).

Date	Event
1991	Introduction of Formation micro-imaging tool. Introduction of Linux® operating system. Introduction of LWD array processing.
1992	Introduction of Array Induction logs. Introduction of in-cab color plotters.
1993	Introduction of cross-dipole acoustic source. Introduction of LWD focused resistivity logs. Introduction of Intel Pentium® microprocessor.
1994	Introduction of LWD azimuthal gamma ray measurements. Introduction of LWD acoustic imaging. Introduction of LWD resistivity imaging.
1995	Introduction of Platform Express®, a compact complete measurement package. Introduction of resistivity through casing measurements. Introduction of LWD azimuthal density measurements.
1996	Introduction of 3-Dimensional induction tools Introduction of improved NMR (CMR) tools.
1997	Introduction of multi-radius NMR tools.
1998	Introduction of high resolution laterolog array tools.
1999	Introduction of horizontal production logging.
2000	Introduction of LWD NMR tools.
2002	Introduction of NMR diffusion tools. Introduction of LWD spectral gamma ray tool.
2004	Introduction of LWD wired pipe
2005	Introduction of pulsed neutron elemental capture gamma ray spectroscopy tools.
2006	Introduction of LWD directional resistivity tools.

could be determined from the AB current, I, the MN potential difference, ΔV, and the electrode array geometry, G, as:

$$R = G\frac{\Delta V}{I} \tag{A.1}$$

In the case of a homogeneous, isotropic, half-space (Figure A.1), Eq. 1 yields the true resistivity, R_t. If the subsurface is not homogeneous and/or isotropic, Eq. A.1 yields the "apparent" resistivity, R_a.

Resistivities and layer thicknesses of a horizontally layered subsurface could be inferred by varying the AB and MN separations and comparing the resulting R_a to those of theoretical *Type Curves*, based on a *Layered-Earth Model*. This technique is often called *Expanding Array Electrical Sounding*. Horizontal discontinuities could be inferred, by keeping fixed AB and MN spacings and shifting the array along the surface, as shown in Figure A.2. This technique is often called *Fixed Array (Electrical) Resistivity Profiling*.

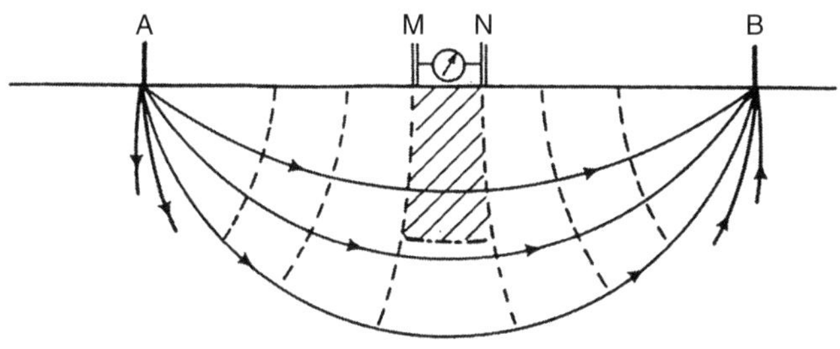

Figure A.1 Schlumberger surface resistivity electrode array (after Ross *et al.*, 1979).

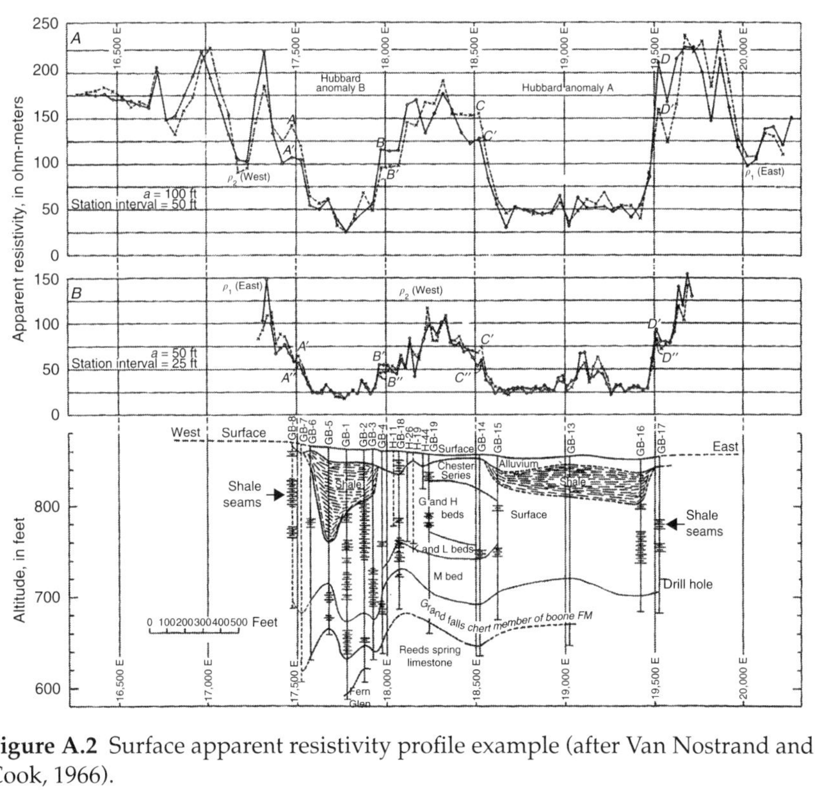

Figure A.2 Surface apparent resistivity profile example (after Van Nostrand and Cook, 1966).

The purpose of the Pechelbronn log was to see if the Schlumberger brothers' surface resistivity profiling technique, dropped down a borehole, could be used to correlate subsurface geological formations between wells, much the same way that the English canal builder, William Smith was doing by visually examining the walls of canals and quarries, road cuts, (new) building excavations, and surface outcrops. Whereas the Pechelbronn log does not look very impressive (Figure A.3), compared to modern logging vendor products, it did demonstrate that wireline techniques could be used for subsurface geological correlation.

In 1929, E. G. Leonardon, manager of Schlumberger Electrical Prospecting Methods, New York, convinced the Schlumberger brothers to publicize their well logging techniques via technical publications (Schlumberger, 1982). The first wireline paper, "Electrical Coring: A Method of Determining Bottom-Hole Data", presented in 1929 and published as A.I.M.E. Technical Publication 462 (Schlumberger *et al.*, 1932), established a tradition of stimulating technical disclosure via publication (by Schlumberger). This tradition has since been followed by virtually all wireline vendors. The Schlumberger brothers quickly published their borehole profiling results in other scientific journals (Schlumberger and Schlumberger, 1929; Schlumberger *et al.*, 1933) as this turned out to be an inexpensive form of promotion. Within three years, Conrad and Marcel were running a multinational service company, Schlumberger Electrical Coring, and logging wells worldwide (Figure A.4).

None of the three men involved in logging the Pechelbronn well, or the Schlumberger brothers, could have predicted that they were about to establish one of the world's largest petroleum service companies. They only knew that someone was crazy enough to pay them to drop their surface resistivity array down a borehole and produce a resistivity profile. *The Schlumberger resistivity profiling technology already existed, but it was the vision of the Pechelbronn Oil Field management that led to it being used in a borehole,* to provide a quick and inexpensive method of correlating geological formations between wells. Even today, with all of the exotic wireline measurements available, structural and stratigraphic correlation are still the most common usages of wireline logs (Figure A.5).

Having found their niche market, the Schlumberger brothers added an additional electrochemical measurement, Spontaneous Potential (SP) to resistivity. Conrad Schlumberger first observed measured electrical potentials in the absence of any applied

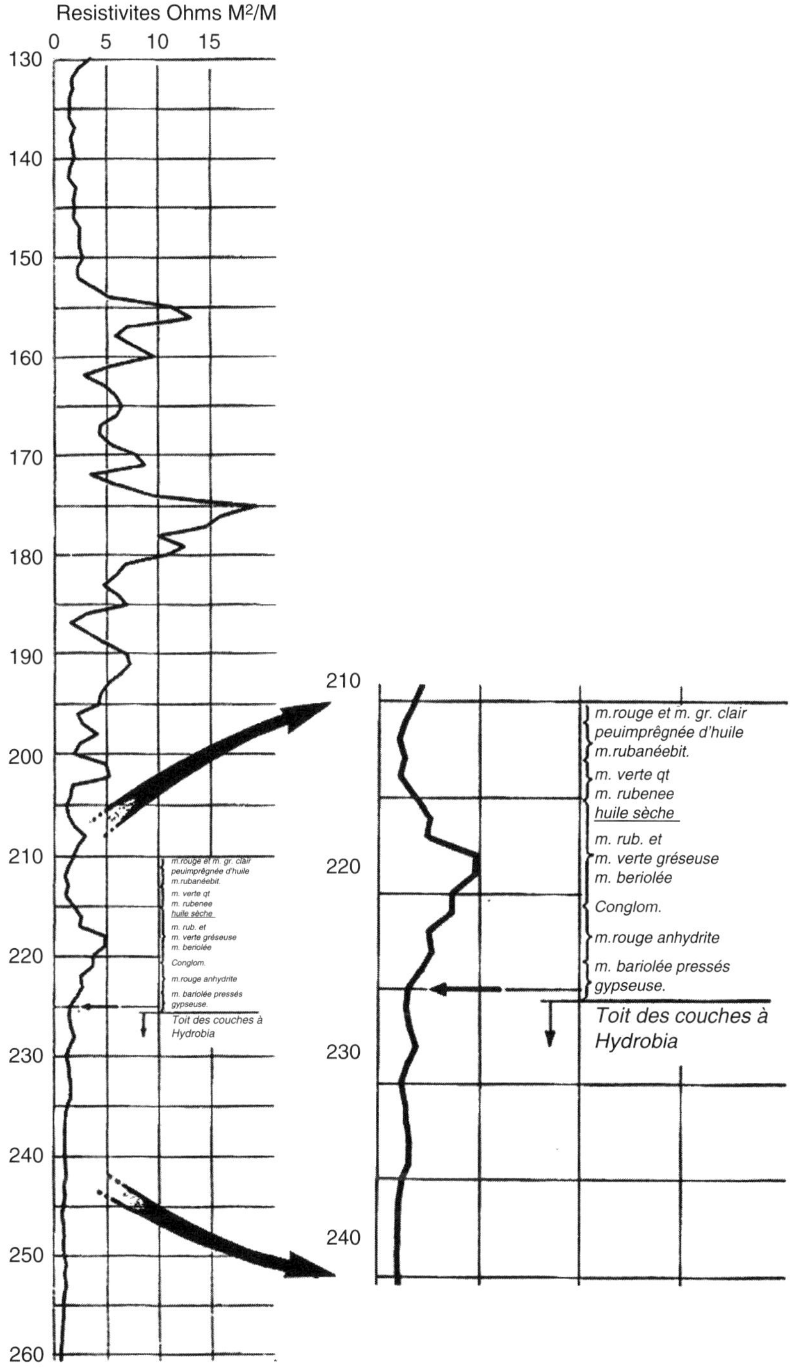

Figure A.3 First well log: Pechelbronn Field, Alsace-Lorraine, France (after Schlumberger, 1982).

Figure A.4 Logging a well near Bakersfield, California, in 1933 (after Ross *et al.*, 1979; Anon., 2006).

currents, over a buried pyrite ore body, in 1913 (Schlumberger, 1982). Because these observed potential anomalies existed in the absence of any applied current, Conrad Schlumberger named the phenomena *Polarisation Spontané*, the English translation of which is Spontaneous Polarization, or SP. The Schlumberger brothers also observed these potential anomalies in their oil well wireline measurements. The anomalies were not constant, but could be correlated to variations in geology. By 1931, SP (often called the "Permeability Log") was being included with resistivity, as part of the Schlumberger Electrical Logs (E-Logs) or Electrical Surveys (ES®). Along with this new measurement came new interpretation theories and techniques (Doll, 1949a).

The brothers mastered unbalanced bridge recording, which allowed continuous logging, and modified their surface resistivity profiling electrode array to one (Normal Array) which was better suited to borehole operations. They also added multiple spacing resistivity curves, to provide information on resistivity variations away from the borehole (ala surface electrical sounding), developed new electrode (Lateral and Laterolog) arrays in attempts to overcome the thin bed resolution/depth of investigation dilemma, and developed optical lever film recording cameras. Through the end of the Second World War, wireline geophysical logs meant Schlumberger Electrical Logs (E-Logs).

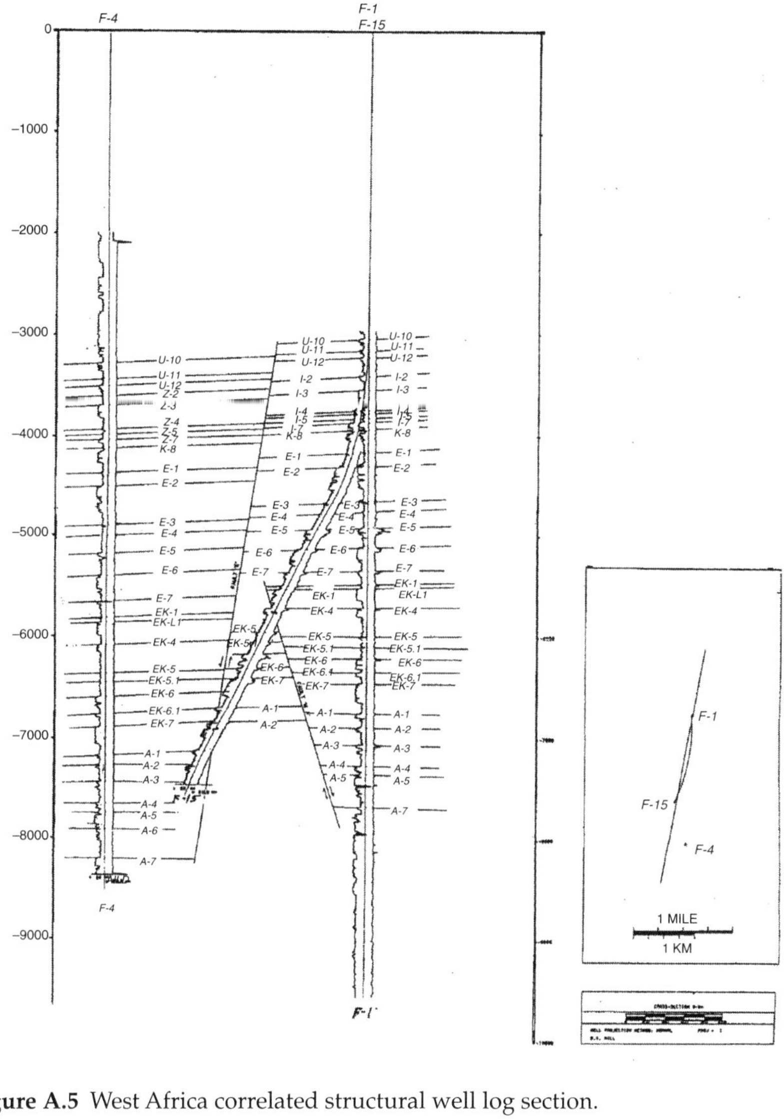

Figure A.5 West Africa correlated structural well log section.

A.3 Gus Archie's Equations and the Dawn of Quantitative Petrophysics

During the period of Schlumberger global expansion, uses of the technology were not static. Geologists and engineers began making observations suggesting that the wireline measurements could be used for more than just subsurface correlation.

An early observation was that ratio of the resistivity of a rock, with pore spaces completely filled with water, R_o, to the resistivity of the water filling the pore spaces, R_w, was relatively constant for any given reservoir rock, and appeared to be related to porosity, ϕ. This ratio, called the *Formation Factor, F*, is defined as:

$$F = \frac{R_o}{R_w}. \tag{A.2}$$

A second observation was that ratio of the resistivity of a rock, with pore spaces only partially filled with water, R_t, to R_o, was also relatively constant, for any given reservoir rock, and appeared to be related to the water saturation (the portion of the pore space filled with water), S_w. This ratio, called the *Formation Resistivity Index, I*, is defined as:

$$I = \frac{R_t}{R_o}. \tag{A.3}$$

In 1942, G. E. (Gus) Archie, of Shell Oil Company, published the results of his laboratory investigations, which related measured resistivities to porosity, ϕ, and water saturation, S_w (Archie, 1942):

$$F = \frac{R_o}{R_w} = a\phi^{-m}, \tag{A.4}$$

and:

$$I = \frac{R_t}{R_o} = S_w^{-n} = \frac{R_t}{FR_w} = \frac{R_t}{R_w a\phi^{-m}} = \frac{R_t \phi^m}{aR_w}. \tag{A.5}$$

The empirical coefficient, *m*, is sometimes called the "*cementation exponent*", as it appears to have some relationship to the degree of rock consolidation and/or cementation. The empirical coefficient, *a*, is sometimes called the "*tortuosity coefficient*". It appears to have some relationship to the ratio of the length of the circuitous path followed by fluids and electrical current, through a reservoir rock, to a direct (straight-line) distance. The empirical coefficient, *n*, is sometimes called the "*saturation exponent*", as it is the exponent of water saturation, S_w, in Archie's Second Equation (A.5). Archie's original values, for a, m, and n, were $a = 1$. $m = 2$, and $n = 2$. Over time, other values have appeared in the literature. Bassiouni (1994) listed the following ranges: $0.35 \le a \le 4.78$, $1.14 \le m \le 2.52$, and $1 \le n \le 2.5$. Individual values of a, m, and n appear to be reservoir specific

Archie's equations (Eq. 4 and 5) offered the opportunity to obtain quantitative reserves information from wireline measurements. To do this routinely, however, required wireline porosity tools.

Archie's equations are empirical relationships. He simply cross-plotted (probably the first petrophysical usage of cross-plotting) on different grids until he found one (bi-logarithmic) on which he could draw a straight line through the data. To raise Archie's equations to the level of a physical "Law", as some non-petrophysicists have done, would probably be rather amusing and embarrassing to their originator. The failure of Archie's equations, under certain circumstances, illustrate their empirical nature and have required the development of more complex petrophysical models.

Because of the relationships between reservoir rock resistivity and water saturation, logs that measure resistivity are often called *Saturation Tools*.

A.4 Air-Filled Boreholes, Oil-Based Muds, and Induction Logs

The SP log and all resistivity logs, discussed so far, require galvanic coupling to the formation, which means that the fluid filling the borehole must be electrically conductive. The borehole fluids in wells drilled with air and oil based muds, however, are non- conductive. The need for saturation tools, for use in wells drilled with air and/or oil-based muds, led to the development of induction logs (Doll, 1949b).

It used to be said that the way to get ahead in Schlumberger was to be (1) French, (2) a graduate of *l'École Polytechnique*, or to (3) marry a Schlumberger heir. Doll Henri took no chances, touching all three bases by marrying Conrad's daughter Anne. He was also a brilliant scientist and engineer, as well as an excellent technical manager. The Schlumberger-Doll Research Center, recently relocated from Ridgefield Connecticut to Cambridge Massachusetts, is the only Schlumberger facility worldwide to have any name other than Schlumberger.

Of Doll's many contributions to wireline technology, the Induction Log (Doll, 1949b) may have been his most significant. Electromagnetic (EM) induction was a well-established surface geophysical technique for locating buried metallic ore bodies, even before Doll invented the first Induction logging tool. An alternating electrical current (AC) through a transmitter coil induces an alternating (primary) magnetic dipole field. This alternating magnetic dipole, in turn, induces AC eddy currents in conductive materials in the subsurface. These induced eddy currents, in turn, induce secondary alternating magnetic dipole fields, which can be measured by a receiver coil. Doll was able to take the principles of surface EM technology, miniaturize it to fit into a logging sonde, and calibrate the signals in terms of apparent resistivity of the material surrounding the borehole. This new tool answered an immediate need, as many wells were being air drilled in arid regions with no ready water supplies. In the case of this breakthrough, the vision came from Schlumberger management (including, by this time, Doll).

A.5 World War II Technology Legacy

Many technologies developed for military applications, during the Second World War (WW-II), found critical civilian applications in minerals and petroleum exploration when they were declassified after the war. Aeromagnetics and *Induced Polarization* (IP), developed as anti-submarine measures, have become common geophysical exploration techniques. Specialized analogue computer technology (developed for navigation, gunnery targeting, and cryptography), became a mainstay for analogue wireline surface units, analogue seismic recording trucks and analogue seismic play-back systems. Digital computers, developed for rapid code breaking, ushered in the information age, accelerated by minerals and petroleum exploration and production (E&P) data processing. Nuclear physics, developed to end the war, allowed estimation of porosity and lithology, via density and neutron logs. In all of these cases, though the technologies were developed for one purpose, the vision of minerals and petroleum industry managers applied them to solve needs in their own industries. In some instances, this vision came from service companies, but in many others this vision came from mining and petroleum company management and in-house mining and petroleum E&P research laboratories.

A.6 Cased-Hole Correlation and Natural Gamma Ray Logs

Many wells, still in service today, were drilled before the use of wireline measurements and, consequently, have no open-hole logs for correlation. The presence of steel casing eliminates the use of SP logs and, until recently at least, resistivity logs. With the primary correlation logs unavailable, another wireline measurement correlation logs must be used for these wells. The simplest and least expensive option is the use of natural *Gamma Ray* (GR) logs.

Natural gamma radiation occurs from three sources: (1) ^{40}K to ^{40}A decay, (2) ^{238}U decay series, and (3) ^{232}Th decay series. Humble Oil and Refining (Now EXXON-Mobil Exploration & Production Research, or EPR) first evaluated natural gamma ray logging by utilizing ionization chambers and, later, Geiger-Müler tube detectors in logging sondes (Howell and Forsch, 1939). The first commercial natural gamma ray logging services were, provided by Well Surveys, Inc., of Tulsa, Oklahoma (Russell, 1944). This firm was later acquired by Lane Wells, a Dresser Industries Company, and became the foundation of a second major wireline vendor (now Baker-Atlas Wireline Services).

Interest in uranium exploration stimulated the broad usage of natural gamma radiation logging tools. During the Uranium Boom, following WW-II, thousands of would-be millionaires bought themselves Geiger-Mueller tube detectors (Geiger Counters) and set off for the wilds of Western US and the Pre-Cambrian shield of Canada, looking for "*Atomic-Age Gold*". The involvement of large number of amateurs doing uranium prospecting, led to many new approaches, most of which were not successful. However, the concept of putting radiation detectors in a pressure housing and lowering them in boreholes, to record a log of radioactivity vs. depth, was very successful. Later, natural gamma ray logs were also used for thorium and potash (Dewan and Greenwood, 1955) mineral exploration, as well. A side effect of the use of natural gamma ray logs for commercial radioactive mineral deposits was that these logs also appeared to make good sand/shale indicators and could be used as correlation logs in cased wells.

In this case the need was for cased-hole correlation tools. The vision for the solution to this need, was provided by wireline industry management, who recognized that gamma ray measurements could answer that need, and oil company management, who recognized the utility of using gamma ray logs as cased-hole correlation tools.

A.7 Seismic Velocities, Acoustic Logs, and Jessie Wylie's Time Average Equation

Interest in better seismic velocity information led to the development of acoustic logs. The California Research Company (now part of Chevron Energy Technology Company, or ETC), Magnolia Petroleum (now part of EXXON-Mobil EPR), and Shell Development Company (now part of Shell Upstream Technology) all developed operational acoustic logging tools at about the same time. The first commercial acoustic tools (called Continuous Velocity Logs, CVL), were marketed by Seismograph Service (Anon., 1953).

In 1956, M. R. J. (Jessie) Wyllie, and associates, at Gulf Research and Development Company (now part of Chevron ETC), developed what became called the Wyllie *Time Average Equation* (Wylie *et al.*, 1956, 1958):

$$\varphi_s = \frac{\Delta t - \Delta t_{ma}}{\Delta t_f - \Delta t_{ma}} \cdot \frac{1}{c}, \tag{A.6}$$

where: ϕ_s is the (Wyllie) sonic porosity, Δt is the observed rock interval acoustic transit time (in μsec/ft), Δt_{ma} is the rock matrix interval acoustic transit time (in μsec/ft), Δt_f is the fluid interval acoustic transit time (in μsec/ft), and c is a coefficient, determined by the data.

The empirical coefficient, c, is sometimes called a "compaction correction", as it appears to be loosely related to the compaction of unconsolidated clastic reservoir rocks. It is not part of the original formulation, but was added later as the $\Delta t/\varphi$ relationship is not linear. This empirical relationship and the CVL proved to be one of the earliest reliable wireline porosity tools. A subsequent empirical relationship, by Raymer *et al.* (1980), of Schlumberger, proved to be more useful in unconsolidated sandstone reservoirs. The simplest form of this relationship, now called the *Raymer-Hunt-Gardner (RHG) Equation* is provided by Dewan (1983):

$$\phi_{RHG} = RHG\left(1 - \frac{\Delta t_{ma}}{\Delta t}\right), \tag{A.7}$$

where: $4 < RHG < 8$ is an arbitrary constant, and: $\varnothing_{RHG}$ is the *Raymer-Hunt-Gardner* sonic porosity, Δt is the observed interval transit time, and Δt_{ma} is the matrix interval transit time (note: there is no Δt_f, in the RHG relationship).

In this case, the need was for a wireline porosity tool. The initial vision was supplied by Jessie Wylie, and his associates at GRDC, who developed the *Time Average Equation* to estimate porosity from interval transit time measurements acquired from the acoustic logging tool, which was originally designed to provide detailed seismic velocity information for seismic inversion.

A.8 The Manhattan Project and Nuclear Logging

The Manhattan Project was directly responsible for ending WW-II, with the dropping of Atomic bombs on Hiroshima and Nagasaki. After WW-II, the declassified nuclear technology that helped to end it (and unemployed nuclear engineers who implemented this technology) led to the development of modern density and neutron logs. Both of these logging tools are based on absorption and/or attenuation of atomic radiation (gamma rays and neutrons, respectively) and can be used to estimate porosity. The latest generation density tools also generate a curve dependent only on the rock matrix and, as such, can be used to estimate rock type.

The need, in this case was for reliable wireline porosity measurements. The vision was provided, by wireline vendors like Lane Wells (now Baker-Atlas), and the unemployed nuclear engineers, needed to modify a weapons development technology and convert it to peaceful uses.

A.9 Space Program Technology Legacy

Most new logging techniques based upon applications of different branches of physics had occurred by the late 1960's. The development since then has been primarily in the areas of increased measurement accuracy, equipment reliability, data acquisition, data transmission, and data processing techniques. In many ways, these last technological advances have been even more impressive than the initial meaurement technique developments.

The race for outer space, following the October 4, 1957, Soviet launch of the first artificial satellite Sputnik-1 (Smithsonian Air and Space Museum Sputknik-1 URL) produced tremendous innovation in the development of low-energy drain miniaturized microcircuits. These developments led not only to laptop computers, cellular phones, personal digital assistants (PDA), and MP3 players, but also to microcomputers in the logging surface units and more compact and complex logging sonde tool stacks.

Gearhart Industries introduced the first use of a general-purpose mini-computer in a field logging system in 1975 (Burgen, 1975). The use of an on-board mini-computer allowed, not only digital recording of the data, but also some processing and analysis, in the field (Head, 1977; Head and Gearhart, 1975). Other major wireline vendors quickly followed suit and the ensuing competition greatly expanded the on-board processing and display capabilities. Current on-board petrophysical analysis capabilities essentially duplicate what previously could only be done at centralized data processing centers.

A.10 SANDIA Geothermal Log Program and Hardened Microcircuits

The final piece of the digital revolution involved replacing downhole analog electronics in the logging sonde, with miniaturized analog to digital (A/D) conversion and microprocessor chips. This development was aided by two unrelated programs:

1. Miniaturization was stimulated by the space program (see above). Logging tools, however, required circuitry, which could withstand the high temperatures and pressures of the borehole environment, which often also contained caustic fluids.
2. In the late 1970's, SANDIA National Laboratory sponsored a high-temperature/high-pressure *Hostile Environment Logging* (HEL) tool component development program, which completed the transition.

Down-hole A/D conversion, sampling, storage, and processing meant more stable logging sondes, and greater (higher volume) data acquisition, as well as faster data acquisition, encoding, and transmission (*i.e.*, broadband) speed and volumes to the surface. This allowed the development of more complex *Sondes*, such as micro-resistivity borehole scanning tools, array acoustic, array induction, and array Laterolog *Sondes*, as well as shorter multi-tool stacks. For example, tool stacks, which used to be over 100 ft long have been replaced by more complex tool stacks which are less than 50 ft long. These newer tool stacks feature downhole A/D conversion and preprocessing, with multiplexed data transmission to the surface, resulting in fewer logging trips/log suite and more information/logging trip.

None of these developments, however, could occur without the development of hardened micro-circuitry, which came about as a result of the SANDIA HEL project. Not only did the HEL project stimulate the development of commercial HEL components, but Anthony Veneruso, the program principal investigator (PI), was hired by Gearhart Industries to upgrade their downhole systems at the project's conclusion.

A.11 Extended-Reach Directional Drilling, Horizontal Wells, Deep Water, Ultra Deep Wells and Measurements While Drilling

Without logs, a well is just an expensive hole in the ground. As drilling targets moved further offshore into deeper and deeper waters, drilling and rig rentals escalated. Time spent conditioning wellbores for logging and running the logs is time that is not spent "making hole".

The costs of offshore site preparation have encouraged drilling extended-reach wells from common drillsites and platforms. Urban, valuable agriculture use, and environmentally sensitive locations also encouraged the drilling of extended reach wells from drilling "islands". These extended-reach wells, as well as being costly, also pose higher risk for loosing the well during drilling. Finally, extreme extended-reach and horizontal wells are often difficult to log with wireline measurements, as the logging sondes must be physically pushed to the bottom (or end) of these boreholes.

A recent major technological development has been the introduction of specialized drillcollars, with instrumentation packages, recording, storage, and/or telemetry, which allow formation evaluation measurements during the drilling operation. Since the introduction of the first *Measurements While Drilling* (MWD) subassemblies in 1979, most wireline measurement capabilities have been duplicated by MWD (also called *Logging While Drilling*, or LWD). At present, almost all wireline measurements can be duplicated by MWD/LWD measurements, with nearly the same (or even better) data quality. In addition, some MWD/LWD techniques have been developed which have no wireline analog.

A.12 Data Acquisition, Data Recording, and Data Transmission Developments

The original Pechelbronn log (Figure A.3) was acquired, using station logging techniques. The logging sonde (array) was suspended at a fixed location, while the logging engineer(s) manually balanced electrical measurement bridges to obtain the Figure A.1 AB currents and MN potential differences needed to estimate apparent resistivity. This operational technique essentially transported the surface Resistivity profiling technique (Figure A.2) from the surface to the borehole. While station logging was considered acceptable, for surface profiling, borehole resistivity measurements required very detailed apparent resistivity (and later measurements) measurement spacing. Greater detail translated into more stations per well and, consequently, more time required to acquire the densely spaced data. Figure A.4 shows two logging Engineers simultaneously balancing bridges to obtain resistivity and spontaneous polarization (SP) data, thus doubling the data acquisition rate (but also the required manpower), for the same station spacing density.

The next data acquisition improvement was to do this operation continuously. The Engineer's bridge rheostats were mechanically linked to a chart recorder ink pen stylus, with the chart recorder drive linked to the logging cable depth encoder. The engineers would furiously spin their rheostats, to maintain their bridge balance, as the logging array was slowly retrieved from the bottom of the well, with the chart recorder recording a continuous record of the formation apparent resistivity and SP variations. This greatly increased logging sped and provided continuous coverage. The number of measurements, however, were limited by the number of engineers required to balance a bridge for each measurement. If all measurements were to be recorded on a single chart, the number of engineers and, consequently, the number of measurements, was effectively limited to two, because of the required mechanical link between the bridge rheostat and the chart recorder.

The next step was to replace the manually balanced bridges, using *Unbalanced Bridge* measurement circuitry, to drive the chart recorder pens electronically. This not only allowed more measurements, but now everything could be supervised by a single engineer. It also allowed faster logging speeds at the same, or increased accuracy.

Petroleum companies soon realized that they need multiple log copies. The Paper chart records did not reproduce very well. The next step in data acquisition was to replace the paper charts and ink pens with light lever galvanometers and photographic film. At one point Schlumberger was reported to be the largest single customer of Kodak 8.25 in (21 cm) continuous feed film.

The need, in all of these cases, was for detailed measurement information, more measurements, and for shorter rig time to acquire these measurements. The vision to meet these needs was provided by the logging vendor, which at this time was still only Schlumberger. This single provider situation, however, was to change with the entry of other logging vendors.

A common problem with both paper chart and film recorded logs was that if different log recording parameters were desired, a well had to be re-logged. Because re-logging a well meant more rig time, to acquire the data, this was seldom done. Consequently, the logging engineer's first "guess" for the logging parameters was often his last and "best" guess. Formation evaluation (FE) specialists had also begun to develop increasingly complex well log analysis techniques. These techniques could only be described as tedious, if applied manually. A method of recording the log data, in the field, which could later be played back and reprocessed using different parameters and combined for analysis, became a dream for FE specialists.

Encouraged by the success of seismic FM analog tape recording and play-back systems, Dresser-Atlas is supposed to have evaluated this technology for field recording well logs, in the early 1960's. While the technology was claimed to have been a technical success, lack of client interest and required specialized analog processing equipment resulted in the project being quietly dropped.

Another failed attempt at field recording log data involved paper teletype tape (TT), for direct input into mane-frame computers (Bateman, 2009). This medium proved not only to be an extremely noisy distraction to the engineer and very slow, but also was not very well-suited for the field conditions in which most logging units worked. It was also quietly dropped.

In 1968, Schlumberger introduced their seven track *Truck Tape Recorder* (TTR) transports and became the first logging vender to field a successful digital field recording system. This system, however, could not be played back, in the field and could only be read by Schlumberger in-house processing computers. Any operating company wishing to do their own in-house log processing had to request nine-track binary tapes from a Schlumberger processing center.

With Schlumberger offering to deliver binary nine-track tapes to clients, major petroleum companies began to develop their own complex log analysis software packages, and specialized log analysis service companies, such as Scientific Software Corporation, began offering their own log analysis services to the industry. These developments led, in turn, to all major, and several minor, wireline service companies offering to deliver client binary tapes to those who requested them. The only problem with all of these different binary tape sources was that each vendor used its own unique binary tape format.

This library tape format confusion ended with the establishment of the Petrotechnical Open Software Corporation (POSC, now Energistics) and standardized library (binary) tape formats, which everyone could write and read (Bateman, 2009). Two of the more common library tape formats are Library Information Standard (LIS), developed by Schlumberger, and Library ASCII Standard (LAS), developed by the Canadian Society of Well Log Analysists. Because of its simplicity, essentially all wireline vendors now routinely deliver well log data in LAS format.

In 1975, Gearhart Industries introduced their Direct Digital Logging System (DDS), which featured mini computers in the logging units, which could do on-site log play-back and analysis. Schlumberger followed with their Cyber Service Unit (CSU) logging units in 1977, and the logging industry was in the digital age to stay.

A.13 Log Analysis Developments

Formation Evaluation (Well Log Analysis) sophistication has kept pace with wireline and MWD/LWD hardware developments. Before calculators were readily available, log analysts utilized slide rules (Figure A.6), chart book nomographs (Hoelscher *et al.*, 1952), and pre-calculated tables. The first chart-book nomographs were calculated from closed form mathematical boundary value problem solutions (remember that Conrad Schlumberger was a physics professor). As logging tools and petrophysical solutions became more complex specialized analog computers were developed to generate the appropriate nomographs (Bateman, 2009).

Worksheets were originally utilized to organize the log analysis processes. Programmable calculators simplified this process and made more complex petrophysical models practical. Microcomputer spreadsheet calculations (Figure A.7) increased the speed and accuracy of the process. Specialized formation evaluation software for main-frame and mini-computers (Figure A.8) made detailed well log analysis and impressive displays possible. Inexpensive to modestly priced micro-computer formation evaluation (FE) software (Figure A.9), made this software portable, allowing petrophysicists to conduct formation evaluation from the drillsite, office, or just about any other location with Internet access.

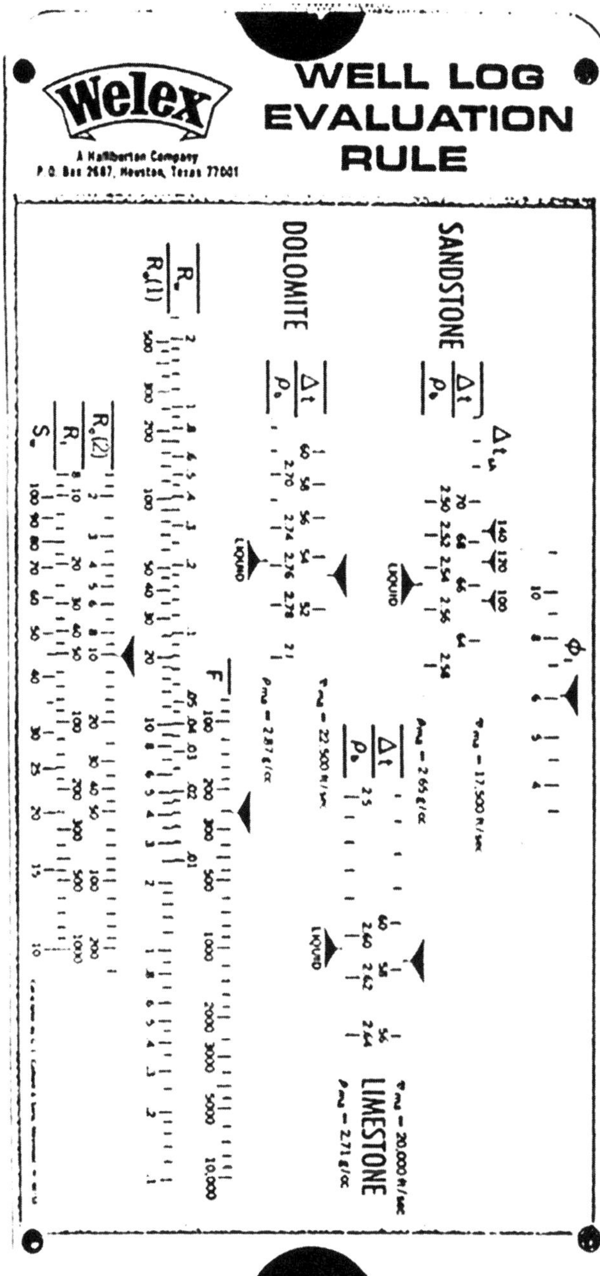

Figure A.6 Specialized formation evaluation slide rule (courtesy of WELEX).

Well:
Field:
Location:
Reference: Rotary table 30.2 m

Microcomputer spreadsheet formation evaluation analysis
a TRWLAS product

Revised: 07.10.88, DGH

Pay Sand	Level	Top MD-KB (m)	Bottom MD-KB (m)	Top TVD-SS (m)	Bottom TVD-SS (m)	Formation temperature (°F)	Formation Rw (ohm-m)	Vsh	Ør	Rt (ohm-m)	Net pay (TVD) (m)	Sw
	1	1115.75	1116.5	-1085.55	-1086.3	166.7	0.0740	0.183	0.223	144.24	0.750	0.030
	2	1116.5	1117.5	-1086.3	-1087.3	166.7	0.0740	0.021	0.227	488.86	1.000	0.022
	3	1117.5	1118.5	-1087.3	-1088.3	166.8	0.0740	0.030	0.269	108.98	1.000	0.055
	4	1118.5	1119.5	-1088.3	-1089.3	166.9	0.0739	0.026	0.252	103.36	1.000	0.061
	5	1119.5	1120	-1089.3	-1089.8	166.9	0.0739	0.026	0.244	88.30	0.500	0.069
	6	1120	1120.5	-1089.8	-1090.3	167.0	0.0739	0.059	0.239	69.39	0.500	0.074
	7	1120.5	1121.5	-1090.3	-1091.3	167.0	0.0739	0.039	0.243	50.09	1.000	0.094
	8	1121.5	1122.5	-1091.3	-1092.3	167.1	0.0738	0.035	0.228	133.15	1.000	0.056
	9	1122.5	1123	-1092.3	-1092.8	167.2	0.0738	0.021	0.271	108.97	0.500	0.056
	10	1123	1123.5	-1092.8	-1093.3	167.2	0.0738	0.044	0.255	26.10	0.500	0.130
	11	1125	1126	-1094.8	-1095.8	167.4	0.0737	0.097	0.232	20.39	1.000	0.149
	12	1126	1127	-1095.8	-1096.8	167.5	0.0737	0.059	0.209	14.42	1.000	0.217
	13	1127	1127.5	-1096.8	-1097.3	167.5	0.0736	0.147	0.218	8.70	0.500	0.246
	14	1128.5	1129.5	-1098.3	-1199.3	167.7	0.0736	0.154	0.233	8.70	1.000	0.229
	15	1130.5	1131.5	-1100.3	-1101.3	167.8	0.0735	0.525	0.146	2.80	1.000	0.429
	16	1132.5	1133.5	-1102.3	-1103.3	168.0	0.0735	0.259	0.174	2.80	0.000	0.526

Logs used: doop latorolog resistivity (RLLD), density (RHOB), neutron (NPHI), sonic (DT), gamma ray (GR), and caliper.
Schlumberger enironmental corrections applied to: RLLD, RHOB, & NPHI.

Rmf @ 75 F:	0.0671 ohm-m	Salinity:	112328.9 ppm (NaCl eq)	Density:	1.0820 gm/cc
Rwa @ 75 F:	0.157 ohm-m	Salinity:	40638.1 ppm (NaCl eq)		
Rm @ 75 F:	0.776 ohm-m				
Mud weight:	9.5 lb/gal				
Mud Temperature:	71.6 +				
Formation Temperature:	80 -				
Bit size:	8.5 in				

Vsh: Estimated from GR, using stelber equation and from ss desity/neutron cross-over.
GR(sd): 33 API
GR(sh): 115 API
Neutron sand line bais: 0.11

PHID: Estimated from RHOB, using the density porosity equation.
Rho(ma): 2.65 gm/cc
Rho(f): 1.0820 gm/cc

PHIN: Environmentally corrected.

PHIS: Estimated from sonic DT, using the Wylie time-average equation.
DT(ma): 55.5 μsec/ft
DT(f): 189 μsec/ft
c: 1.2

PHIE: Estimated from Vsh corrected PHID and PHIN, using the Gaymard equation.
PHID(sh): 0.12
PHIN(sh): 0.46

PHIR: PHIE corrected to reservoir conditions by:
Ør - 0.94 Øθ -0.006

Rt: Environmentally corrected RLLD.

Sw: Estimated from PHIE, AND Rt, using the Simadoux equation.
a: 0.91
m: 1.91
n: 1.8
Rsh: 1.7 ohm-m

Net pay cut-off:
Sw: 0.5

MD-KB: From log prints.
TVD-SS: From WAPET.

Interval summary	
Gross pay (TVD):	17.75 m
Net pay (TVD):	12.25 m
Net Gas (TVD):	0 m
Net Oil (TVD):	12.25 m
Average Ø:	0.234
Average Sw:	0.109
Average Vsh:	0.123
Hydrocarbon thickness:	2.557 m

Figure A.7 EXCEL® formation evaluation spreadsheet.

At this point, a disclaimer is in order. All of these data processing and analysis developments did not, in and of themselves, make the results more accurate, just faster to achieve. In the hands of an inexperienced petrophysicist, these data processing aids just increased the speed and quantity of questionable results, over what could be obtained by hand analysis. In the hands of an experienced petrophysicist, however, these aids seemed to be "like magic".

Modern logging field units can perform most log analysis activities in the logging unit and deliver LAS digital files, via satellite from the wellsite to processing centers, or to client offices worldwide. In some cases this can be done, as the well is being logged. This capability allows detailed digital well log analysis within hours of running the logs.

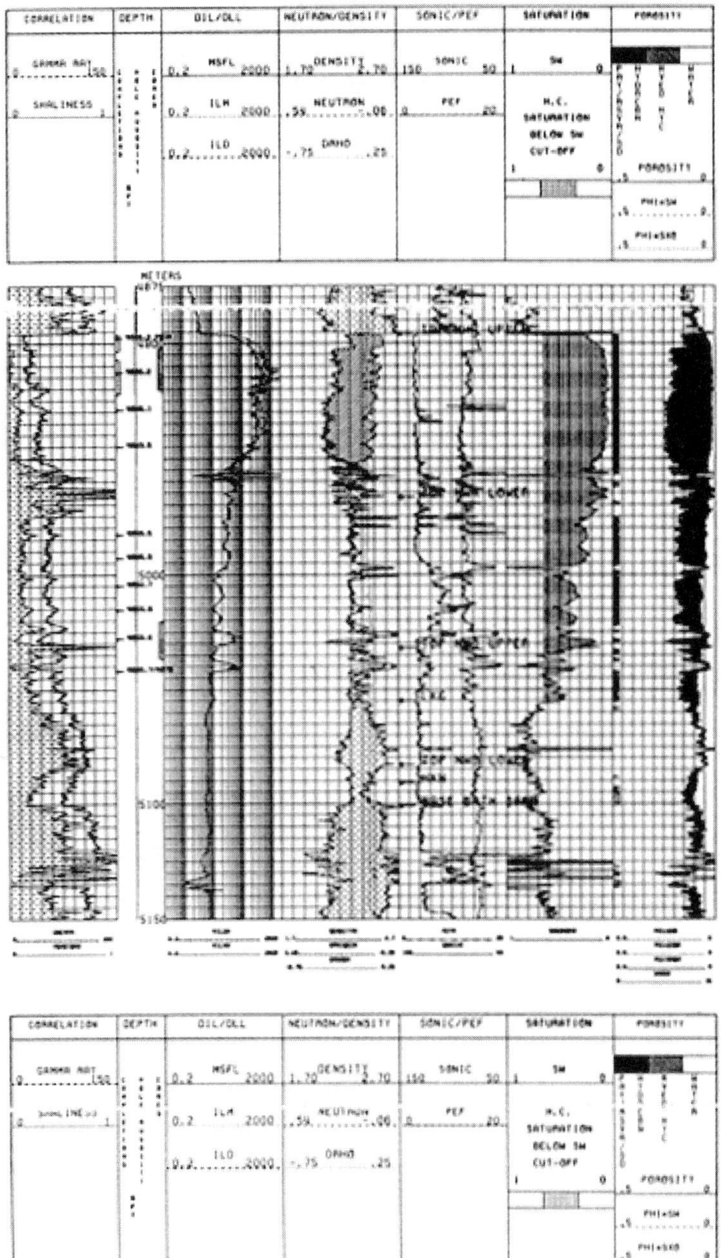

Figure A.8 Computerized multiple log formation evaluation analysis.

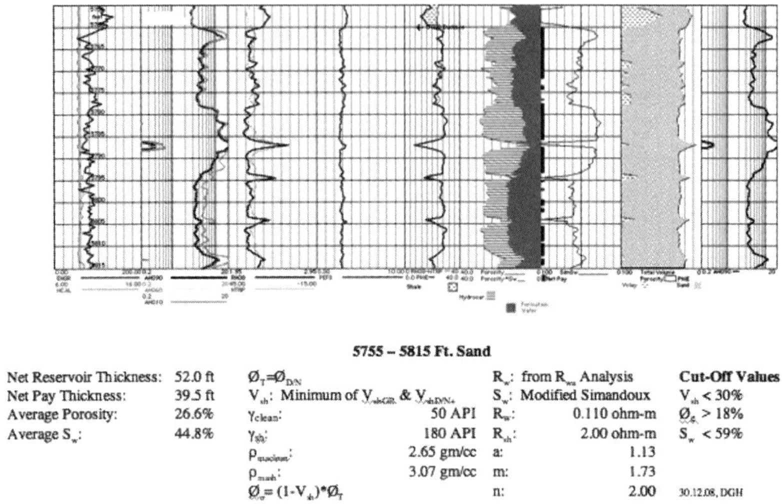

Figure A.9 Micro-computer formation evaluation product.

Wireline vendors can also download well log data, at the wellsite, to USB memory sticks or floppy disks. This allows detailed digital well log analysis, on laptop microcomputers at the wellsite, utilizing the same software that is available in vendor and client offices. As long as there is a broad-band or dial up connection available, a petrophysicist can work from anywhere: at the well site, in the client's home or remote office, at the petrophysicist's home base, or even in remote vacation cabins.

A.14 Formation True Resistivity, R_t, Flushed Zone Resistivity, R_{xo}, Water Saturation, S_w, and Flushed Zone Saturation, S_{xo}

Archie's Equations (Eq. A.4 and A.5) allowed production and reservoir engineers to estimate the volume of hydrocarbons in a reservoir, if they knew the porosity and the formation resistivity. Unfortunately, the apparent resistivity equation (Eq. A.1) assumes that there are no effects of the borehole and borehole fluids, which is not true.

To address this problem, Conrad Schlumberger and his son-in-law, Henri Doll, drew upon their surface resistivity sounding experiences and developed multiple electrode arrays and spacings, which would deliver apparent resistivities with varying degrees of influence from the borehole and borehole fluids. These multiple apparent resistivities could then be inverted to obtain true resistivities for rings of cylindrical borehole environment models (Figure A.10), similar to the type-curves used to invert surface electrical sounding data. Two of the model ring resistivities have become very useful for reservoir engineering. These are the true resistivity, R_t, of the uninvaded zone and the flushed zone resistivity, R_{xo}. They are used with Archie's equations and porosity to estimate the uninvaded zone and flushed zone water saturations, S_w, and S_{xo}, respectively. The former is then used to estimate the total volume of hydrocarbons in

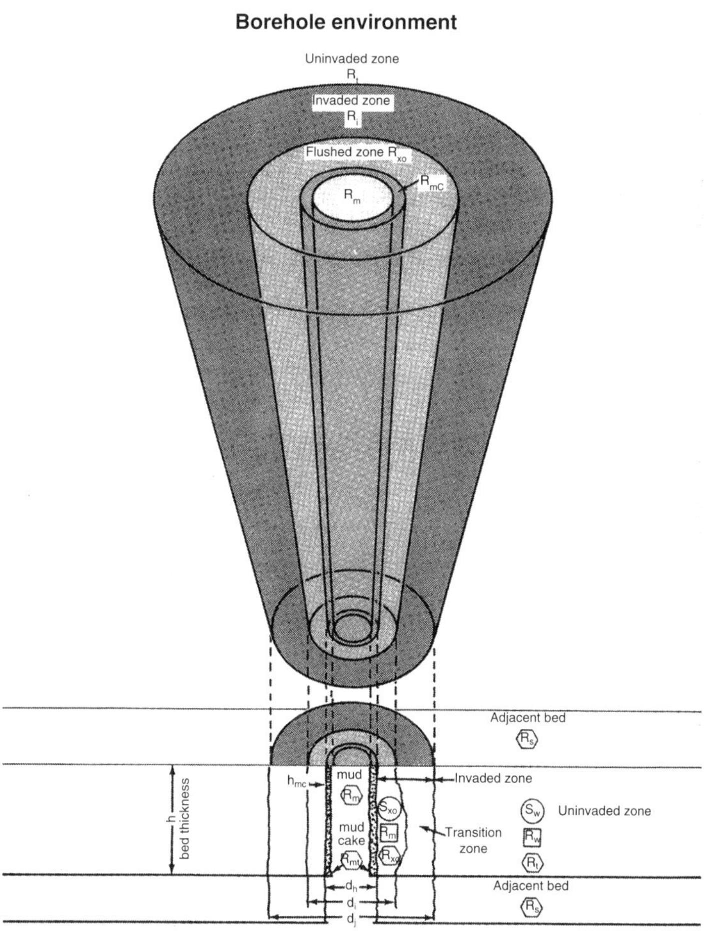

Figure A.10 Generalized borehole environment model (after Anon., 1983).

place, whereas the latter is used to estimate the volume immobile hydrocarbons, with the difference providing the volume of movable (recoverable) hydrocarbons. Essentially all resistivity (conductivity) tool developments, since the initial Schlumberger resistivity tool, have been aimed at providing more accurate estimates of R_t and R_{xo}, under varying borehole conditions.

A.15 Rat Holes, Bed Resolution, Depth of Investigation, and Laterolog Developments

The electrode array used to log the Pechelbronn well (Figure A.11) was essentially the surface Schlumberger electrode array dropped down the borehole. The array measure point was at the center of the MN potential electrode dipole, centered in the larger AB current electrode dipole. This meant that the deepest point, which could be measured, would be AB/2 from the bottom of the well. To address the half-array length lost data problem and avoid the necessity of drilling an AB/2 *Rat Hole*, below the deepest point of interest, Conrad

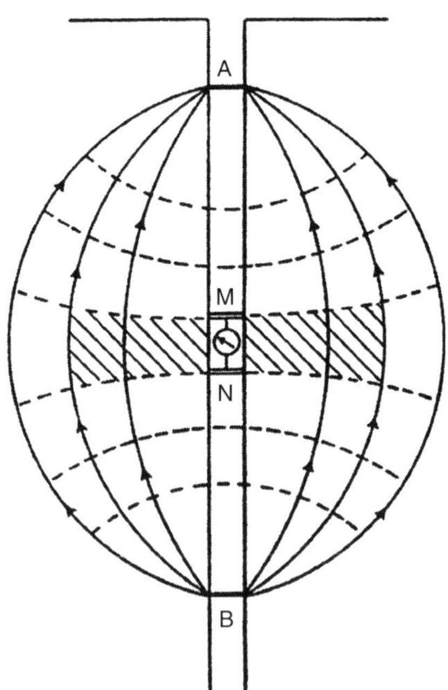

Figure A.11 Pechelbronn well resistivity log electrode array (after Ross *et al.*, 1979; Schlumberger, 1982).

Schlumberger developed the *Normal Array* (Figure A.12), which was essentially half of the Schlumberger surface electrode array, with the B current and N potential electrode moved to the surface (or at least very far up the borehole from the A current electrode and M potential electrode). The normal array required a much smaller rat hole, than the Pechelbronn array, because the AM spacing of the normal array was much shorter than the AB/2, for the Pechelbronn array.

The tool "Span" (effective measured thickness) of the Pechelbronn array was the MN potential dipole separation. For the normal array, it became the AM separation. Almost immediately, problems arose. Geologists wanted to be able to correlate very thin beds, which meant that AM spacing (AM) should be small. The geometrical factor, G, in the apparent resistivity equation (Eq. A.1), however, assumed that the borehole and formation had the same electrical properties. If the borehole was large enough, compared

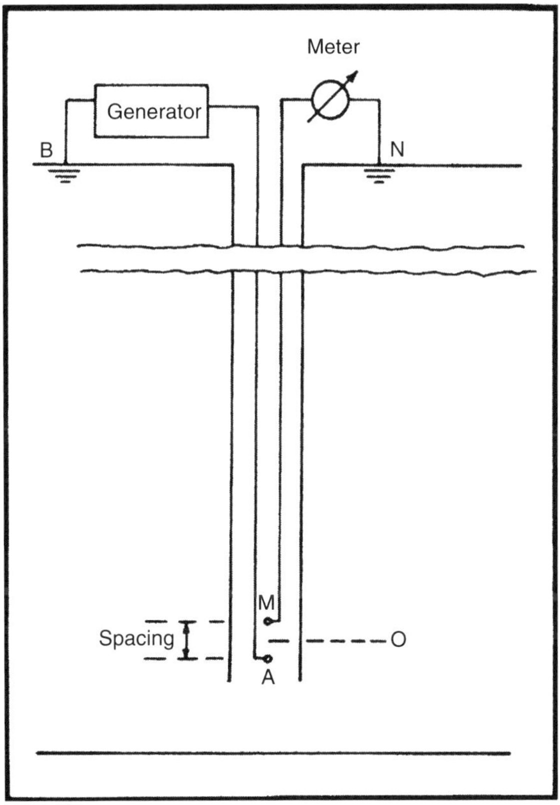

Figure A.12 Normal resistivity log electrode array (after Anon., 1972).

to AM and the mud resistivities differed greatly from those of the formation, the apparent resistivity, given by Eq. A.1 could be quite different from the true formation resistivity. These borehole effects were greatest for short AM spacings and decreased for longer AM spacings (*i.e.*, AM >> than the borehole diameter). Bed boundaries were also much sharper, however, with short AM spacings than they were for longer AM spacings.

The goals of sharp bed boundaries, thin bed resolution, and accurate R_t gave rise to a bed resolution/depth of investigation dilemma: You could either design your normal array with a short AM to get sharp bed boundaries and thin bed resolution, or with large AM to reduce the borehole effects on R_a, but you could not accomplish both goals with a single normal electrode array.

Conrad Schlumberger's first attempt to deal with this bed resolution/depth of investigation dilemma was to fall back on his surface electrical sounding experience and use sondes with multiple AM spacings. Over time, two normal array AM spacings: 16 inch *Short Normal* (SN) and 64 inch *Long Normal* (LN) became the standard normal array spacings, although some slim-hole E-Logs use a 48 inch LN.

Conrad Sclumberger's second attempt at resolving the bed resolution/depth of investigation dilemma was to design a new electrode array, which he called the *Lateral Array* (Figure A.13). The lateral array was essentially ¾ of the Pechelbronn array, with only the B current electrode removed from the sonde. The tool measure point was midway between the M and N potential electrodes and the tool span was the MN dipole spacing, whereas the depth of investigation was governed by the separation of the downhole current electrode (A) and the center of the MN potential dipole (AO). The lateral array was a perfect example of something that looked much better on paper than it performed in practice. The short MN dipole gave fairly good thin bed resolution and required a very small Rat -Hole, while the long AO spacing did reduce the borehole effects. Unfortunately, the Lateral array bed boundary response was so asymmetrical that it became almost impossible for all but a few *Lateral Log Specialists* to interpret. Schlumberger attempted to address this problem with catalogues of *Lateral Log Type Curve* case history books, but the measurements were still viewed with suspicion, by many in the petroleum industry. As a result, the lateral log never became popular and can only be found, today, on some slim-hole E-Log sondes.

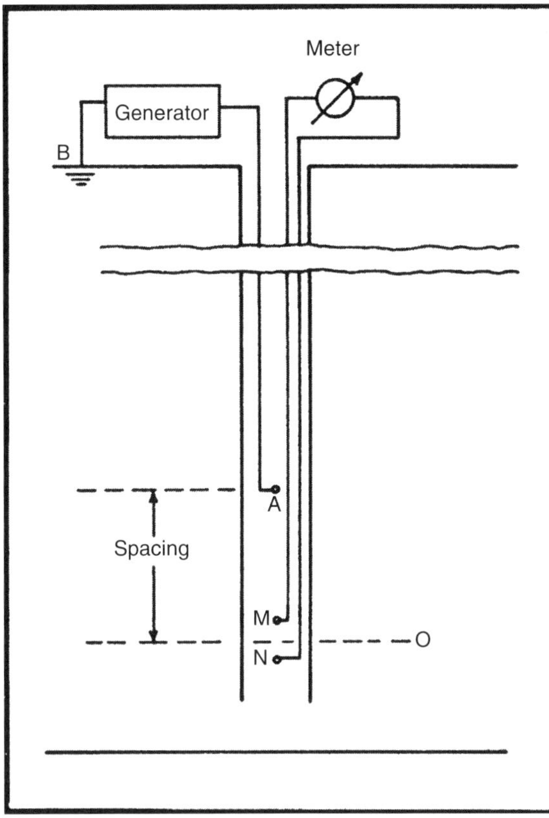

Figure A.13 Lateral resistivity log electrode array (after Anon., 1972).

The next attempt at resolving the bed resolution/depth of investigation dilemma was the guarded electrode (Figure A.14), called *Laterolog-3®* (*LL3*), by Schlumberger and *Guard Resistivity*, by all other vendors. There is some uncertainty as just who originated the guarded resistivity tool and just when they did it. The earliest published papers apparently were by George Keller (1949, 1950), followed by that of Owen and Greer (1951). During the discussion of his introductory *Laterolog-7®* paper, Doll (1951) claimed that Conrad Schlumberger first developed the guarded electrode configuration, in 1927, and disclosed his invention during 1940–41 litigation. Schlumberger did not begin offering LL3 services, however, until after other wireline vendors proved guard resistivity to be of commercial value.

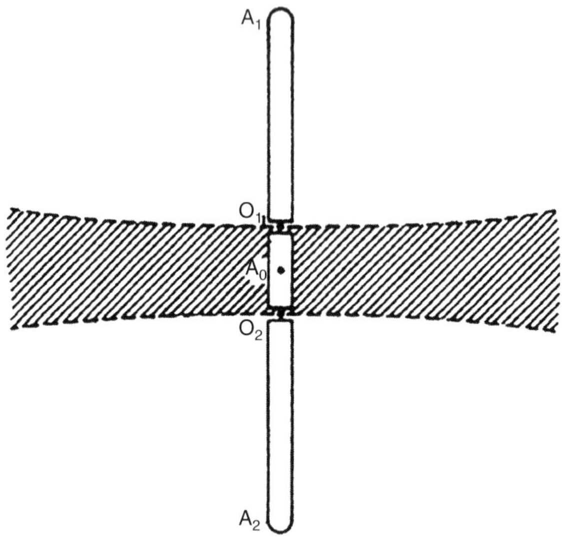

Figure A.14 Current distribution from guarded, or Laterolog-3® (after Anon., 1972).

The guarded or LL3 Resistivity tool utilizes two elongated *Guard Electrodes* maintained at the same potential as an isolated center *Current Electrode*. Resistivity is determined by variations in current from the current electrode. The net effect of the electrode guards is to force the current from the current electrode into the formation, as shown in Figure A.14, which minimizes the effects of the borehole fluids. This appears to be quite successful, when the apparent resistivity, R_a, is greater than the mud resistivity, R_m, but breaks down, when the reverse is true (Figure A.15). The guarded electrode also precluded measuring meaningful SP signals, because of the massive guard and current electrodes, in the well, as well as the high current densities emanating from them.

In spite of its obvious shortcomings, guarded electrode resistivity logs proved to be very popular for carbonate and clastic reservoirs, drilled with conductive muds. Gulf Oil Company <Nigeria>, Ltd., for example, used LL3/GR and side wall cores (SWC) as their primary development drilling log suite through the mid 1970's. This rudimentary log suite allowed the drillsite geologists and completion engineers to discriminate sands from shales, using the *Gamma Ray* (GR) logs; pay from non-pay sands, using the LL3; and oil sands (fluorescence) from gas sands (no fluorescence), using the SWC. Unfortunately, this log suite was useless for depletion planning,

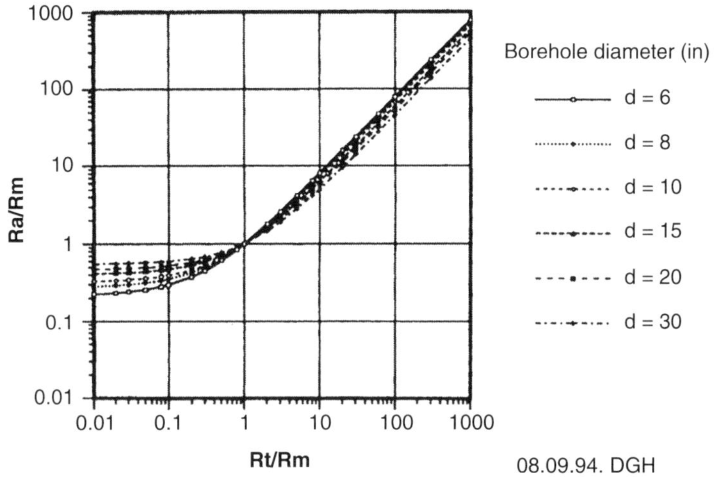

Figure A.15 Slim-Hole guarded electrode departure curves, showing low borehole effects, for $R_a > R_m$, but large borehole effects for $R_a < R_m$.

because it had no porosity tool. High-permeability zircon-rich beach sands were often missed, because they were radioactive. High-gravity light Niger Delta crudes were also often missed, because they evaporated easily and/or had only faint fluorescence. All of these undesirable consequences could have been avoided by using more modern, though costlier, log suites available at the time.

Many vendors (e.g., Dresser-Atlas, Gearhart, and Halliburton) continued to use guarded electrodes as the shallow resistivity device (instead of the SN array) on their induction and dual induction log sondes, even after stand-alone guarded electrode tools were phased-out of use. Guarded electrode resistivity tools continue to be offered by many slim-hole logging vendors, for mining applications where $R_a >> R_m$, because of the excellent bed boundary and thin-bed resolution offered by the guarded resistivity tools.

Whereas Schlumberger may, or may not, have invented the guarded electrode resistivity tool, Henri Doll definitely did invent the *Laterolog-7®* (LL7) tool (Doll, 1951). This tool used a combination of seven current, potential, and dynamic (servo controlled) bucking current electrodes to focus the measuring current into the formation, much like the guarded electrode, but (hopefully) without its drawbacks (see Figure A.16). Over time another pair of electrodes was added, to aid in the focusing, and two signal frequencies were used, allowing measurement of a deep and shallow

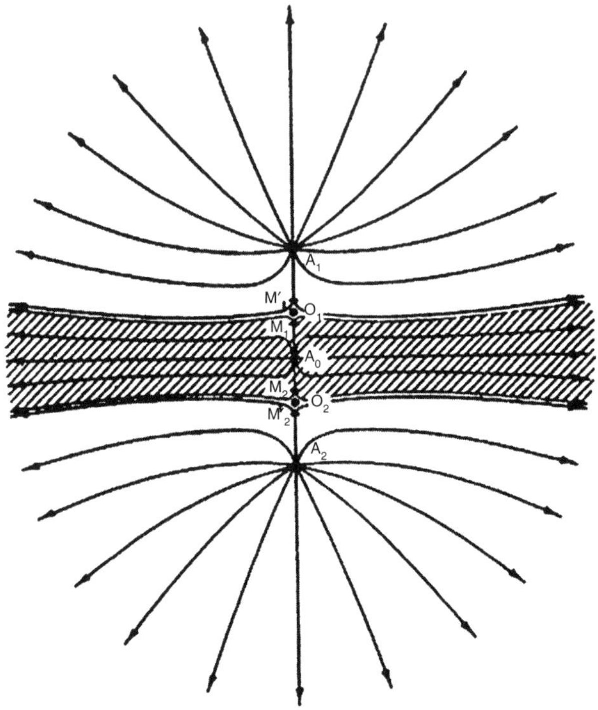

Figure A.16 Schematic current flow streamlines for Laterolog-7® (LL7) tool (after Anon., 1972).

apparent resistivity (see Figure A.17). The resulting tool, called the *Dual Laterolog®* (DLL) is in use today by essentially all major logging vendors.

Although not called a laterolog, the Spherically *Focused Log®* (SFL) is a focused current resistivity tool based on the laterolog type design (Figure A.18). The SFL was designed to provide shallow resistivity information, for use with induction logs, as an alternative to the LL3 or SN (Anon, 1984a).

All of the resistivity (conductivity) tools discussed, to this point had remote current return (B) electrodes. These remote current return electrodes were placed either at the surface or used the logging cable armor, separated by a 36-foot long insulated member, called a *Bridle*, from the logging tool. In certain situations, notably below thick high-resistivity beds, the remote current return electrode caused erroneous apparent resistivities because of distortion in the assumed current density streamlines.

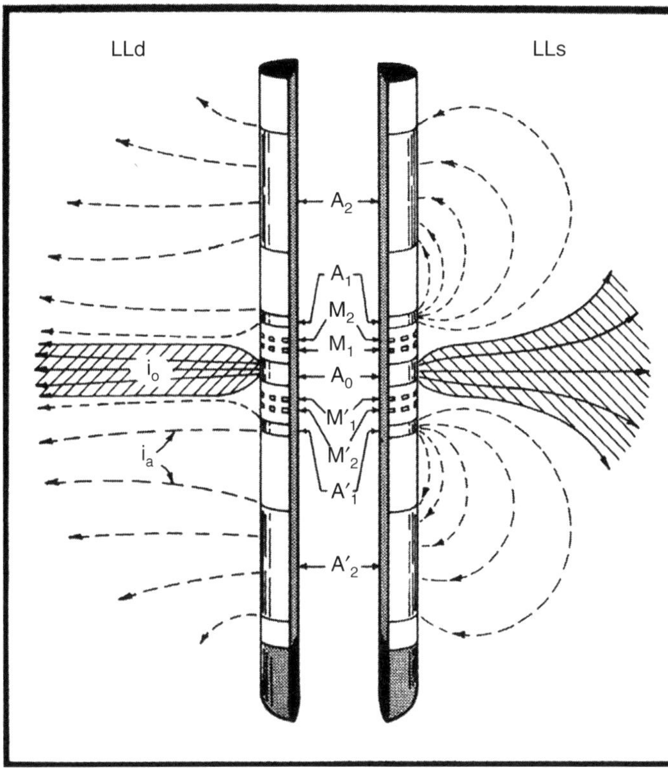

Figure A.17 Schematic current flow streamlines for Dual Laterolog® (DLL) tool (after Anon., 1972).

The resistivity tools discussed, thus far, were also analog devices, with down-hole potentials and currents actually measured at the surface. The finite number of logging cable conductors (9, or less) limited the number of measurements. Self and mutual inductances in the long logging cables precluded meaningful measurements of signal phase shifts.

The latest incarnations of laterolog tools are array laterologs. The Schlumberger *High-Resolution Laterolog Array®* (HRLA) tool, introduced in 1998, uses downhole microprocessors, multiple electrode arrays, multiple frequencies, and records both amplitude (impedance) and phase to provide 6 apparent resistivities, which can be inverted to provide two-dimensional (cylindrical) resistivity models (Anon, 2000). This particular tool also has all current electrodes on the sonde, eliminating problems associated with a surface, or

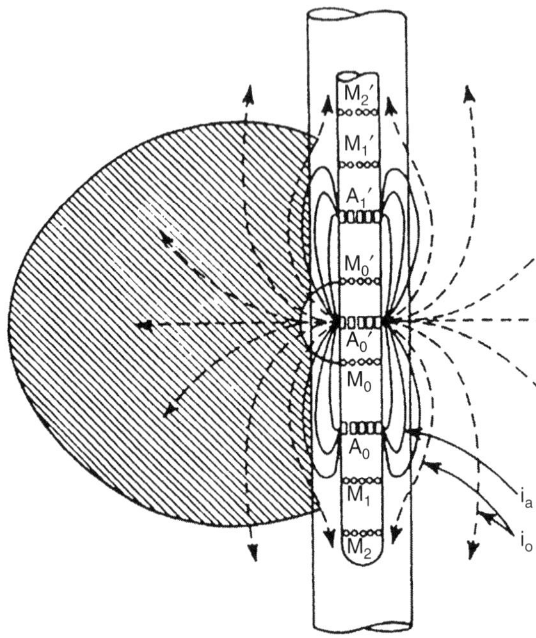

Figure A.18 Schematic current flow streamlines for Spherically Focused Log® (SFL) tool (after Anon., 1972).

bridle, current return electrode. Figure A.19 shows schematic equipotential shells and current streamlines for the 6 HRLA apparent resistivities (modes 0–5). Mode 0 is assumed to provide a downhole mud resistivity, R_m, measurement, whereas mode 5 is assumed to be close to true formation resistivity, R_t.

A.16 Air, Mist and Oil-Based Muds: Induction Log Developments

All of the resistivity (conductivity) measurement tools, described in the last section require electrically conductive fluids in the borehole so that electrical current can pass from the tool electrodes to the formation. The use of air and mist (to drill in areas where water supplies are scarce or thief zones are present) and oil-based muds (used to minimize water invasion and/or reduce swelling clays) required resistivity measurement tools, which did not need conductive muds. The solution was the *Induction Log*, developed by Henri Doll and introduced by Schlumberger in 1949.

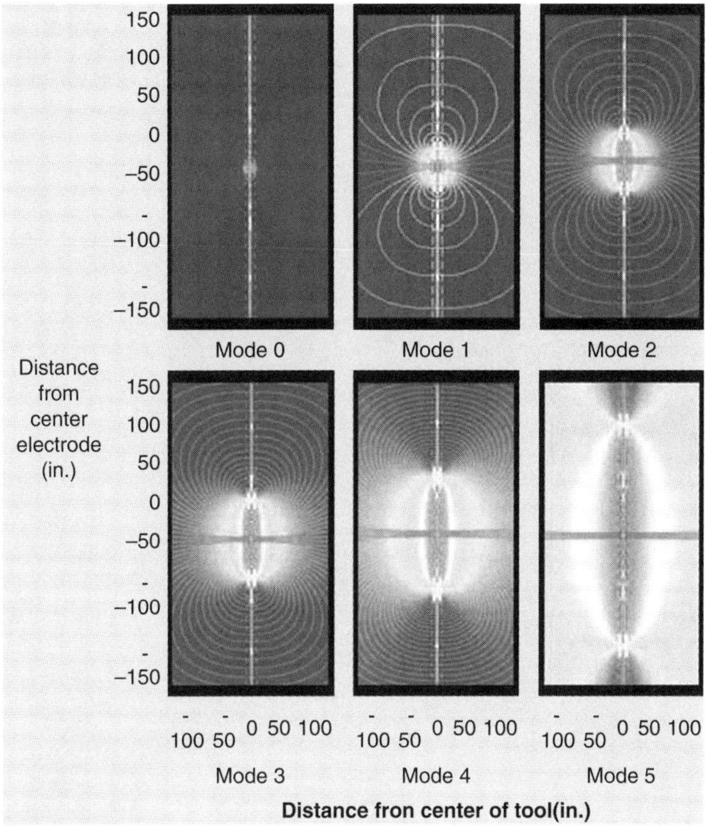

Figure A.19 Schematic equipotential shells and current flow streamlines for High-Resolution Laterolog Array® (HRLA) tool (after Anon., 2000).

Doll (1949b) took existing ground and airborne *Electro-Magnetic* (EM) geophysical prospecting technology used for metallic ore body exploration, in the minerals industry, and redesigned it for use in boreholes drilled in conductive sediments (see Figure A.20). In the process, he defined how induction log theory would be treated in the logging literature (see Figure A.21). Moran and Kunz (1962) refined and expanded Doll's original induction log theory. Philip Nelson's (Hearst and Nelson, 1985; Hearst *et al.*, 2000) induction log development was done from the traditional minerals industry approach. These different developments reach the same conclusion and complement one another, nicely.

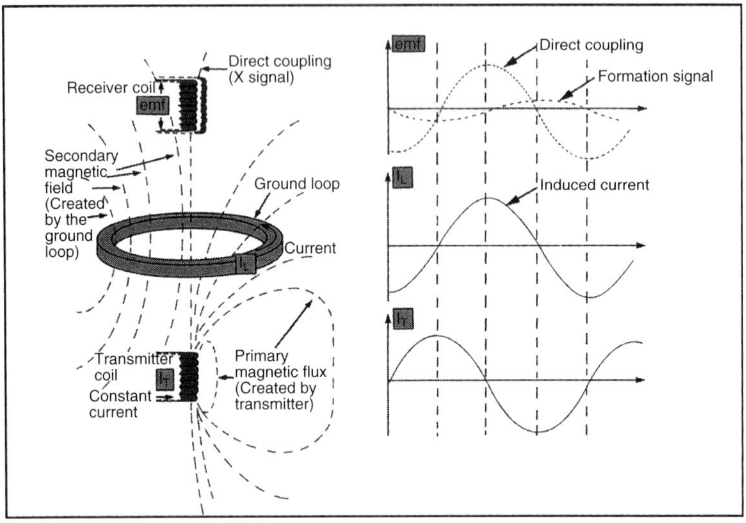

Figure A.20 Schematic Induction Log® principle (after Anon., 1984a).

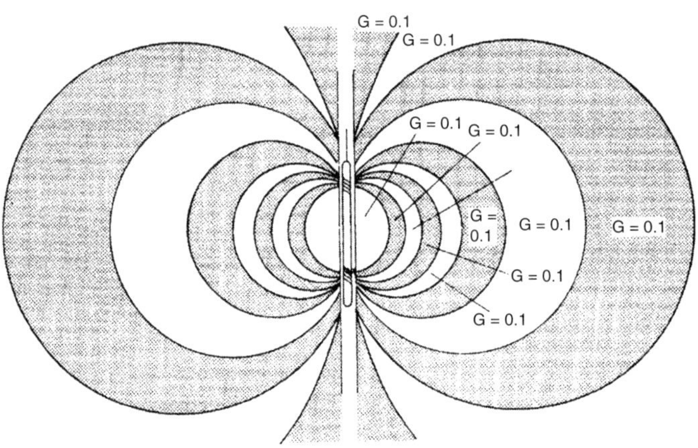

Figure A.21 Schematic illustration of Induction Log® regions of equal geometric factor (after Doll, 1949a).

Doll's original (1949b) paper implied the use of a single transmitter/receiver coil pair. In fact, all Schlumberger induction tools have utilized multiple receiver coils to improve tuning, even if only a single conductivity (resistivity) curve was produced. For example, the 6FF40® tool uses 6 coils, but produces only a single conductivity/resistivity curve.

Once the induction tool proved to be successful in air and oil-based drilling fluid environments, the next development was to use it in fresh-water mud environments (Dumanoir *et al.*, 1957). To add thin-bed resolution and apparent resistivities (conductivities) with varying amounts of invasion effects, a 16-inch SN measurement was added to the sonde.

The next innovation was to use the multiple transmitter/receiver coil separations to develop apparent conductivities (resistivities) with different amounts of invasion influence. This Dual Induction tool also included shallow electrode array resistivity (conductivity) measurements. Schlumberger introduced a new shallow laterolog called the *Laterolog-8®* (LL8) and later, the *Spherically Focused Log®* (SFL), while Dresser-Atlas (now Baker-Atlas), Gearhart (now Halliburton), and Halliburton used short guarded logs (LL3), to obtain shallow resistivity/conductivity measurements.

All of the above induction log devices were analog tools, with transmitter voltages and currents generated at the surface and the downhole receiver coil voltages and currents measured at the surface. Only total impedances (conductances) were measured, even though EM phase measurements were common in the minerals industry and James Moran (1964) had patented an impedance and phase interpretation method for induction logging tools.

The reason, for the lack of phase measurement, was because the self and mutual inductances of the long logging cables and short transmitter-Receiver coil spacing precluded monitoring the small signal phase shifts occurring between the transmitted and received signals. This all changed with the introduction of (downhole) digital and array induction tools. BPB Wireline Services (now Weatherford) was the first to introduce a downhole digital array induction tool (Martin *et al.*, 1984), followed closely by Schlumberger's *Phasor®*, or DITE, digital dual induction tool (Barber, 1985).

The BPB tool was quickly promoted as an array induction tool, while Schlumberger took a more conservative approach, using their Phasor® to resolve the details of phase measurement and inversion before introducing their own *Array Induction Tool®* (AIT) (Barber, 1985). The original Schlumberger AIT tool (Figure A.22) used 28 independent signals from eight transmitter/Receiver coil arrays, using at lest two different frequencies. The resulting 28 signals are either deconvolved or inverted in the field to provide up to 15 (borehole) environmentally corrected apparent resistivities (three groups of five), representing different vertical resolutions and

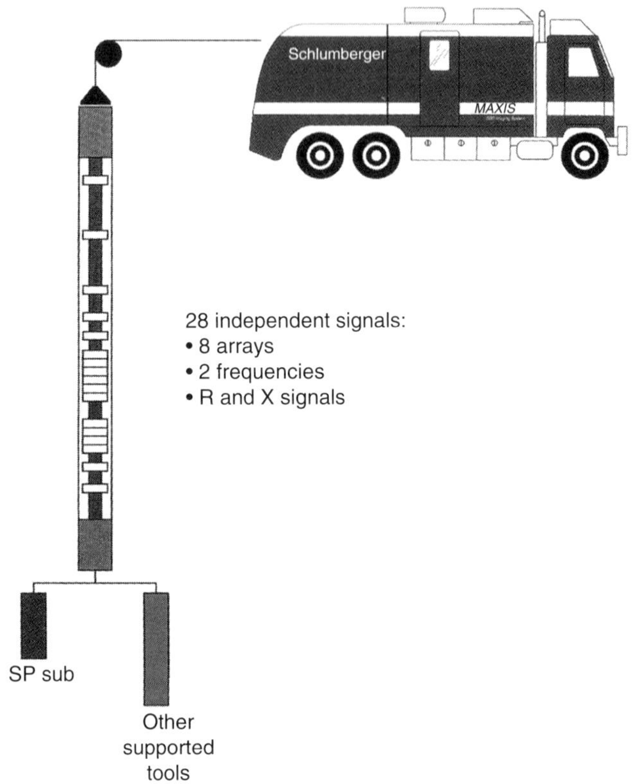

Figure A.22 Schlumberger Array Induction Tool® (AIT) schematic illustration (after Anon., 1992c).

focusing depths. These apparent resistivities are then be inverted to provide concentric cylindrical ring resistivity models (Figure A.23). All major logging vendors now offer their own versions of array induction logs and tool complexity has increased beyond the original numbers of signals, used for inversion.

The digital induction and array induction logs, with their multiple frequency/multiple coil spacing and phase measurement capabilities are possible, only because of the hardened microcircuit advances described earlier. These hardened microcircuits allowed the transmitter signal generators and receiver coil signal analyzers to be moved from the surface to the logging sondes. Downhole analog to digital (A/D) conversion, data processing, data encoding, and data transmission circuits, eliminated the problems caused by cable inductance. All of this has added up to increased measurement capabilities and greater information for analysis.

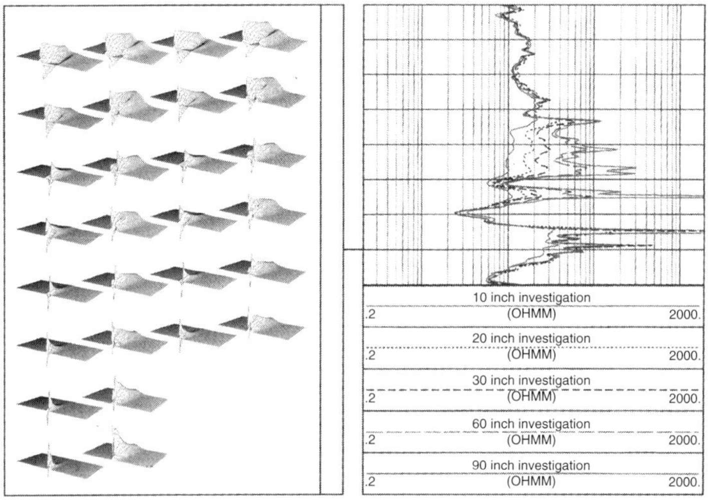

Figure A.23 Schlumberger Array Induction Tool® signals and resulting resistivity curves (after Anon., 1992c).

A.17 Departure Curves, Tornado Charts and Inversion

All of the analog resistivity and induction tools provided *apparent* resistivities, R_a, and/or conductivities, C_a, based on tool geometry and measured voltages and currents. Petrophysicists wishing to estimate S_w and S_{x0} had to first correct the measured R_a for borehole effects using departure curves (e.g., Figure A.15, and A.24) and then invert the multiple borehole corrected apparent resistivities, using *Tornado Charts* (Figure A.25).

The digital tools (e.g., DITE, AIT, HRLA) changed all of that. For these tools, borehole correction and inversion is done "in the cab", during the logging run, with the resulting data being both borehole corrected and inverted into model "Ring Resistivities". This is both a blessing and a curse. The petrophysicists' job is made easier as they do not have to make these corrections and inversions, themselves. However, the logging Engineer has made decisions, during the logging job, which effect the inverted log results. Setting up the inversion/deconvolution model for the array tools involve multiple assumptions and model parameter settings. The opportunity for creating errors "on the fly" is extremely large and the petrophysicist must be extremely vigilant in running quality control

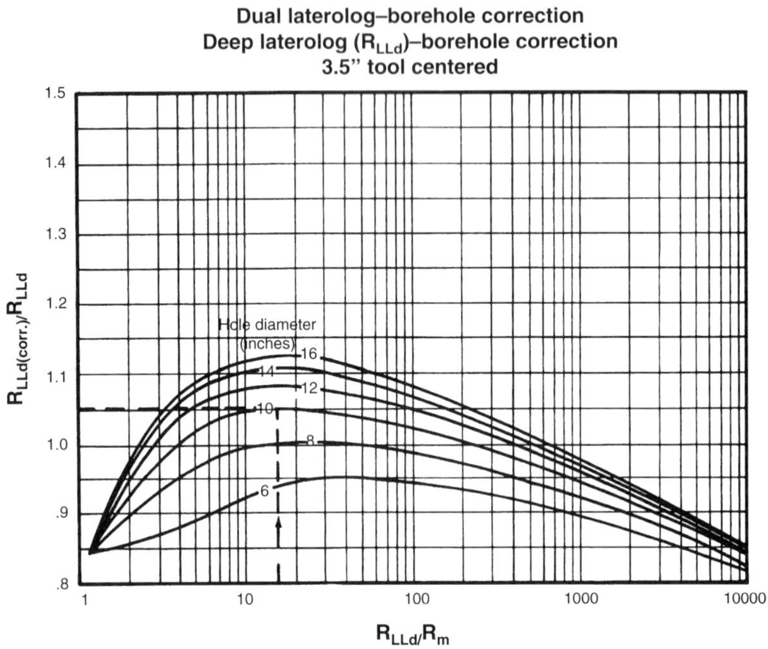

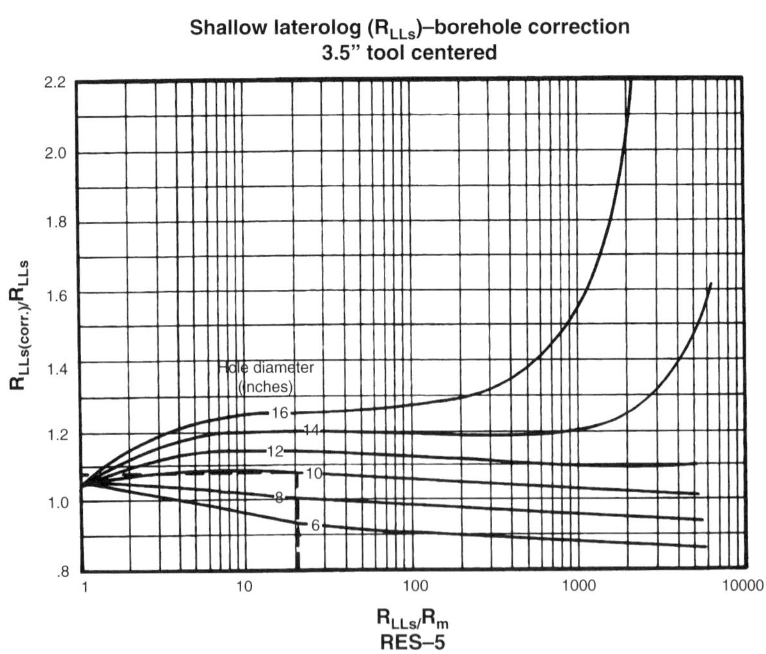

Figure A.24 Gearhart Dual Laterolog® (DLL) departure curves (after Anon., 1983).

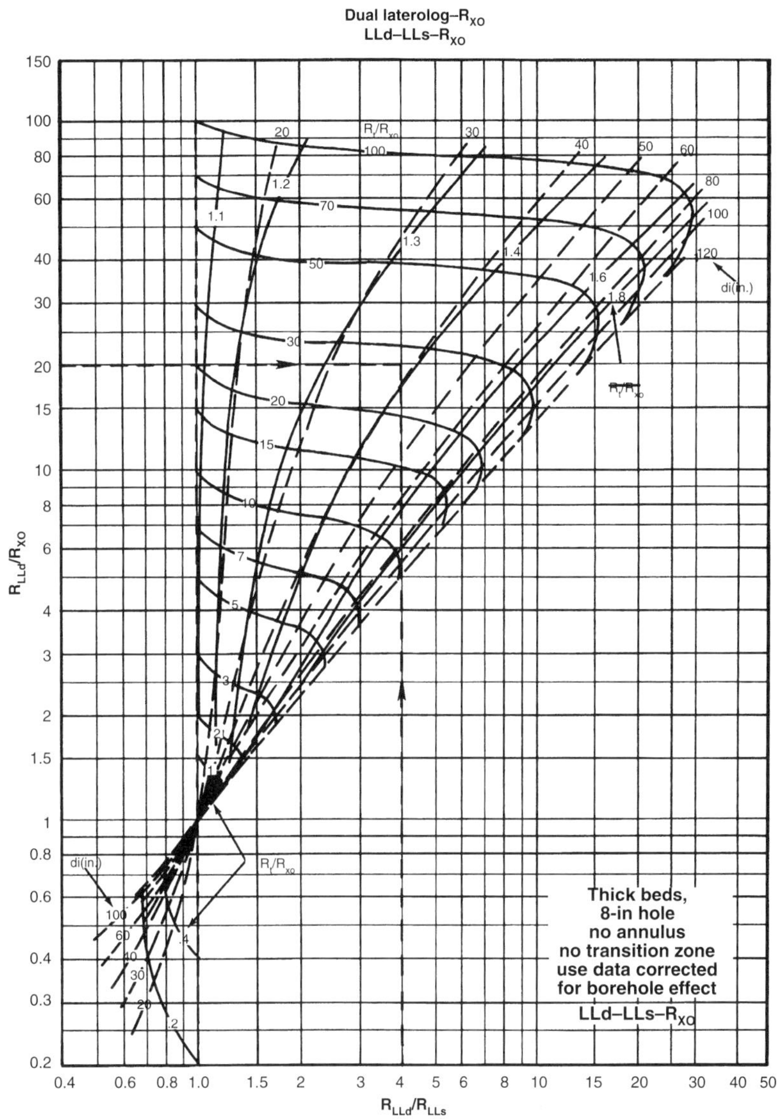

Figure A.25 Gearhart Dual Laterolog® (DLL)/Micro Spherically Focused Log® (MSFL) tornado chart curves (after Anon., 1983).

of the resulting inverted data. This often involves requesting the logging vendor to reprocess the array tool data from the engineer's *Data Acquisition Tape* (DAT) at the vendor's expense, as it involves a data quality issue. Since the petrophysicist never sees the original (proprietary) DAT data, this request must be made on the basis of the appearance of the resulting inverted data: An action which requires considerable persistence, insistence, and multiple calls to the logging vendor's home engineering center for support of the request.

A.18 Acoustic Log – The Accidental Porosity Tool

Acoustic logs are a favorite porosity tool for independent gas producers, for example in California's Sacramento Valley, because they can be used as a single porosity tool and do not have chemical radioactive sources. The original purpose of acoustic logs, however, was not to determine porosity. The original purpose was to provide detailed seismic velocity information for surface seismic data interpretation.

Seismic reflection exploration methods measure reflected acoustic echoes from lithology contrasts in the subsurface. The resulting echo reflection times are converted to depths via subsurface seismic velocity models. The common methods of developing these models were originally via seismic refraction surveys and/or *Well Shots*, where seismic detectors and/or sources were lowered in an open well, with sources near the surface and other detectors on the surface. Figure A.26 shows a typical Well-Shot analysis display. Both

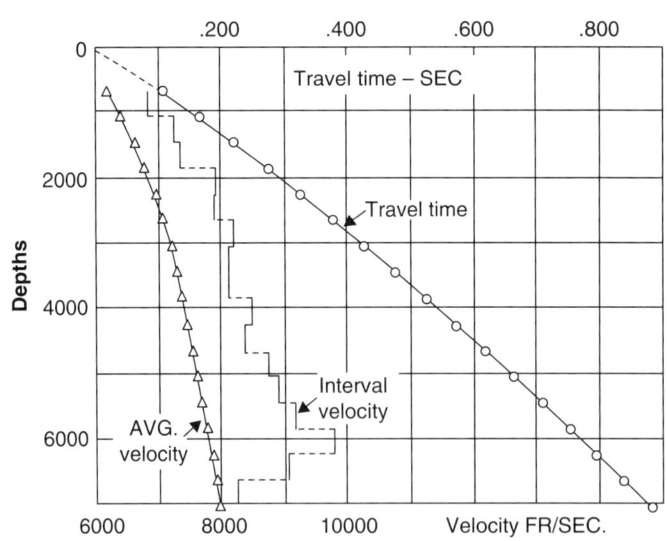

Figure A.26 Well shot velocity analysis display, showing: (one-way) travel-time, interval velocity, and average (reflection) velocity (after Kokesh, 1956).

of the above two methods of obtaining subsurface velocity information delivered only *interval* (blocks of constant) velocity models. More sophisticated seismic interpretation techniques required more continuous velocity models than were practical from either refraction or well shots.

Magnolia Field Research Laboratory (now part of EXXON Production Research Co., or EPR) and Shell Development Co. (now Shell E&P Technology Co.) independently and almost simultaneously developed prototype continuous velocity logging tools (Summers and Broding, 1952; Vogel, 1952). Seismograph Service Company (SSC) began offering commercial *Continuous Velocity Log*® (CVL) service the following year, under license from Magnolia Field Research Laboratory (see Figure 27).

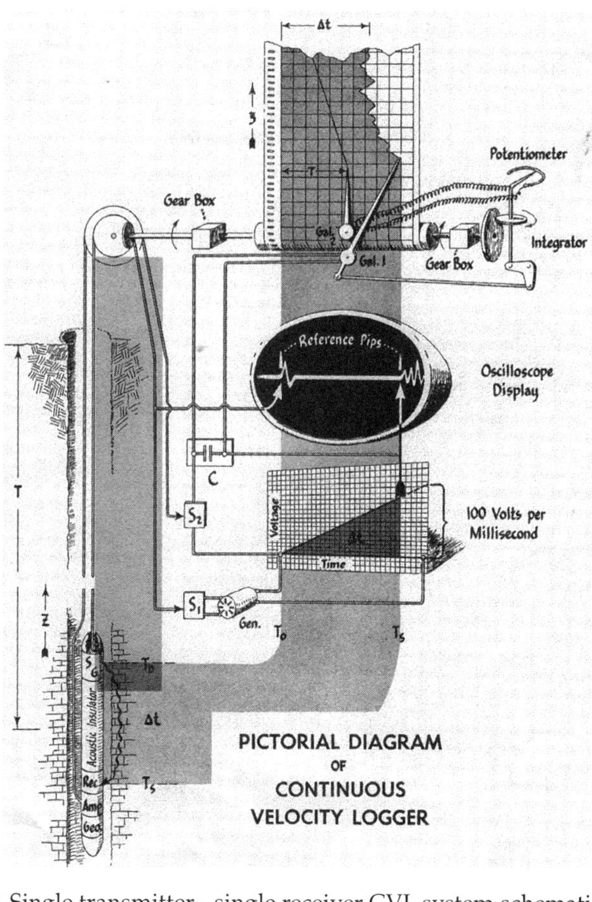

Figure A.27 Single transmitter - single receiver CVL system schematic (after Anon., 1953).

Acoustic logs are based on seismic refraction theory (Dobrin, 1960). The transmitter and receiver are placed far enough apart so that the acoustic signal that: (1) travels from the transmitter through the mud column to the borehole wall; (2) is critically refracted along the borehole wall at the speed of sound in the wall rock; (3) is critically refracted back into the mud column to be detected at the receiver; and (4) arrives ahead of the signal that travels along the sonde housing or through the mud column. The desired information is the wall rock velocity, not the total travel time of the above three-segment route. Because of this, single-receiver acoustic sondes were quickly replaced by dual-receiver sondes, with only the receiver arrival time differences recorded. The latter could easily be converted to interval (inverse velocities) transit times or inverted into velocities. The success of the CVL prompted SSC to start its own logging company, Birdwell. This success also stimulated all of the major logging vendors to provide their own acoustic logs under such names as Acoustilog® and Sonic Log®.

The original stimulation, for the acoustic log, was the desire for more detailed seismic velocity information. It was the vision and persistence by major petroleum research centers (in this case, Magnolia and Shell), which led to the development of prototype tools. Support from all of the major petroleum companies also encouraged wireline vendors to provide the service.

With the seismic velocity application of acoustic logs finalized, major oil companies began looking for other applications. Jessie Wylie and associates at Gulf Research and Development Co. (now Chevron ETC) developed the Wylie time-average equation (Eq. A.6), which allowed estimation of porosities from acoustic interval (inverse velocity) transit time. The original Wylie time-average equation did not include the compaction coefficient, *c*, because Wylie and his associates worked only with consolidated rocks (carbonates and well-cemented sandstones). The arbitrary compaction coefficient was added, so that the time-average equation could be used for unconsolidated clastic rocks. Raymer *et al.* (1980) introduced an alternative unconsolidated clastic rock porosity estimator (Eq. A.7).

Acoustic logs also have been used for two other purposes: (1) evaluation of casing/cement bond quality; and (2) estimation of rock mechanical properties. Casing *not* well bonded to its cement sheath will ring like a bell, resulting in very high amplitude acoustic waveforms, whereas casing which *is* well bonded, will not. This simple usage of full waveform acoustic logs in cased wells allows production engineers to identify locations of fluid leaks between casing and cement and design remedial "squeeze" repairs. Acoustic logs, in combination with density logs, allow production engineers to estimate formation strength and drilling engineers to select appropriate drill bits.

Four recent developments have greatly increased acoustic log utility: (1) multiple receivers (array tools) with full waveform recording; (2) specialized shear wave rich sources; (3) shear and compressional wave transit time (inverse velocity) processing and analysis of full waveform logs; and (4) successful 360° acoustic borehole imaging tools. The first and third developments were made possible because of the hardened micro-circuitry developments (described above); downhole A/D sampling, storage, processing, encoding, and transmission, as well as waveform semblance processing. The second development was required to enhance the third development. The last development ushered in borehole imagery and brought geology back into log evaluation (see below).

There are two types of seismic (acoustic) waves. Compressional waves involve propagation of expansion and compression of volumes, with particle motion in line with wave propagation direction. Shear waves involve propagation of shape distortion with particle motion transverse to the direction of propagation. Early acoustic tools, primarily utilized only compressional wave information, because the first arrivals at all receivers were compressional waves. Analogue acoustic tools usually delivered waveforms from only two receivers, which made it very difficult to correctly identify the (later) shear wave arrivals. Compressional wave velocities could be used to determine both seismic velocities and estimate porosities. As a result, there was little incentive to pursue shear wave identification.

Recent advances in seismic interpretation have shown the advantages of using both compressional and shear wave reflection interpretations, which required detailed shear wave velocity models. Mechanical properties estimated, using both compressional and shear wave velocities are more reliable than those estimated using compressional wave velocities alone. These new applications created the need for shear wave acoustic log measurements.

Multiple receiver array acoustic tools made it much easier to distinguish the shear wave arrival from the later Stoneley boundary wave arrival (see Figure A.28). The use of down-hole A/D conversion, processing, encryption, digital data transmission and surface processing, using semblance logic, made it possible to deliver continuous log curves of both compressional and shear wave (inverse velocity) interval transit times.

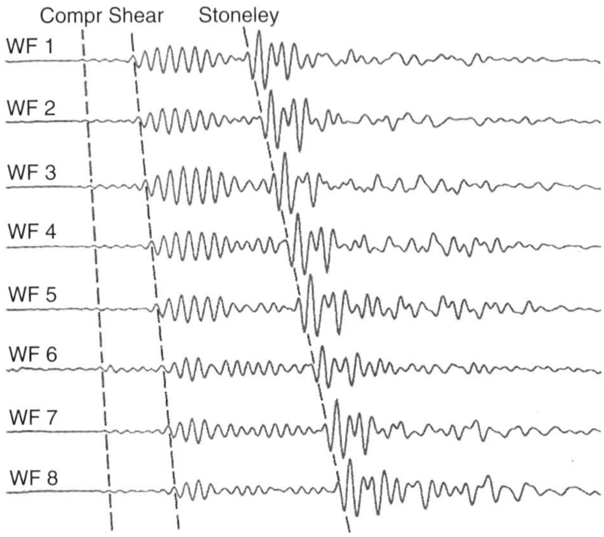

Figure A.28 Eight-receiver acoustic log full waveforms with compressional, shear, and Stoneley wave arrivals identified (after Anon., 1985).

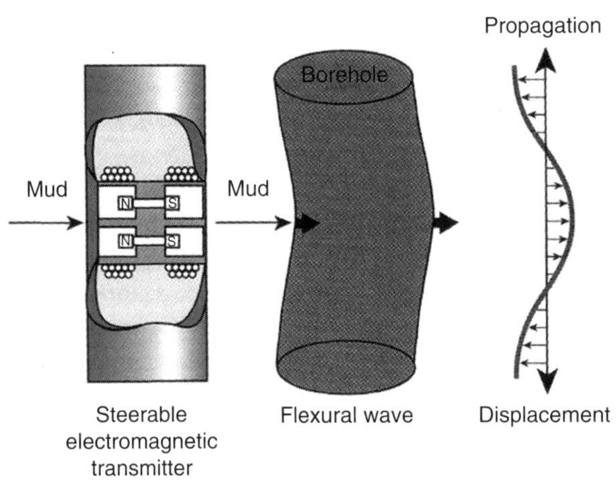

Figure A.29 Dipole type acoustic source schematic (after Anon., 1997).

Whereas compressional waves propagate through both solids and fluids, shear waves propagate *only* through solids. This means that shear wave acoustic logs must use signals, which undergo two compressional/shear wave mode conversions at the borehole wall. This reduces the resulting signal amplitudes at the receivers. The recent development of dipole acoustic sources (see Figure A.29) has increased the shear wave energy component in the borehole wall rock, resulting in stronger shear wave arrival signals at the receivers.

A.19 Neutron Log – The First True Porosity Tool

Archie's Equations (Eqs. A.4 and A.5) offered the opportunity for production and reservoir engineers to estimate hydrocarbon reserve volumes, *if they knew the reservoir rock porosities*. Porosities could be obtained from core measurements, but that involved the time and expense of cutting cores and making laboratory measurements. Initial attempts using resistivity and micro-resistivity logs were less than satisfactory for a variety of reasons.

The neutron log was the first wireline measurement developed to estimate porosity. Gulf Research and Development Co. (now Chevron ETC) and Magnolia Field Research Co. (Now ExxonMobil EPR) conducted early evaluations of neutron logs (Caldwell, 1958; Faul and Tittle, 1951; Tittle *et al.*, 1951; Wyllie, 1952). Well Surveys (now Baker-Atlas) introduced the first commercial neutron logging system based on field trials dating back to 1938 (Anon, 1974).

Fast (high energy) neutrons are attenuated (slowed), and eventually captured via interactions with the nuclei of the atoms and ions of the material through which they pass. The *Neutron Capture Cross-Section* (ability to attenuate and/or capture neutrons) is inversely related to the mass of the target atom/ion nucleus. Hydrogen ions, which have atomic masses closest to that of neutrons, are excellent thermal neutron absorbers and have a much higher thermal neutron capture cross-section than most other ions in the borehole environment. When hydrogen ions capture thermal neutrons, they emit gamma rays of capture. Many of the other neutron-target interactions also release capture gamma rays, as well.

Neutron logs bombard the borehole wall with fast neutrons, which moderate to thermal energies via collisions with dense ions in the borehole wall. Once reaching thermal energies, they can then be captured by hydrogen ions. Water, oil, and clay minerals are the major source of hydrogen ions in the borehole environment. For clean (*i.e.*, low clay mineral content) reservoir rocks, monitoring either the remaining thermal neutron or capture gamma ray flux can be related to water/hydrogen content of the rocks. Because the hydrogen content (often called *Hydrogen Index*) of oil and water molecules is similar, it can be related to the rock porosity, assuming no gas, or clay minerals are present.

Early neutron log sondes (Figure A.30) contained chemical (usually AmBe), or neutron generator (deuterium/tritium), fast neutron sources and a single gamma ray or thermal neutron detector. These tools (often called GNT tools) delivered normalized count rate (API) units and were usually calibrated using core measured porosities (Swulius, 1986). Different calibrations were required for different reservoirs, tools, and borehole diameters. This approach worked quite well for carbonates, where boreholes tend to be in gage, but was difficult to use in unconsolidated clastics, which were prone to washouts. In spite of this limitation, many operators (e.g., Gulf Oil Company <Nigeria>, Ltd.) continued to use GNT neutron tools into the late 1970's, even after successive (and improved) neutron tool types were available.

The first attempt to improve on the GNT type neutron sondes was the Sidewall Neutron Porosity (SNP) type tools. Schlumberger introduced their SNP tool in 1965 (Tittman *et al.*, 1965), followed quickly by the other logging vendors. These tools used chemical

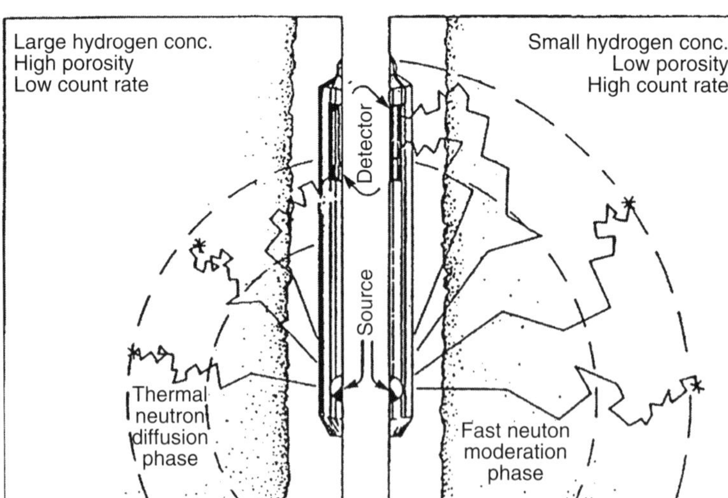

Figure A.30 Schematic of single detector neutron logging sonde (after Ellis, 1987).

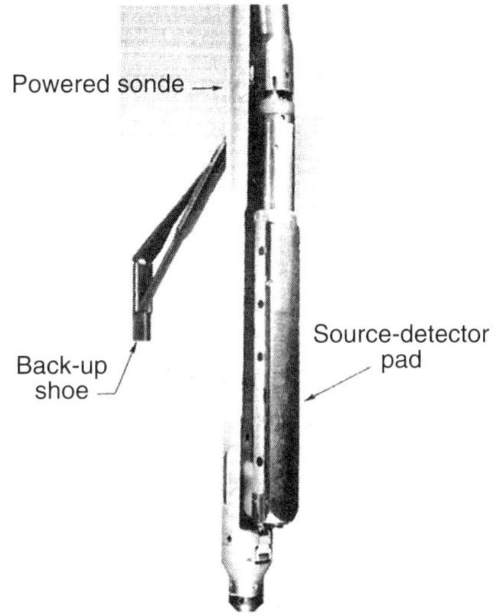

Figure A.31 Photo of single detector sidewall neutron logging sonde (after Tittman *et al.*, 1965).

or neutron generator fast neutron sources and a single epithermal neutron detector located on a mandrel skid, with a back-up single arm caliper (Schlumberger/Halliburton, see Figure A.31) or on one arm of a two-arm caliper (Gearhart/Dresser). These neutron tools were calibrated directly in terms of (limestone, sandstone, or dolomite) lithology-porosity. The SNP tools were even more sensitive to borehole washouts than GNT tools, but it was hoped that by forcing the source and detector against the borehole wall, that problem could be overcome. Unfortunately, this was not the case. Neither SNP tool design could reach the borehole wall in the case of deep washouts. The mandrel design tended to bridge over thin washouts. The two-arm caliper-pad design could be damaged by hard borehole ledges, in spite of its leading edge skid rail.

The final neutron log sonde design improvement, called a Compensated Neutron Log (CNL) involved a chemical source with two thermal and/or epithermal detectors (see Figure A.32). Schlumberger introduced their CNL tool in 1971 (Alger *et al.*, 1971), followed quickly by the other logging vendors. While not strictly a side-wall device, the CNL design is held against the borehole wall by a back-up bow-spring, which limits borehole wall offset. It is much less sensitive to washouts, than the either the GNT or SNP design. With modifications (including down-hole counting, data

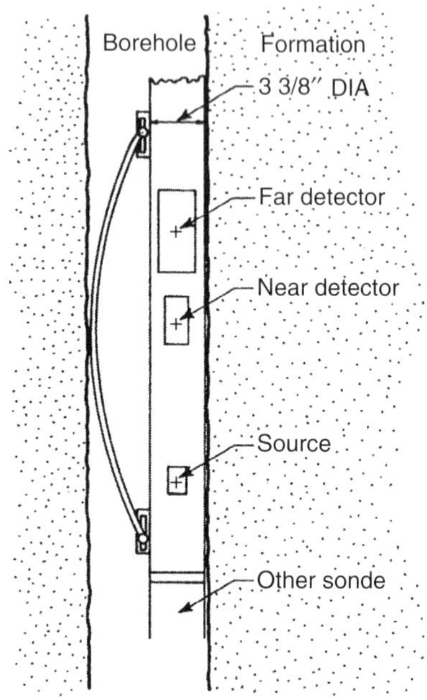

Figure A.32 Dual detector compensated neutron log (CNL) schematic (after Bassiouni, 1994).

processing, encoding and transmission micro-circuitry), this is the neutron tool, currently in use by most wireline and MWD vendors. With the exception of clay minerals (anomalously high apparent porosity) and gas (anomalously low apparent porosity), the CNL tool has proven to be a reliable porosity tool. Fortunately, the neutron/density combination has proven to be an excellent porosity tool, with each tool complementing the other's shortcomings.

The most recent advance, in neutron logging tool development, involved putting four CNL tools on fins of MWD sub-assemblies, to produce four quadrant neutron curves during MWD operations. This is significant, because MWD devices cannot use back-up bow-springs, like wireline tools and must rely on gravity and drill string rotation to place the MWD sub-assembly against the bottom and/or right hand side of an inclined or horizontal borehole. Because the directional/horizontal well drill pipe rotates during drilling, even with downhole drill motors, the same side of the drill assembly is not always against the borehole wall. Using four neutron tools on MWD sub-assemblies increases the chances that one of the four, usually the right-hand (RH) quadrant is against the borehole wall at all times.

Recently, several vendors have attempted to use fuzzy logic to produce borehole images, from the rotating quadrant data. These displays, however, are more qualitative than quantitative, but do improve insight about the condition of the borehole.

A.20 Density Log – The Porosity Tool that almost did not Make It

The last of the open-hole porosity logs is the (Gamma-Gamma) *Density Log*. Gamma-gamma density logs use a collimated chemical gamma ray source (usually ^{60}Co or ^{137}Cs) pressed against the borehole wall and monitor the resulting *Compton Scattering Energy Window* gamma ray flux at one, or more distances from the source.

Gamma rays in the Compton energy window loose energy upon colliding with the orbital electrons in a material. The reduction in Compton energy window gamma ray flux, at some distance from an intermediate energy gamma ray source, can be related to the *Electron Density*, ρ_e, of the intervening material. The *Bulk Density*, ρ_b, of most petroleum reservoir rock minerals and liquids can be simply related to the electron density.

For a simple reservoir rock model, with a matrix of density, ρ_{ma}, porosity, ϕ, and fluid of density, ρ_f, the bulk density, ρ_b, is equal to:

$$\rho_B = RHOB = \frac{m}{V_B} = \phi\rho_f + (1-\phi)\rho_{ma}, \tag{A.8}$$

where: *RHOB* is the bulk density, V_B is the bulk volume and m is the mass.

Eq. A.8 can be inverted as:

$$\phi_D = \frac{(\rho_{ma} - \rho_B)}{(\rho_{ma} - \rho_f)}. \tag{A.9}$$

The first technical papers, discussing density logs, were by members of major oil company research facilities (e.g., Baker, 1957; Pickell and Heacock, 1960). Lane Wells (now Baker-Atlas) introduced the first commercial density logging system, the *DensiLog®*, in 1954 (Anon, 1974; Schneider and Watt, 1964). The first design (Figure A.33) featured a single detector and delivered normalized count rates, like the GNT tool. It was extremely sensitive to borehole diameter and mud weight. This single detector tool was so

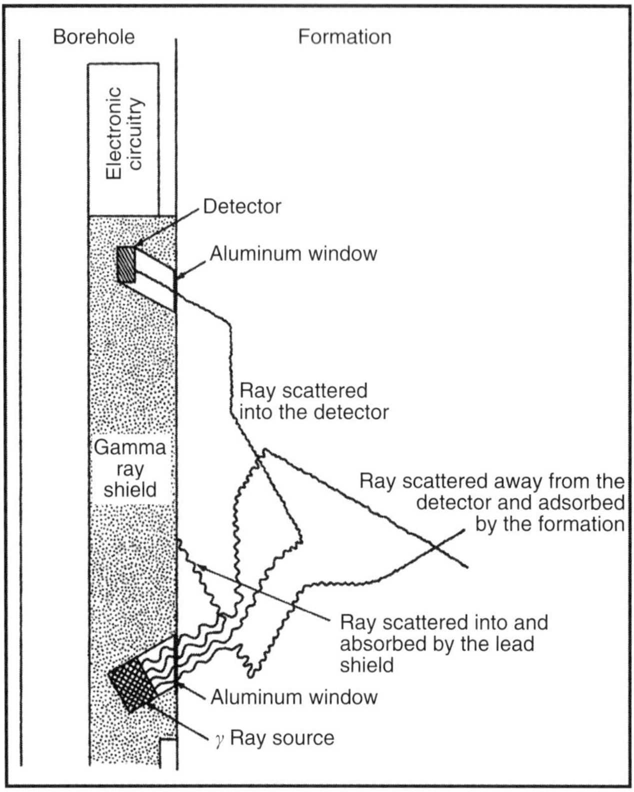

Figure A.33 Schematic of single detector density log sonde (after Bassiouni, 1994).

difficult to use that it, alone, nearly torpedoed the entire borehole density measurement concept. Only the insistence by the major oil companies (AMOCO, BP, ESSO, Gulf, Mobil, Shell, and SOCAL), that forced wireline vendors to deliver a serviceable tool kept density log development alive.

The solution to the problems of the single-detector density log design problems, like that of the GNT, was to measure gamma ray flux at two distances from the source. This compensated density log (CDL) design allowed the tool to self compensate for the effects of mud cake and Barite mud (Figure A.34). Like the SNP tool, the CDL was a pad tool forced against the borehole wall by a back-up caliper. Also, like the SNP tool, two different design types emerged. Figure A.35 shows the centered housing design, used by Atlas and

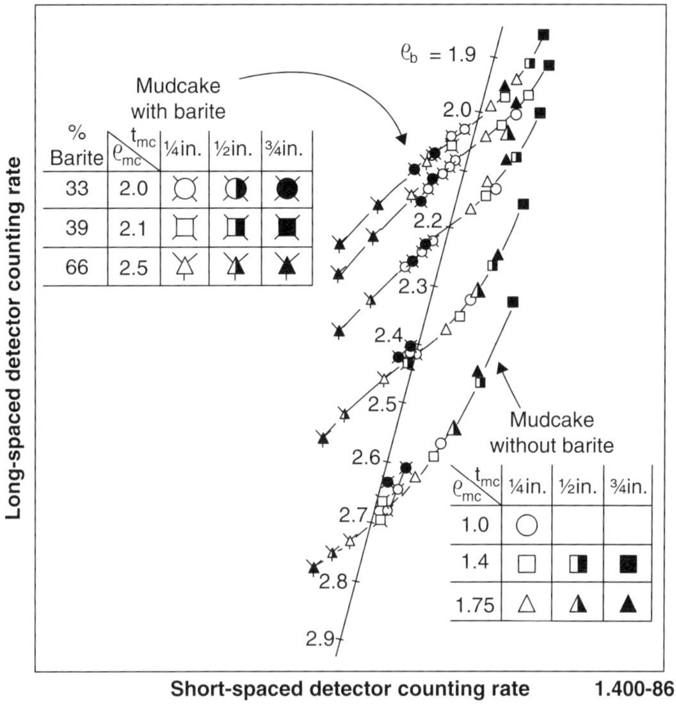

Figure A.34 Schlumberger FDC® rib-spine calibration (after Anon., 1987).

Gearhart, whereas Figure A.36 shows the mandrel tool design used by Schlumberger and Halliburton. The centered housing tool pad is shorter than the mandrel tool and does not bridge short borehole washouts, like the mandrel tool. It may, however, be more easily damaged by ledges, in the borehole than the mandrel tool.

Compensated density tools proved to be quite serviceable and most of the significant wildcat and development wells, drilled since 1975, have used them. They have proven to be particularly useful when run in combination with the CNL log. Eq. A.9 assumes no clay minerals, in the matrix, and that the formation fluid is water or oil. The presence of gas, in the formation, will give anomalously high ϕ_D, while clay minerals (at least for sandstones) will give anomalously low ϕ_D. The reverse is true for neutron log response. Cross-plotting and/or over-plotting ϕ_N vs. ϕ_D, or ϕ_N vs. ρ_B allows the interpreter to quickly identify these situations and interpret them correctly.

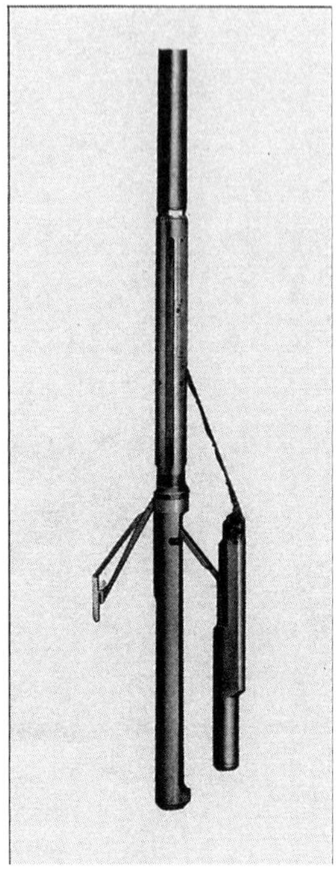

Figure A.35 Atlas & Gearhart centered-housing density log sonde design (after Anon., 1974).

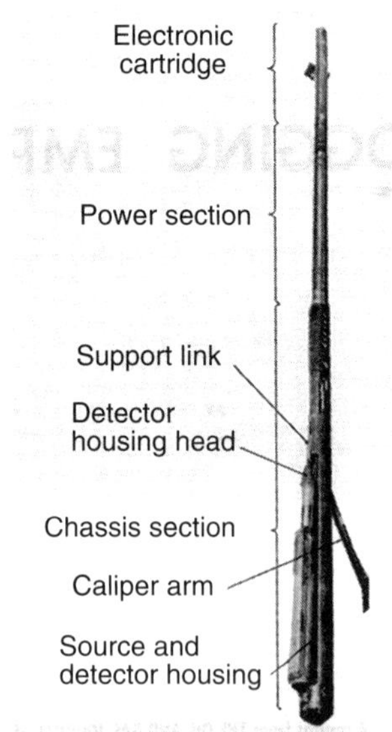

Figure A.36 Schlumberger & Halliburton mandrel density log sonde design (after Rodermund, *et al.*, 1961).

The next advance, for density logs, involved better identification of lithology. Figure A.37 is a cross-plot of gamma ray incident energy vs. mass absorption coefficients, showing the energy regions for *Pair Production*, *Compton Scattering* and the *Photo-Electric Effect*. The CDL type tools use Compton energy window gamma ray flux changes to estimate bulk density of the reservoir.

There is another, lower energy, gamma ray/electron interaction, which is also of great utility: the *Photo-Electric Effect* (PE). The PE capture cross section of an atom or ion is directly proportional to it atomic number, with the exception of the hydrogen ion. Inasmuch as the hydrogen ion has no electrons, it has a very low PE. Thus, water and oil, which have large numbers of hydrogen ions, have much lower PE than the matrix minerals in a reservoir. Consequently a tool, which measures changes in the flux of PE window gamma rays, becomes a lithology tool. This measurement has not been that widely accepted for simple (*i.e.*, sand-shale) clastic reservoirs. For complex lithologies, however, PE type measurements have proved to be extremely useful. Examples of these environments are evaporite sequences and interbeded clastics and carbonates, such as the Upper Cretaceous rocks of the Congo Basin. Another application

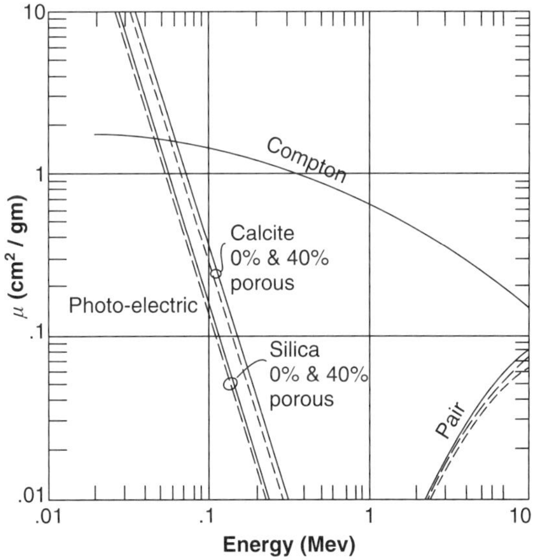

Figure A.37 Gamma ray incident energy vs. mass absorption coefficients (after Tittman and Wahl, 1965).

of the PE curve is the identification of arkosic sands, as the PE of potassium Feldspars is much different than that of quartz.

The final density log advancement involves quadrant type MWD measurements. The same arguments applied to the quadrant neutron measurements also apply to density measurements. Again, these multiple measurements would not be possible without the development of low-power drain, hardened microprocessors, which can be encased in MWD subassembly fins.

A.21 Pulsed Neutron Capture Logs – The All Purpose Tool

Pulsed Neutron Capture (PNC) Logs involve two technologies:

1. Neutron generator sources.
2. Gamma ray spectral measurements.

Neutron generator sources are nearly as old as neutron logs. The McCullough Tool Company (a name, which seems more closely associated with chain saws than logging services) was the first vendor to introduce a pulsed-neutron logging source (Martin, 1956; Stroud and Schaller, 1958). A neutron generator has two components:

1. A downhole sealed tube particle accelerator, which accelerates deuterium ions through a large potential drop to strike a tritium target, discharging 14 MeV neutrons (Figure A.38).

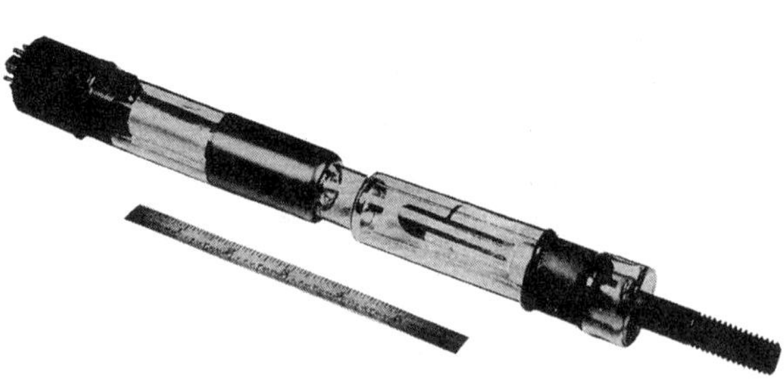

Figure A.38 Down-hole sealed particle accelerator tube (after Tittman and Nelligan, 1959).

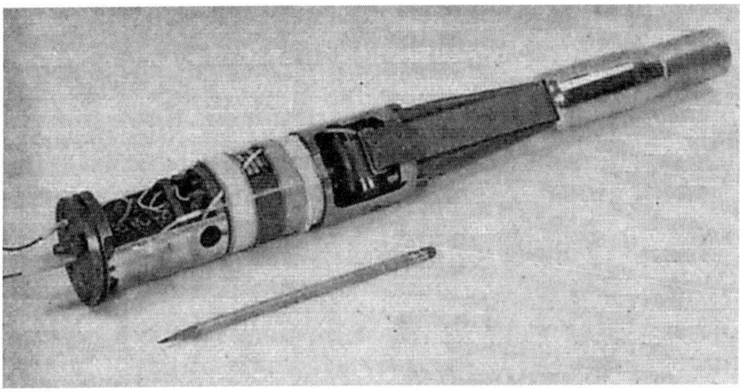

Figure A.39 Down-hole Van de Graaff generator (after Little and Youmans, 1964).

2. A down-hole Van de Graaff generator to develop the high voltages needed to power the particle accelerator (Figure A.39).

The first applications of pulsed neutron generators, was for conventional neutron logs. Neutron generators could be turned off, while the AmBe chemical sources could not. Whereas pulsed neutron generators were safer and easier to handle than chemical sources, they were also much more expensive, much less stable, and much less reliable than chemical sources. As a result, pulsed neutron generators are no longer used for most neutron porosity tools. They remain the source of choice, however, for PNC tools, because many of the desired measurements, such as inelastic scattering measurements, require an "OFF" period, with no neutron bombardment (Hearst *et al.*, 2000).

The second technology involved with PNC logs is gamma ray spectroscopy. Thermal neutron capture and other neutron nucleus interactions release gamma rays unique to each target element. Sorting and counting gamma rays by energy allows qualitative and quantitative estimates of the concentrations of the various elements present in the borehole environment.

One of the first applications of PNC logs was for detecting the chlorine. The thermal neutron capture cross-section of chlorine is about 100 times that of hydrogen (Stroud and Schaller, 1958). The chlorine capture gamma ray flux provides a qualitative, if not quantitative, picture of oil saturation, in those reservoirs with high-salinity brines.

The primary minerals in sandstones are quartz (SiO_2), whereas clay minerals usually contain aluminum (Al). A second application of PNC tools was to estimate the Si/Al ratio, of materials in the borehole wall, as an aid in determining sand/shale lithology (Wichmann and Webb, 1969).

Both Lane-Wells (now Baker-Atlas) and Schlumberger introduced *PNC Neutron Die-Away* tools in the 1960's and early 1970's (Wahl *et al.*, 1970; Youmans *et al.*, 1964). These tools monitor the speed at which thermal neutrons are captured within the borehole wall rock, monitoring thermal neutron flux at multiple times after the generator has been turned off. The technique is particularly useful for obtaining both porosity and saturation in cased wells.

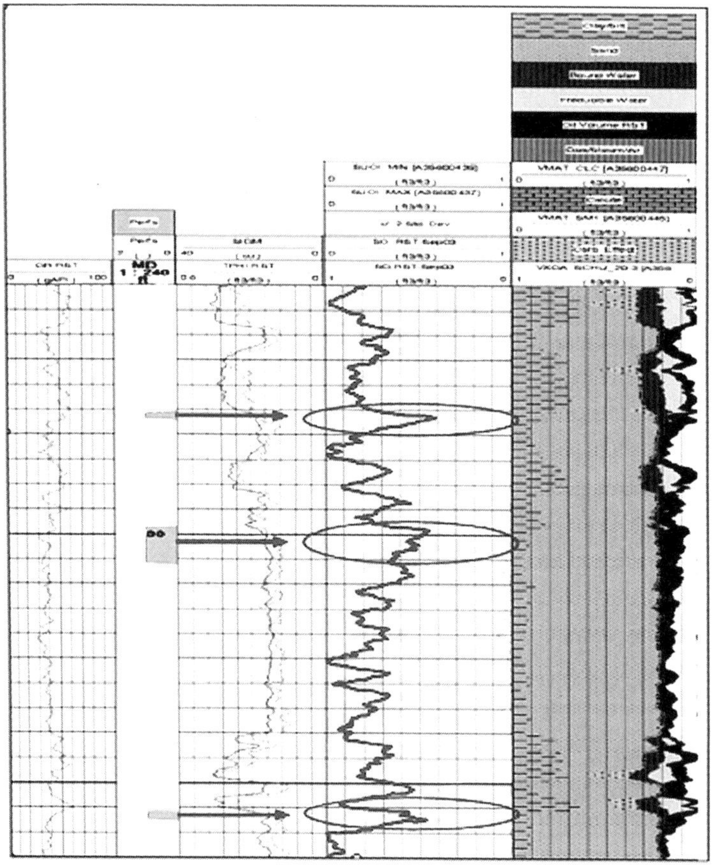

Figure A.40 Example of cased-hole PNC run in C/O mode, indicating three intervals of bypassed oil (courtesy of Schlumberger).

Monitoring inelastic neutron scattering gamma rays has proven useful and now represents one of the prime applications of modern PNC tools, such as Schlumberger's *Reservoir Saturation Tool®*, RST. Prime PNC targets are to monitor the 4.44 MeV carbon, 6.1 MeV oxygen, 3.7 MeV calcium and 1.78 MeV silicon inelastic scattering gamma rays, to provide C/O and Ca/Si ratios, for identification of hydrocarbons (Hearst *et al.*, 2000). These modern C/O and Ca/Si PNC applications differ from the earlier generation C/O and Al/Si tools with the use of high resolution spectral gamma ray detectors, providing more accurate discrimination. Figure A.40 shows an example of a cased-hole PNC log run in C/O mode to locate bypassed oil zones.

This PNC/Spectral gamma array technique has also been used for a variety of other elements. Hoyer and Lock (1972), Nargolwalla (1973), Nargolwalla *et al.* (1976), and Nargolwalla and Seigel (1977) attempted to use this technique for estimating grades of metallic ores. The resolution of the spectral scintillation detectors and stripping algorithms then available, however, was not sufficient to meet the low-grade ore resolution targets needed. Modern BaGe and GeI detectors have allowed much finer spectral gamma ray measurements, leading to the development of such geochemical logging tools as Schlumberger's *Gamma Ray Spectral Tool®* (GST) and *Geochemical Logging Tool®* (GLT).

A.22 Through Casing Resistivity Measurements – Well Logging's Holy Grail

The sagas of the fabled English King Arthur and the Knights of the Round Table include many references to Quests in search of the Holy Grail, "holiest of Christian religious relics" to be discovered only by "the best knight in the land, the only man capable of sitting in the mysterious 'Siege Perilous' whose heart and life were pure beyond question" (see website references).

There are many oilfields in the world, which were discovered and developed before the arrival of modern logs. There are also many existing fields, with pay zones that were not considered economic to place on production, at the time of discovery. With facilities available on site and declining production from the original reservoirs, many of these marginal and/or mised pay zones may now be very attractive to develop. Unfortunately, the wells in most of these fields do not have modern logs and many have no wireline measurements at all.

Many of the logging options available, when a well is open, are no longer available once casing has been run. The most significant option loss, once casing has been run, is electrical resistivity/conductivity logs. Steel casing is a very good electrical conductor, which will short-out electrode logging tools and shield induction logging tools from formation resistivities.

Because of the casing electrical-shielding problem, the development of a method of measuring formation resistivities behind casing has seemed to be an unreachable goal, much like the Arthurian Quests in search of the Holy Grail.

In the late 1980's, a start-up firm, called Magnetic Pulse, Inc., developed a pulsed induction tool, with very sensitive decay measurements. Whereas some of their field demonstration results looked encouraging (Gill *et al.*, 1991), the company never captured enough market with their specialized logging tool and eventually ceased operations.

Recently, Schlumberger (Aulia *et al.*, 2001) and Weatherford (Geldmacher and Jonkers, 2007) have developed through-casing electrode array systems, which appear to offer great promise. Through-casing resistivity measurements may really be more than an "impossible dream".

Figure A.41 is a schematic drawing for a through-casing resistivity survey tool. Source current electrodes are located at the top and bottom of a long sonde, with a current return electrode at the surface. Multiple micro-potentiometer electrodes are located along the body of the sonde. All of the downhole electrodes are located on sets of three hydraulically controlled pads, which force them against (some implementations actually cut through any corrosion) the inside of the casing wall. Measurements are made in "Station Logging" mode, taking several minutes at each station (this must be a very low signal/noise measurement). The downhole micro-potentiometer circuitry monitors current leakage through the casing into the formation and is capable of measuring nano-volt (nV) potential differences between the downhole electrode triplets.

Figure A.42 shows an example of a through-casing resistivity log overlain on the original (open-hole) resistivity log, for a California reservoir undergoing waterflood. There is considerable separation between the open-hole and cased-hole resistivity measurements, for the deeper sands, indicating that the waterflood front has already

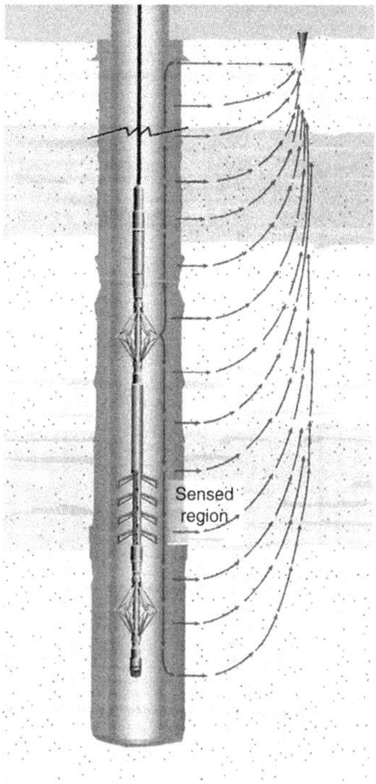

Figure A.41 Through-casing resistivity tool schematic (courtesy of Schlumberger).

swept past this well. The shallowest sand, near the depth of 6700 ft, however, shows little curve separation, indicating that this sand still contains significant oil reserves.

A Schlumberger Regional Sales manager told me, however, "The Jury's still out", on this measurement. There have been several claimed through casing resistivity measurement successes, such in California's San Joaquin Valley, and Central Sumatra, Indonesia, where the measurements are being used to monitor water flood operations. The technique appears to be most successful with fresh water reservoirs and fairly recent casing. This may be an exciting technology to watch.

The ability to measure resistivity behind casing has been a dream of production engineers, for years. The development of micro-circuitry, a fall out of the space program, and hardened electronics, a fall out from the SANDIA high-temperature, high-pressure logging circuitry program, made the sensitive nano-volt measurements possible, whereas the ingenuity of logging tool designers made it work.

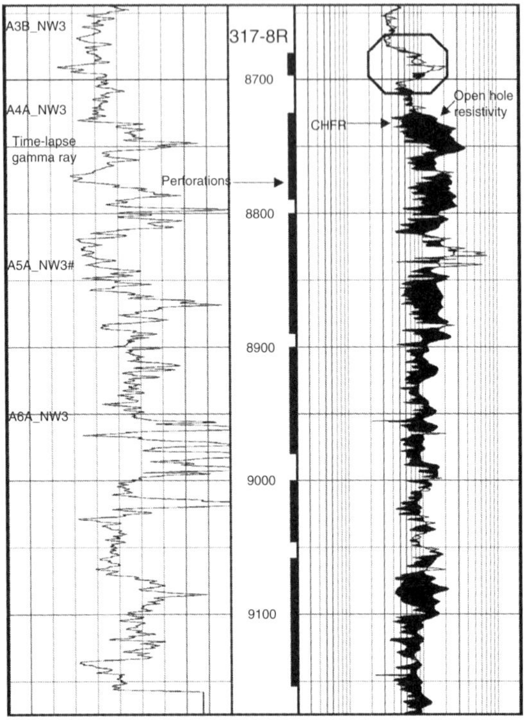

Figure A.42 Through-casing resistivity log showing swept and unswept waterflood intervals (after: Starcher, *et al.*, 2002).

A.23 Nuclear Magnetic Resonance Log – Patience and Persistence

The development of *Nuclear Magnetic Resonance* (NMR) logs illustrates just how difficult it can be to develop an entirely new logging technology. The potential of oilfield NMR measurements was recognized in the 1950's and 1960's (Brown and Fatt, 1956; Brown and Gamson, 1960; Seevers, 1966), but reliable wireline and MWD tools were not developed until the 1990's (Coates *et al.*, 1991a; 1991b).

The difficulties of transforming laboratory NMR measurements to the borehole proved to be daunting. The market for this measurement, at least initially, appeared to be limited to heavy oil reservoirs with fresh waters. Vendors dismissed the technology as not being worth the effort. Only sustained pressure by major oil companies (primarily, Chevron and Shell) kept the technology alive. Technology developments, outside of the petroleum industry, made viable borehole measurements possible. Departure of a key logging tool design engineer (Coates), from a major logging vendor to a start-up venture, made it happen. Once a reliable logging tool was available and a market established, the major logging vendors introduced their own versions and developed additional applications.

All of the measurements described, so far, provide viable estimates of volumetrics (e.g., porosities and gas/oil/water saturations) only. Another long-sought log-derived property, however, was pore size distribution and/or permeability. Seevers (1966) established that NMR measurements could provide this information.

One of the atomic particle quantum numbers relates to the particle's spin, which generates a magnetic moment. Protons, neutrons and electrons all have quantum spin numbers. Protons, however, have the largest NMR response (Coates *et al.*, 1999), as described below.

For ions with multiple protons (*i.e.*, $Z > 1$), protons pair up with protons of the opposite (spun) quantum number. Ions with odd atomic numbers have one unpaired proton and have greater NMR responses than those with even atomic numbers. The ion with the greatest NMR response is hydrogen, with only one proton and no neutrons or electrons. The oxygen ion in water, by contrast, has only paired protons, neutrons, and electrons, and a very low NMR response. Consequently, the NMR response of water is primarily due to the hydrogen ions.

In the absence of any external magnetic field, proton pair spins have random orientations. In the presences of an external magnetic field, proton spin moment can be oriented either parallel or antiparallel to the external field. If a larger external magnetic field is imposed, protons will reorient, with their axes precessing (rotating) around the new net magnetic field, much like a gyroscope processing around the earth's gravitational field (see Figure A.43). The frequency of this rotation is proportional to the external magnetic field strength. This proton spin precession about external magnetic field vectors is called *Proton Precession*, or *Nuclear Magnetic Resonance* (NMR). In the mid 1950's, Varian Associates utilized this concept to develop a *Proton Precession Magnetometer®*, which monitored the precession frequency of a fixed volume of water (Dobrin, 1960).

Over time, the proton spins of an assemblage of protons will align parallel and anti-parallel to the external magnetic field (see Figure A.44). This polarization alignment follows an exponential

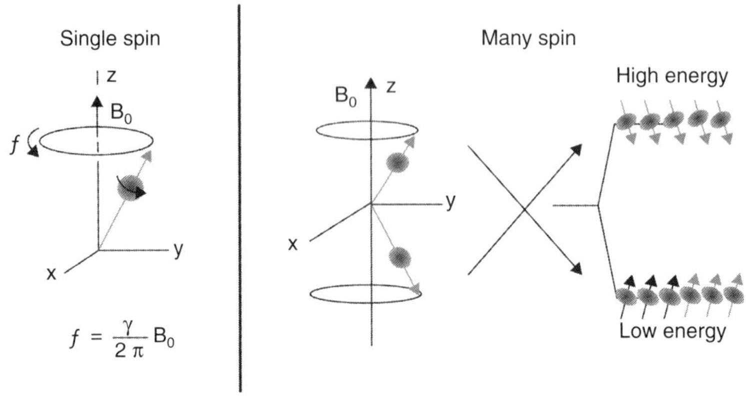

Figure A.43 Proton(s) precessing about an external magnetic field (after Coates, *et al.*, 1999).

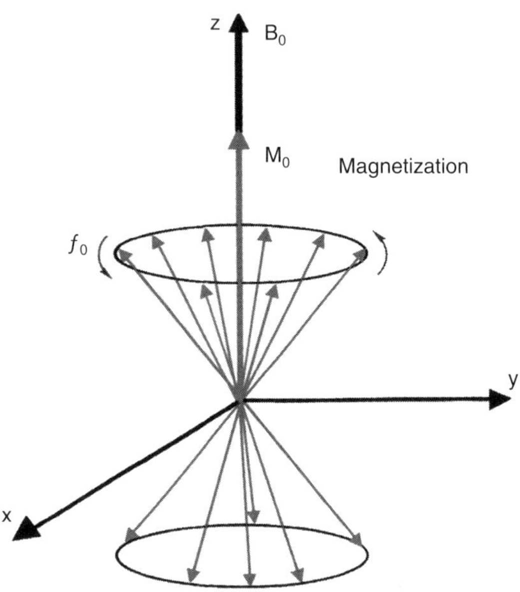

Figure A.44 Macroscopic (net) magnetization vector precessing about an external magnetic field (after Coates, *et al.*, 1999).

polarization model (Coates *et al.*, 1999). If this external field is removed, the proton spins will precess around the original magnetic field, again following an exponential decay curve. The time required for the proton spins to reach 63% alignment with the external field is called the relaxation time, T_r. The exponential model means that it takes $3T_r$ to reach 95% alignment.

The bulk relaxation times for fluids are much longer than for the same fluids, when they fill the pore spaces of reservoir rocks. Fine-grained rocks, which have narrow pore throats with large surface areas, have shorter relaxation times than coarse-grained rocks of the same mineralogy (Seevers, 1966). This difference is even more pronounced when the fluid is the preferential wetting fluid in the reservoir (Hurst *et al.*, 2000). The obvious conclusion is that NMR relaxation is related to specific surface area and can be used to estimate the effective porosity and permeability.

The California Research Company (now Chevron ETC) and Byron Jackson (BJ) Tools introduced the first NMR logging tool in the 1950's (Brown and Fatt, 1956; Brown and Gamson, 1960). This analogue tool was really not much more than a prototype and suffered the reliability and interpretation problems associated with most prototypes. One of the most endearing features of this tool was the requirement that the mud be "poisoned" with iron filings to "kill" the borehole signal. In spite of its shortcomings, the BJ tool did demonstrate that formation NMR could be measured in boreholes and that, under the right circumstances, it could be used to estimate porosity and permeability.

Armed with BJ NMR tool results, Standard Oil Company of California (now Chevron), joined by Shell Oil Co., approached the major logging vendors about establishing NMR (now called NML) logging services. Both Schlumberger and Dresser-Atlas (now Baker-Atlas) built a few prototype tools and offered limited NML services. These tools, like the BJ tool, also experienced reliability and interpretation problems. In as much as the only market appeared to be heavy oil reservoirs with fresh waters, these logging vendors saw little incentive to develop more sophisticated and robust NML tools.

Little changed for over twenty years. The major logging vendors would have dropped the technology, altogether, if not for the insistence by Chevron, Shell, and (later) EXXON. In the late 1980's, Numar Corporation, a small Israeli start-up, adopted powerful new ferrite magnet technology and lured several expert NMR log tool design engineers (including George Coates) from the major logging vendors, and developed their own NMR tool, called the *Magnetic Resonance Imaging Log®* (MRIL). This tool incorporated: (1) state-of-the-art ferrite technology; (2) down-hole digital circuitry, A/D sampling, data processing, encoding, and transmission; and (3) up-hole data processing and interpretation. Numar Corporation also worked closely with those oil companies attempting to obtain NMR logging services to meet their needs. In less than 15 years, Numar

Corporation, established NMR logging as a viable commercial market and themselves, as a viable NMR logging vendor. With the NMR log market established, Schlumberger introduced its own version of new NMR tools and Halliburton bought Numar Corporation.

The original application of NMR log was to estimate saturations in heavy oil reservoirs with fresh waters. A second application of NMR logs as to estimate the reservoir pore-size distribution. Coates *et al.* (1999) listed the following applications for NMR logs:

1. Presence and quantities of different fluids (water, oil, and gas).
2. Porosity and pore size distribution.
3. Bound and free water saturations.
4. Effective porosity and permeability.
5. Flushed zone saturation, S_{x0}, for wells drilled with oil based mud.

Potential applications for NMR log data have been known, since the 1950's, based on laboratory investigations at major oil companies (primarily Chevron and Shell) research facilities. The need, for well-bore NMR measurements, was kept alive by E&P managers, from these major oil companies, with personal knowledge of the laboratory and field data. Technology (ferrite and hardened microcircuitry) developments outside the petroleum industry provided the breakthroughs needed to build viable commercial NMR tools. A high-tech start-up company (Numar) took the risks to develop viable NMR logging tools and establish a NMR logging market. Established logging vendors moved into the NMR logging market, once it was established. With the added competition, new applications for NMR logs have been demonstrated.

All of the above elements were required to arrive at the current technical state of NMR logs.

A.24 S_{xo} Tool Developments

Specialized logging tools are needed to estimate saturation within the flushed zone. These measurements are used to determine: (1) near borehole apparent resistivity for resistivity model interpretation inversion; and (2) flushed zone water and hydrocarbon saturation estimates, for estimating movable (and therefore recoverable) hydrocarbon volumes.

Figure A.45 shows several (different) micro-resistivity electrode arrays, which are employed on hydraulic levered pads, pressed

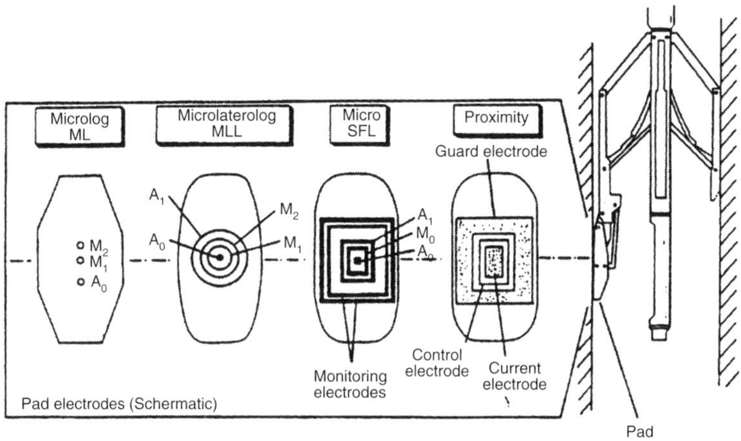

Figure A.45 Micro--resistivity pad designs (after Ballengee, 1990).

against the borehole wall, to obtain R_{xo}, from which S_{xo} could be estimated. Schlumberger recently introduced yet a new micro-resistivity tool: the *Micro-Cylindrically Focused Log*® (MCFL®). This new pad-type micro-resistivity measurement is included on the pad of one arm of a two-arm caliper, of the three-detector *Litho-Density Tool*® (LDT®), as part of their *Platform Express*® tool stack (Figure A.46). This tool delivers pseudo-microlog curves, which can be utilized for qualitative identification of permeable formations.

A.25 Dielectric Tool Developments

Formation (electrical) dielectric constant also, has a petrophysical utility (Hoyer and Rumble, 1976). In three dimensions, the current density vector, **J**, for alternating current (AC) signals can be written as (von Hippel, 1954a, 1954b):

$$\mathbf{J} = \sigma\mathbf{E} + j\omega\varepsilon\mathbf{E} = \left(\sigma + j\omega\varepsilon\right)\mathbf{E}, \tag{A.10}$$

where: σ is the electrical conductivity of the media.
$\omega = 2\pi f$ I the angular (signal) frequency.
ε is the electrical (dielectric) permittivity of the media.
$j = \sqrt{-1}$, indicates the out of phase (quadrature) signal component.
E is the electrical field strength vector.

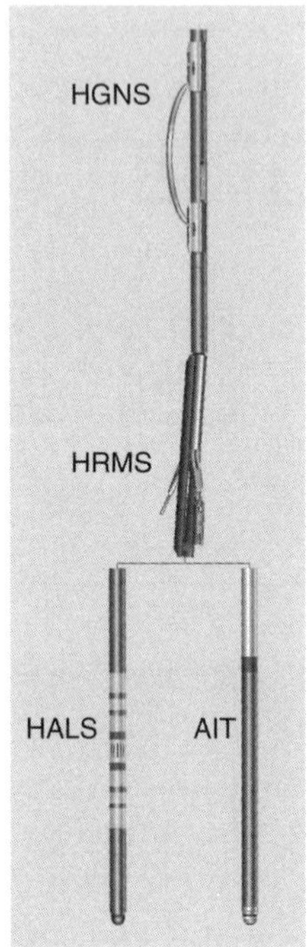

Figure A.46 Schlumberger Platform Express®, showing: High Resolution Azimuthal Laterolog (HALS®) or Array Induction Tool (AIT®); High Resolution Mechanical Sonde (HRMS®), containing the three- detector Litho-Density Tool (LDT®) and Micro-Cylindrically Focused Log (MCFL®); and the Highly Integrated Gamma Ray Neutron Sonde (HGNS®) components (courtesy of Schlumberger).

The dielectric constant, K_e, is the relative permittivity:

$$K_e = \frac{\varepsilon}{\varepsilon_0},$$

where: ε_0 is the permittivity of free space.

The phase angle, ϕ, between the in-phase and quadrature portions of Eq. A.10 is given by:

$$\phi = \tan^{-1}\left(\frac{\mathrm{Im}}{\mathrm{Re}}\right) = \tan^{-1}\left(\frac{\omega\varepsilon}{\sigma}\right). \quad \text{(A.11)}$$

The electromagnetic (EM) wave phase velocity, v, is given by (von Hippel, 1054a, 1054b):

$$\upsilon = \frac{1}{\sqrt{\varepsilon\mu}}, \quad \text{(A.12)}$$

where: μ is the magnetic permeability of the media.

There have been several studies of the dielectric constants of reservoir materials (e.g., Howell and Licastro, 1961; Keller and Licastro, 1959; Scott *et al.*, 1967). It is important to note that the dielectric constant of water is essentially an order of magnitude greater than any other component of reservoir rocks (Freedman and Vogiatzis, 1979) and varies little with salinity (Anon, 1984b, 1988). Both of these properties prove to be extremely valuable, for heavy oil reservoirs with fresh waters.

Whereas the induction logs operate at kHz frequencies, dielectric logs operate at MHz to GHz frequencies. Logging vendors have developed a considerable variety of dielectric logging tools to cover this signal frequency range, but they can be grouped into the following two basic types.

1. Low (MHz) frequency dielectric tools are essentially very high frequency induction tools, which measure the EM signal phase change between two sets of receiver coils on a borehole-centered mandrel tool. Interpretation of data from these tools is based on Eq. A.11. These tools have a greater depth of investigation than the high-frequency dielectric tools, but do not have the thin-bed resolution of the high-frequency tools.
2. High (GHz) frequency dielectric tools are pad devices using microwave transmitters at either end of a two-microwave receiver array. These tools measure the interval transit times of GHz impulse signals, similar to the compensated acoustic log measures acoustic interval transit times and attenuation of the signal pulse between receivers. Interpretation of data from these tools is based on Eq. A.12. These tools have much better thin-bed resolution than the low-frequency dielectric tools, but are definitely S_{xo} tools.

Dielectric tools have proven to be quite popular for evaluating heavy oil reservoirs with fresh waters and for in-field wells in fields undergoing steam or waterflood, where the formation water salinities are not known. Dielectric tools may have even delayed the development of viable NMR tools, as long as logging vendors thought the only NMR logging market was heavy oil with fresh waters.

A.26 Dipmeters to Borehole Imaging

Formation marker tops in a well provide only one dimension, unless combined with those from many other nearby wells or seismic reflection data, to form a structural/stratigraphic picture of an oil/gas field. For many purposes, however, it is desirable to get more detailed information than is available from surface geophysics or correlation of well logs between wells.

One option to obtain this detailed stratigraphic and/or structural information is to cut oriented whole cores, while drilling a well. This has often been done, in cases of thick reservoir rocks with fracture permeability; thin sands; multiple interconnected pay intervals; and/or vuggy intervals. Examples of these reservoirs are the Green River Formation in NE Utah and the Pinda Formation in the Congo Basin. This is an extremely expensive undertaking, however, which many operators are unwilling to accept. Consequently, a log option had to be devised.

An initial, and partial, solution to this need for detailed in-well structural and stratigraphic information was the development of dipmeter tools. Schlumberger introduced the first dipmeter in 1941, by adding SP electrodes to pads on the arms of a three-arm caliper. Expanded (10 in = 100 ft, or 1:100) depth scale plots of the three SP curves were manually correlated to solve the three-point geometrical problems, in order to obtain bedding and fault structural dips and strike. Later dipmeters used micro-resistivity electrode arrays, which provided more detailed correlation curves. Eventually, four and six-arm dipmeters were introduced, in attempts to improve full 360° coverage of the borehole.

Manual correlation of the 3–6 dipmeter curves was a tedious and labor-intensive task performed by only a few *Dipmeter Specialists*. Figure A.47 shows an image of a manual goniometer, used to solve 3–4 point dip problems for correlations if the individual dipmeter curves. Figure A.48 shows the meridian stereonet, utilized by structural geologists to solve the same problem. Figure A.49 shows the stereographic projection of an inclined plane, providing dip and strike, from a stereonet solution.

Digital recording of dipmeter data and data processing, using statistical correlation and semblance algorithms, speeded up the process. These data processing advances, however, added another layer of mystique, because the results were very sensitive to data processing parameters. Inasmuch as dipmeter processing

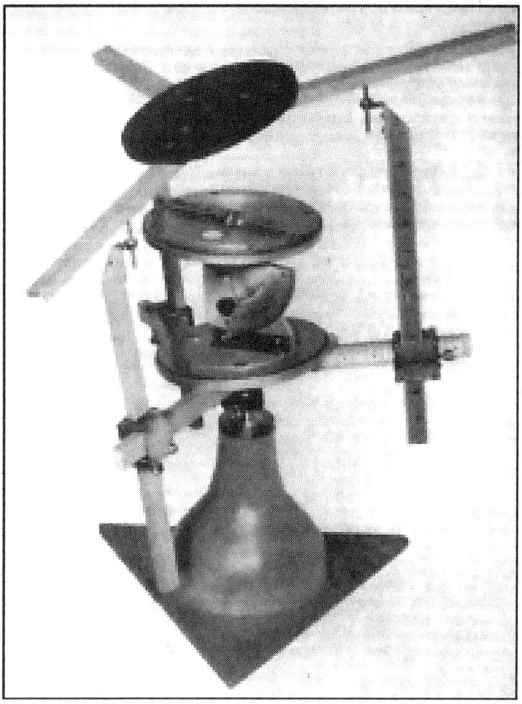

Figure A.47 Manual dipmeter goniometer computer (after Bateman, 2009).

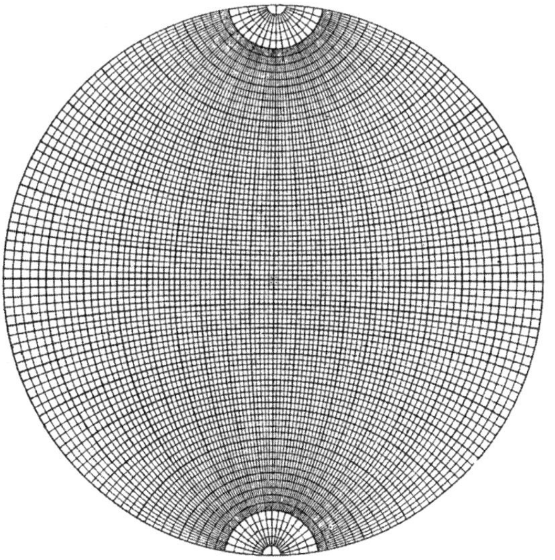

Figure A.48 Meridian stereonet (after Bucher, 1944).

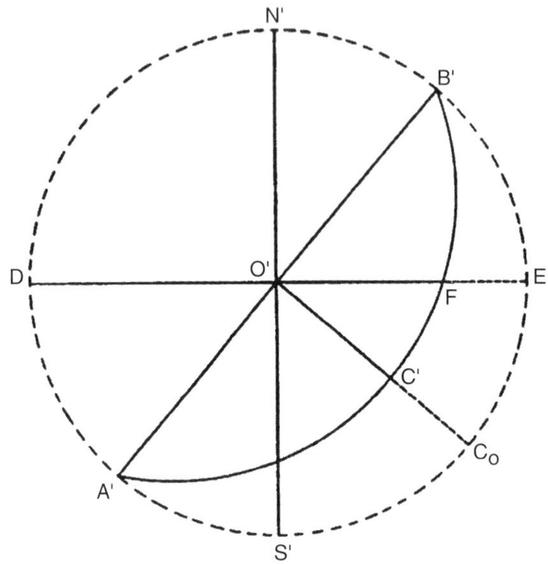

Figure A.49 Stereographic projection of inclined plane (after Bucher, 1944).

specialists usually were not geologists, the end users (geologists, geophysicists, and petrophysical engineers) tended to regard the entire dipmeter data acquisition and data processing as a necessary *black art* evil.

The advent of borehole imaging tools and interpretation workstations, improved the detailed structural and stratigraphic situation, greatly. Geological end users (geologists, geophysicists, and petrophysicists) could review the resulting borehole images, as if they were looking at whole cores, to identify and record structural and stratigraphic dips and strikes electronically in computer databases, for future usage.

The first borehole-imaging tool was called the *Borehole Televiewer* (BHTV), developed by Mobil Field Laboratories (now ExxonMobil EPR) in 1969. It was licensed to all of the major logging vendors, as well as a few specialty companies (Zemanek *et al.*, 1970). In spite of its name, this was really a SONAR tool, using acoustic reflections from the borehole wall to map its reflectivity and the borehole shape. Like the BJ NMR tool, the Mobil BHTV tool was really more of a prototype than production sonde. As such, it suffered considerable performance and reliability problems.

Because it was considered a specialty tool with only a niche market, the major logging vendors did not spend much effort developing improved versions. The tool languished in logging tool purgatory until Gearhart Industries revived it, in the mid 1980's, using new design concepts and state-of-the-art downhole components. The first descriptions the new tool and it capabilities (Edmiston *et al.*, 1990; Goetz *et al.*, 1990), however, were not published until after Gearhart had been acquired by Halliburton. The new Gearhart/Halliburton BHTV design, called *Circumferential Acoustic Imaging Tool*® (CAST) was much more robust and reliable than the earlier BHTV tools and provided crisp, high-resolution images. Work station image processing, introduced with the CAST tool, provided the interpreter with hands-on geological interpretation tools, allowing three-dimensional manipulation of the borehole images, as if they were cores. Full, 360° images of the borehole wall could be transposed to a pseudo-core image and rotated on the monitor screen, much like a geologist might view a physical core. Structural and stratigraphic features could be selected, using an on-screen curser, driven by an interpreter controlled joy-stick, track-ball, or mouse. Individual features could be selected at several locations around the borehole image circumference and converted to dips and strikes, using a virtual goniometer, and the resulting data stored in a pick file for later usage. Picks could be superimposed on the borehole and pseudo core images, for later display and/or presentation. The combination of the borehole imager and the work station, brought the geologist back into the center of the interpretation operation, with imaging and data process specialists assisting only in image enhancement, image manipulation, and data base management.

Figure A.50 and A.51 illustrate the value of borehole imagery. Figure A.50 shows a cross plot of routine (PKS) core analysis helium porosities vs. wireline Density/Neutron Cross-Plot porosities for a Congo Basin well. The dashed line in Figure A.50 is the 1:1 line, along which the data should plot if the two sources of information agree. In the case of Figure A.50, the log porosities are all significantly greater than the core porosities for the same depths. Whereas core/log porosity disagreements are quite common, it is the core porosities, which are usually much higher than the log porosities. A CAST® False Core Image display for this reservoir (Figure A.51) shows the reason for this discrepancy. The reservoir has numerous inter-connected small-diameter (pea to grape-sized) vugs, which were detected by the wireline measurements but not

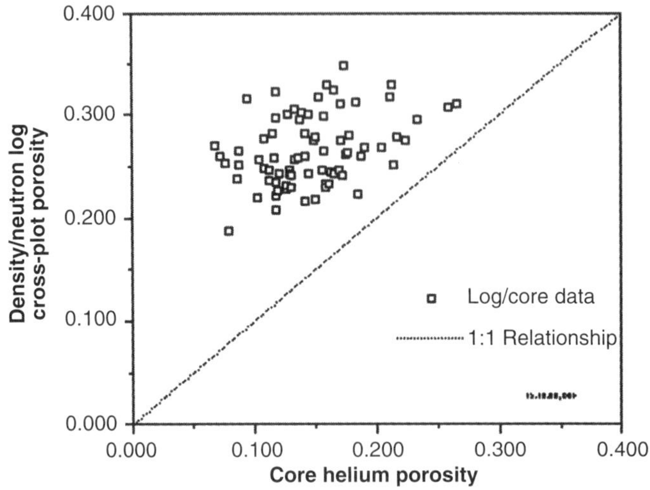

Figure A.50 Congo Basin reservoir cross-plot of routine core analysis helium porosities vs. density/neutron log cross-plot porosities.

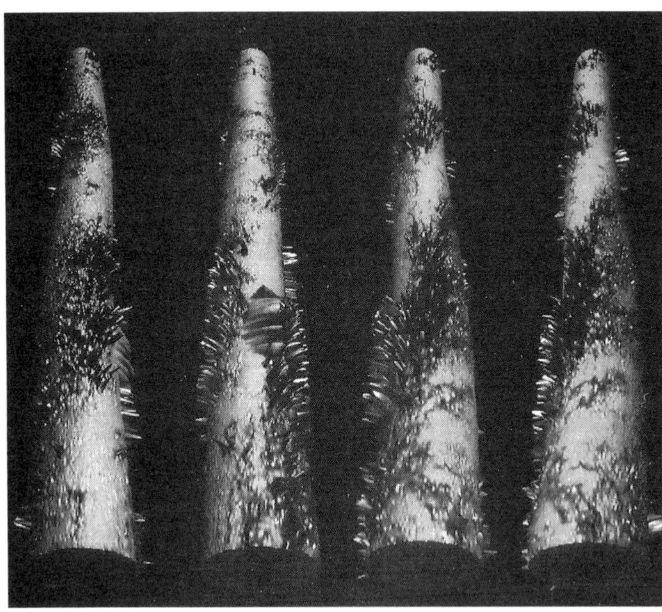

Figure A.51 Congo Basin reservoir CAST® false core images, Through the Figure A.48 reservoir, showing extensive small vug systems.

the core measurements. Discussions with two core measurement laboratories which had worked on the core for this field confirmed that neither vendor's protocol (they were different) would have measured the small interconnected vugs, as pore space.

Prior to obtaining the CAST® images of Figure A.51, production and reservoir engineers, working this field, insisted that the wireline measurements should agree with the core measurements. Inasmuch as the vuggy zone was near the top of a reservoir slated for periphery water flooding, the CAST® images may have saved the field from premature water breakthrough and potential reservoir damage. With the success of the Gearhart/Halliburton CAST® tool, both Schlumberger and Baker Atlas introduced their own versions of improved BHTV tools.

A second type of borehole imager involves detailed micro-resistivity images of the borehole wall. The Schlumberger Formation MicroScanner (FMS)/Fullbore MicroImager (FMI) utilizes multi-button microresistivity electrodes on each of the arms of a four-arm caliper. To increase the circumferential coverage in large boreholes, each pad has an auxiliary flap, which can be opened and forced into contact with the borehole wall (Figure A.52). The Halliburton

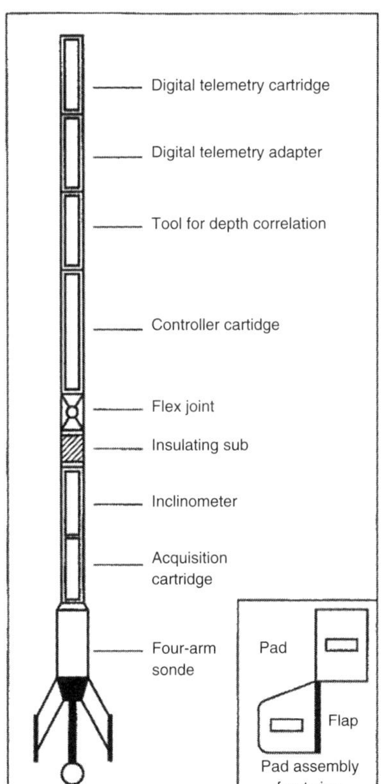

Figure A.52 Schlumberger Fullbore MicroImager® (FMI) schematic (after Anon, 1992b).

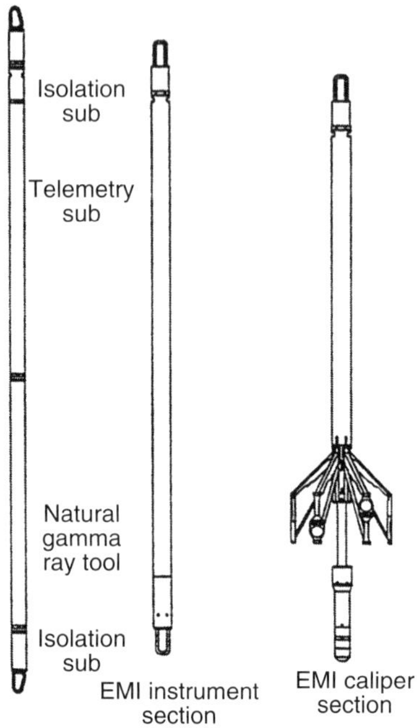

Figure A.53 Halliburton Electric Micro Imager® (EMI) (after Anon, 1994).

Electro Micro Imager (EMI), utilizes multi-button micro-resistivity electrodes on each arm of a Gearhart/Halliburton six-arm caliper (Figure A.53).

Neither one of these micro-resistivity imaging tools provides full 360° borehole wall coverage, like acoustic televiewers do. There are gaps between the imaging strips, which become larger for larger-diameter boreholes. Each of the 4 image strips (from the Schlumberger tool) cover more of the borehole circumference than the individual Halliburton strips. The gaps between the Schlumberger tool strips, however, are also greater than those between the Halliburton tool strips. Both of these micro-resistivity imaging tools provide comparable detail resolution. Both are supported by excellent proprietary interpretative work-stations. Figure A.54 shows a Schlumberger FMS interpretation, while Figure A.55 shows a Halliburton EMI interpretation. The choice between the two services is really an aesthetic one, by the user.

Figure A.54 and A.55 also illustrate the dip-picking algorithms available for all borehole imaging tools. Dipping surfaces will show-up on the two-dimensional displays as sinusoids. The

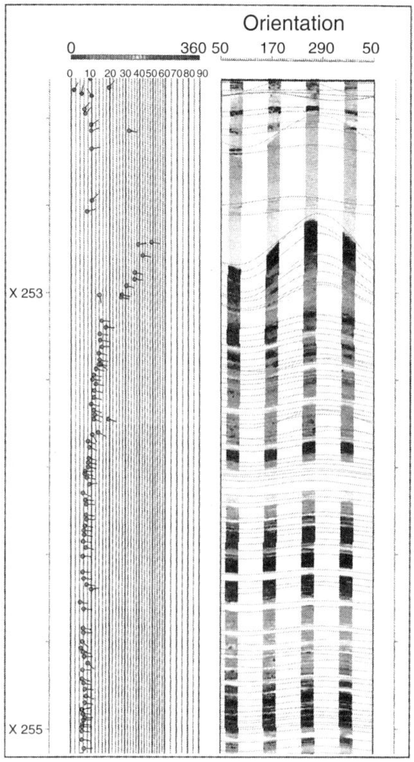

Figure A.54 Schlumberger Formation MicroScanner (FMSI) false core image interpretation display, showing bedding correlations and dip pick tadpoles. (after Serra, 1989).

interpreter needs only to make 4–6 detections on an interpreted bed and the work-station will: (1) fit a sinusoid to the picks; (2) fit a surface to the sinusoid; (3) compensate for borehole enlargements; (4) calculate the dip and strike of the pick; (5) store the pick information in a pick file; (6) and post the pick on a "tadpole plot" log. Borehole imager workstation display and interpretation software is interactive. This removes the mystique, which surrounded dipmeter processing and interpretation. The interpreter immediately sees their picks on the pseudo core images, and tadpole plots. Borehole imagers and interpretative workstations have, for all practical purposes, replaced dipmeters as structural and stratigraphic interpretation tools.

Other borehole imaging tools have also been introduced. The acoustic and micro-resistivity borehole imagers, however,

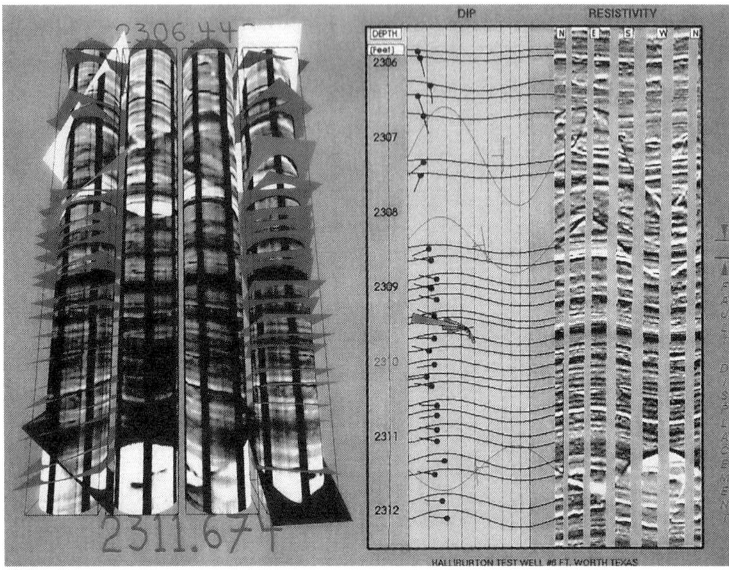

Figure A.55 Halliburton (Six Pad) EMI false core 3-D and 2-D interpretation displays, showing bedding correlations and dip pick tadpoles (after Anon, 1994).

provide the finest resolution. Between these two types of borehole imagers, the micro-resistivity imagers currently provide finer resolution, but the acoustic imagers provide full 360° borehole coverage. Acoustic imagers work well in oil base mud systems, whereas the micro-resistivity images are still of much lower quality, in these situations.

A.27 Wireline Formation Testers

Petrophysicists, as well as production and reservoir engineers, need to have samples of reservoir fluids for formation evaluation, establishing value, and reservoir depletion planning. Originally this had to wait for drillstem or well tests. Wireline formation testers (Moran and Finklea, 1962) represented a significant breakthrough, which allowed formation fluid samples to be acquired while a well was still being drilled, at a lower cost and risk than either drill stem tests or well tests.

The earliest formation tester models, however, contained only a single sample bottle which could not be flushed, downhole. Once a sample had been taken, the tool had to be retrieved to the surface, without knowledge of whether the sampling had been successful or not. Later versions, included downhole dynamic pressure gauges, which monitored the sampling operation, multiple sample bottles, and the ability to purge the sample bottles, if all that was desired was formation pressure. The latter, enabled petrophysicists to develop detailed sub-surface formation pressure models to aid in determining fluid contacts in long transition zones. Figure A.56 illustrates the utilization of formation tester pressure data to differentiate between oil and water legs and the oil/water (O/W) contact in a long transition zone.

Even with repeat samplers, the fluid samples had to be returned to the surface and analyzed before anyone could tell if there was significant mud filtrate contamination. Recent wireline formation samplers, include electrical and optical fluid measurements, which allow surface operators to repeat sample collections, until they are certain that they have uncontaminated formation fluid samples (Anon, 1992d). This capability also allows down-hole fluid identification, which can be extremely useful in the case of stacked reservoirs having multiple hydrocarbon types, as in the Gulf of Thailand.

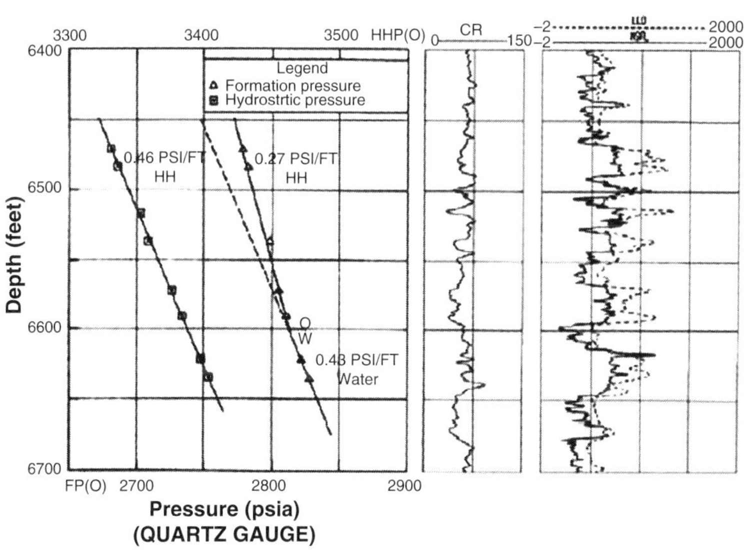

Figure A.56 Use of formation tester pressure measurements to distinguish oil and water gradients and O/W contact in a long transition zone (after Gunter and More, 1987).

A.28 Shaly Sands

Gus Archie's *Formation Factor/Porosity* and *Resistivity Index/Saturation* relationships introduced quantitative formation evaluation and the ability to estimate reserves in clean sands and carbonates. In the late 1940's, however, as more marginal reservoirs were being developed, Archie's equations began to have limitations. If clay minerals were present with fresh formation waters, Archie's equations would underestimate and, in worst case scenarios, could even miss reserves. The problem was that clay minerals exhibit surface electrical conductivity, whereas Archie's equations assume that brines are the only electrical conductors in a reservoir. Several shaly-sand saturation models appeared (e.g., Patnode and Wyllie, 1950; Poupon *et al.*, 1970; Poupon and Leveaux, 1971; Simandoux, 1963; Winsauer and McCardell, 1952). Of the hundreds of shaly-sand algorithms, which have been published (c.f., Brown *et al.*, 1982), they probably all worked, at least once, but none of them have seemed to be both practical and theoretically satisfying.

In the 1960's, Archie hired several electrochemists, for Shell Development Company, and assigned them the task of developing simplified shaly-sand petrophysical models, which production and reservoir engineers could use in the field. Waxman and Smits (1968) met this challenge and developed a saturation model, with the following features:

1. It was based on a parallel electrical circuit model (clay mineral surface conductivity and free fluid conductivity), which was easier to apply using electrical conductivity rather than resistivity.
2. It accounted for the differences in clay mineral species, by incorporating cation exchange capacity (*CEC*).
3. It accommodated the electrochemical (Helmholtz) double layer at the clay mineral/brine interface, and low-conductivity brine responses by incorporating the pore volume normalized *CEC*, Q_v.

The Waxman and Smits (W&S) shaly-sand algorithm was electrochemically correct and simpler than many others in the literature. It did, however, have one draw-back. The critical parameter Q_v was not available from any currently available log measurement. It had to be measured on cores or drill cuttings in the laboratory, which usually required several weeks. Wildcats would have to be completed without well specific Q_v data.

Clavier *et al.* (1977) and Juhasz (1979 and 1981) developed empirical Q_v estimators from log measurements. Worthington (1973) set up an automated Q_v laboratory. These, however, were stop-gap measures until a viable Q_v log would become available. Unfortunately, there has been no commercial Q_v log developed, to date.

A.29 Golden Era and Black Period of Petrophysics

When Gus Archie published his *Formation Factor/Porosity* and *Resistivity Index/Saturation* relationships, he heralded what can probably best be described as "The Golden Age of Petrophysics", which reached its zenith from the mid 1950's through the mid 1970's. All of the major petroleum companies and many large independents had their own Exploration and Production (E&P) research and Development (R&D) facilities, as shown in Table A.8. Some of the most significant breakthroughs in petrophysics, as well as all other aspects of exploration and production, came out of these facilities, which were really the "Pride of the Industry". Because several R&D organizations were often working on the same problems, multiple solutions were common. The "Best" solutions survived, whereas others were forgotten.

Beginning in the 1970's, the corporate parents of these in-house R&D facilities began to transfer the charters of these facilities to "Technology Centers", which monitored other people's technology developments and conducted some technical service work for the operating companies, but did little independent research. Later, corporate officers began to require *all* corporate units, including such traditional services organizations, such as archives, libraries, human resources, and R&D, to become profit centers. Those service units, which could not show *profits* were outsourced, downsized, and/or eliminated. With the industry consolidation of the 1990's, most of the former R&D facilities ceased to exist, all together. The net result was a serious loss of what had been a tremendous E&P innovation engine.

Multiple R&D organizations working on common problems may not have been the most cost-effective method of solving those problems, but the resulting competition produced excellent solutions. The results of this type of competitive environment were evident, following the collapse of the Soviet Union. There were

Table A.8 Major Petroleum E&P Research Centers.

Corporate Parent	Laboratory	Location
British Petroleum, Ltd.	BP Research Centre	Sunbury-on-Thames
Continental Oil Company	CONOCO Research Center	Ponca City, Oklahoma
Gulf Oil Company	Gulf Research and Development Company	Harmerville, Pennsylvania
ESSO/EXXON	Exploration & Production Research Center	Houston, Texas
Phillips Petroleum Company	Frank Phillips/Phillips 66 Research Center	Bartlesville, Oklahoma
Schlumberger	Schlumberger-Doll Research Center	Ridgefield. Connecticut
SOCAL/Chevron	California Research Company/Chevron Oil Field Research Company	La Habra, California
SOCONY/Mobil	Magnolia/Mobil Field Research Laboratory	Irving, Texas
Shell Oil Company	Shell Development Company	Bellare, Texas
Standard Oil Company, of Indiana/AMOCO	AMOCO Production Research Co.	Tulsa, Oklahoma
TEXACO	Texaco Bellare Laboratory	Bellare, Texas
Union Oil of California	Fred L. Hartley Research Center	Brea, California

many excellent research institutes operating within the former Soviet Union. The lack of a competitive environment, however, often resulted in equipment development stopping at what western industry would describe as prototypes, rather than production type equipment.

With the disappearance of oil company E&P research centers, the oilfield service companies have stepped up their own R&D programs. These efforts, however, have been aimed more at improving existing technologies, rather than in basic research. Universities and National Laboratories have also attempted to fill the gap left by the disappearance of oil company E&P research facilities. These efforts have not been as focused as were those of the former oil company E&P R&D centers. The university and national laboratory efforts have also been hampered by reductions in government and private funding.

In many ways the industry seems to have lost much of its innovative edge. Will the current tight world oil market and higher oil prices bring renewed oil company funded E&P R&D? The authors of the current volume certainly hope so.

A.30 The Future

Determining what has and has not been significant developments in the past is skating on thin ice. Attempting to predict what may, or may not happen, in the future is even more so. Anyone who does so is merely stating their opinions, so the reader should take the predictions as food for thought. Stephen Prensky (1994a, 1994b, 2002, 2007, and later) has published excellent periodic "Logging and Petrophysics Advances" updates. These reviews, while interesting reading, do suffer from the problem of deciding what is really significant, "on the fly". Developments, which seem significant, at the time, may not seem to be so, after a few years of applications. Others, which did not seem to be that important, when first introduced (e.g., induction log phase measurement), often turn out to have been very significant, upon reflection. Finally, developments in completely unrelated fields (e.g., micro-circutry), may turn out to have great impact.

In the past, advancements in petrophysics and well logging have been a tandem affair. Basic science petrophysics breakthroughs have tended to come from laboratories run by the major oil companies, whereas logging tool, data processing, and interpretation developments have tended to come from vendors. For the past 25 years, however, this tandem has been severely damaged as oil companies slashed their R&D budgets and mergers eliminated several premiere petroleum E&P laboratories.

This situation may be about to change. Record high commodity prices have oil companies looking for ways to shield profits from taxes and R&D has always been a good "loss leader" which occasionally paid big dividends. Out-sourcing R&D has not been as successful as many of its advocates had thought it would be, as it is more difficult to control direction, security, and costs, with outsourced R&D, than it is with in-house research.

Chemical radioactive sources will probably be replaced sooner, rather than later. It is not because the non-chemical sources are better. It is just that it is becoming more and more difficult to obtain, store, transport, handle, and use chemical radioactive sources than non-chemical sources. All logging major vendors now offer pulsed-neutron generators for some of their tools. Current pulsed-neutron generators are already orders of magnitude more reliable than the earliest models. Some major logging vendors have evaluated using deuterium-deuterium neutron generators, because even the tritium targets, of current pulsed-neutron generators, are considered to be hazardous material. Essentially all major logging vendors have evaluated betatrons (electron generators) and X-Ray tubes as alternatives to chemical gamma ray sources. Some vendors have developed prototype tools, which use pulsed neutron generator fast neutrons to bombard the formation and fast neutron scattering and thermal neutron capture gamma rays as gamma ray sources, for density measurement. *None of these potential replacements to chemical gamma ray and neutron sources, however, have the simplicity, strength, reliability, and/or versitility of the existing chemical sources.* Their biggest selling point is that *they are not chemical radioactive sources.*

Ultra-deep water operations, extended-reach directional drilling, and horizontal wells will probably become even more common. This will increase the demand for MWD improvements, particularly in the area of data transmission to the surface.

The development of MWD capabilities has been explosive, starting with simple directional drilling aids and progressing to provide (nearly) full petrophysical measurements. In some recent instances, MWD tool design has progressed beyond simply duplicating wireline measurements to creating new, unique, MWD measurements (e.g., quadrant density and neutron tools and imaging). Several firms are evaluating what measurements can be made to "see ahead of the bit".

New gas (e.g., tight gas sands, Coal Bed Methane, and Shale Gas) and Oil (e.g., Monterey, Bakkan, and Green River Formations) are stimulating new measurements and interpretation approaches.

"Smart Oilfield" down-hole instrumentation and Supervisory Control And Data Acquisition (SCADA) monitoring is a new and developing field. The down-hole sensors to monitor reservoir properties and production activities, not just production equipment, are being developed. The paper by Adeyemi *et al.* (2008) indicates that these types of monitoring systems are already being implemented.

Some petrophysical properties and logging techniques currently confined to "Petrophysical Purgatory" my eventually get out, just as NMR did. One property that has held promise, but has never been commercialized, is Induced Polarization *(IP)*. Conrad Schlumberger observed this phenomena, in 1911, and named it *Polarisation Provoquée*. Vinegar and Waxman (1986) described laboratory results, which indicated that IP could be simply related to Q_v. Similar unpublished results have been obtained elsewhere. Vinegar *et al.* (1986) described field measurements with a prototype *IP* logging sonde to estimate Q_v from direct log measurements.

In conclusion, the technologies that were unthinkable, just a few years ago, are now commonplace and new ones are on the horizon. This is, indeed, an exciting time to be in the petroleum industry.

Bibliography

Adeyemi, O. S., Shryock, S. G., Sankaran, S. and Hostad, O., 2008. Implementing I-Field Initiatives in a Deepwater Green Field, Offshore, Nigeria, *Soc. Petrol. Engr.*, Pap. No. 115367.

Alger, R. P., Locke, S., Nagel, W. A. and Sherman, H., 1971. Dual Spacing Neutron Log–CNL, *Soc. Petrol. Engr.*, Pap. No. 3565.

Alger, R. R., Raymer, L. L., Hoyle, W. R. and Tixier, M. P., 1963. Formation Density Log Applications in Liquid-Filled Holes, *J. Pet. Tech.*, vol. 20 (12): 300–310.

Anderson, B. I. and Barber, T. D., 1996. *Induction Logging*, Schlumberger Wireline and Testing, Sugar Land, 45 pp.

Anon., 1953. *Continuous Velocity Logging*, Seismographic Service Corp., Tulsa, 7 pp.

Anon., 1972 *Log Interpretation, Volume I - Principles*, Schlumberger Limited, New York, 113 pp.

Anon., 1974. *Log Review 1*, Dresser Atlas Division, Dresser Industries, Houston, 177 pp.

Anon., 1983, *Formation Evaluation Chart Book*, Gearhart Industries, Inc., Fort Worth.

Anon., 1984a. *Resistivity Measurement Tools*, Schlumberger Educational Services, Houston, 39 pp.

Anon., 1984b. *Electromagnetic Propagation Tool*, Schlumberger Well Services, Houston 24 pp.

Anon., 1985b. *Array-Sonic® Service*, Schlumberger Well Services, Houston, 20 pp.

Anon., 1987. *Log Interpretation Principles/Applications*, Schlumberger Educational Services, Houston, 198 pp.

Anon., 1988. *Electromagnetic Propagation Logging*, Schlumberger Well Services, Houston, 37 pp.

Anon., 1991a. *Log Interpretation Charts*, Halliburton Logging Services, Houston.

Anon., 1991b. MWD Systems Expand Capabilities, Petroleum Engineer International, vol. 91, (5): 18–23.

Anon., 1992a. *Elan Users Guide*, Schlumberger Data Services, Houston, 132 pp.

Anon., 1992b. *FMI* Fullbore Formation MicroImager*, Schlumberger Education Services, Houston, 42 pp.

Anon., 1992c. *AIT* Array Induction Imager Tool*, Schlumberger Education Services, Houston, 36 pp.

Anon., 1992d. *MDT* Modular Formation Dynamics Tester*, Schlumberger Education Services, Houston, 32 pp.

Anon., 1993a. *Logging While Drilling*, Schlumberger Educational Services, Houston, 109 pp.

Anon., 1994. *EMI Field Examples and Comparison to Core*, Halliburton Energy Services, Houston.

Anon., 1997. DSI® *Dipole Shear Sonic Imager, Schlumberger Oilfield Marketing Services*, Houston, 36 pp.

Anon., 2000. *HRLA High-Resolution Laterolog Array Tool*, Schlumberger, Houston.

Anon., 2006. *Celebrating 40 Golden Years, Golden Gate Chapter*, Soc. Petrol. Engr.

Archie, G. E., 1942. The Electrical Resistivity Log as an Aid in Determining Some Reservoir Characteristics, *Trans., AIME*, 146: 54–62.

Asquith, G. and Krygowski, D., 2004. *Basic Well Log Analysis, AAPG Methods in Exploration Series*, No. 16, American Association of Petroleum Geologists, Tulsa, Oklahoma, 244 pp.

Aulia, K., Poernomo, B., Richmond, W., Wicaksono, A. H., Béguin, P., Benimeli, D., Dubourg, I., Van der Wal, P., Boyd, A., Farag, S., Ferraris, P., McDougall, A., Rosa, A. and Sharbak, D., 2001. Resistivity Behind Casing, *Oilfield Review*, 13 (1): 2–25.

Baker, P. E., 1957. Density Logging with Gamma Rays, *Trans. AIME*, 210: 289–294.

Ballengee, W. G., 1990. Resistivity Logging, *Formation Evaluation Seminar*, Chevron Corporation.

Barber, T., 1985. Introduction to the Phasor Dual Induction Tool, *J Petrol Techn*, 37, (10): 1699–1706.

Bassiouni, Z., 1994. Theory, Measurement and Interpretation of Well Logs, Society of Petroleum Engineers, Richardson, TX, 372 pp.

Bateman, R. M., 2009. Petrophysical Data Acquisition, Transmission, Recording and Processing: A Brief History of Change from Dots to Digits, *SPWLA 50th Annual Logging Symposium*, Pap. DD.

Bergen, J. G., 1975. Direct Digital Laserlogging, *Soc. of Petrol. Engr*, Pap. 5506-MS.

Brown, A. A., Berry, W. R., Coates, G. R., Fertl, W. H., Hoyer, W. A., Patchett, J. G. and Ransom, R. C., 1982. *SPWLA Reprint Volume–Shaly Sand*, SPWLA, Houston.

Brown, R. J. S. and Fatt, I, 1956. Nuclear Magnetism Logging, *Trans. AIME*, 219: 262–264.

Brown, R. J. S. and Gamson, B. W., 1960. Measurement of Fractional Wettability of Oilfield Rocks by the Nuclear Magnetic Relaxation Method, *Trans. AIME*, 207: 201–209.

Bucher, W. H, 1944. The Stereographic Projection, A handy Tool for the Practical Geologist, J. Geology, 52 (3): 191–212.

Caldwell, R. L., 1958. Using Nuclear Methods in Oil-Well Logging, *Nucleonics*, 16, (12): 58–66.

Clavier, C., Coates, G. and Dumanoir, J., 1977. The Theoretical and Experimental Basis for the Dual Water Model for the Interpretation of Shaly Sands, *Soc. Petrol Engr*, Pap. No. 6859.

Coates, G. R., Miller, M., Gillen, M. and Henderson, G., 1991a. The MRIL in Conoco 33-1: An Investigation of a New Magnetic Resonance Imaging Log, *SPWLA 32nd Annual Logging Symposium*, Pap. DD.

Coates, G. R., Peverarp, R. C. A., Hardwick, A. and Roberts, D., 1991b. The Magnetic Resonance Imaging Log Characterized by Comparison with Petrophysical Properties and Laboratory Core Data, *Soc. Petrol. Engr*, Pap. 22723-MS.

Coates, G. R., Xiao, L. and Prammer, M. G., 1999. *NMR Logging Principles and Applications*, Halliburton Energy Services, Houston, 233 pp.

Dakhnov, V. N., 1962, Geophysical Well Logging (translated by G. B. Keller), *Quarterly, Colorado School of Mines*, 574 (2).

Dewan, J. T., 1983. *Essentials of Modern Open-Hole Log Interpretation*, PenWell Books, Tulsa 361, pp.

Dewan, J. T. and Greenwood, H. M., 1955. Calibration of Gamma Ray and Neutron Equipment for Identification and Evaluation of Potash Deposits, *The Technical Review*, 6: (6) 4–12.

Dobrin, M. B., 1960. Introduction to Geophysical Prospecting, McGraw-Hill Book Co., New York, 445 pp.

Doll, H. G., 1949a. The SP Log: Theoretical Analysis and Principals of Interpretation, *Trans. AIME*, 179: 146–185.

Doll, 1949b. Introduction to Induction Logging and Application to Logging of Wells Drilled with Oil Base Mud, *Trans. AIME*, 186: 148–162.

Doll, 1950. The Microlog–A New Electrical Logging Method for Detailed Determination of Permeable Beds, *Trans. AIME*, 187: 155–164.

Doll, 1951. The Laterolog: A New Resistivity Logging Method with Electrodes Using an Automatic Focusing System, *Trans. AIME*, 192: 305–316.

Doveton, J. H., 1994. Geologic Log Analysis Using Computer Methods, AAPG, Tulsa, 169 pp.

Dumanoir, J. L., Tixier, M. P. and Martin, M., 1957. Interpretation of the Induction-Electrical Log in Fresh Mud, *Trans. AIME*, 210: 202–215

Edmiston, C. S., 1990. Field Performance of a New Borehole Televiewer Tool and Associated Image Processing Techniques, *SPWLA 31st Annual Logging Symposium*, Pap. H.

Ellis, D. V., 1987. *Well Logging for Earth Scientists*, Elsevier, New York, 532 pp.

Ellis, D. V. and Singer, J. M., 2007. *Well Logging for Earth Scientists*, Springer, Dordrecht, 692 pp.

Frank, W. F., 1986. *Prospecting with Old E-Logs*, Schlumberger Educational Services, Houston, 161 pp.

Faul, H. and Tittle, C. W. 1951. Logging of Drill Holes by the Neutron-Gamma Method and Gamma Ray Scattering, *Geophysics*, 16 (2): 260–267.

Freedman, R. and Vogiatzis, J. P., 1979. Theory of Microwave Dielectric Constant Logging Using the Electromagnetic Wave Propagation Method, *Geophysics*, 44: (5): 969–986.

Geldmacher, I. M. and Jonkers, J., 2007. A Through Casing Resistivity Field Trial in Alberta Canada, *SPWLA 48th Annual Logging Symposium*, Pap. Z.

Gill, S., Brink, K., Patterson, D. and Hitchcock, S., 1991. Application of a Layered Resistivity Model for Deriving True Resistivity of Thin Dipping Sands in the Asia-Pacific Region, *Soc. Petrol. Engr*, Pap. No. 22990-MS.

Goetz, J. F., Seiler, D. D. and Edmiston, C. S., 1990. Geological and Borehole Features Described by the Circumferential Acoustic Scanning Tool, *SPWLA 31st Annual Logging Symposium*, Pap. C.

Gunter, J. M. and Moore, C. V., 1987. Improved Use of Wireline Testers for Reservoir Evaluation, *J. Petrol. Techn*, 39 (6): 635–644.

Head, M. P., 1977. Wellsite Computer analysis: A Program for Complex Lithologies, *Soc. Petrol. Engr*, Pap. 6824-MS.

Head, M. P. and Gearhart, M., 1977. Wellsite Formation Analysis using the DDL Computer, Soc. Petrol. Engr, Pap. 6541-MS.

Hearst, J. R. and Nelson, P. H., 1985. *Well Logging for Physical Properties, McGraw-Hill Book Company*, New York, 571 pp.

Hearst, J. R., Nelson, P. H. and Paillett, F. L., 2000. *Well Logging for Physical Properties*, John Wiley & Sons, Ltd., New York, 483 pp.

Hicks, W. G. and Berry, J. E., 1956. Application of Continuous Velocity Logs to Determination of Fluid Saturation of Reservoir Rocks, *Geophysics*, 21 (3): 739–754.

Hoelscher, R. P., Arnold, J. N. and Pierce, S. H., 1952. *Graphic Aids in Engineering*, McGraw-Hill Book Company, New York, 197 pp.

Howell, L. G. and Forsch, A 1939. Gamma-Ray Well-Logging, *Geophysics*, 14 (2): 106–111.

Howell, B. F. and Licastro, P. H., 1961. Dielectric Behavior of Rocks and Minerals, *The Am. Mineralogist*, 46 (2): 269–288.

Hoyer, W. A. and Lock, G. A., 1972, Logging for Copper by In-Situ Neutron Activation Analysis, Society of Mining Engineers Transactions, v. 252, pp. 409–417.

Hoyer, W. A. and Rumble, R. C., 1976, Dielectric Constant of Rocks as a Petrophysical Parameter, 17th Annual Well Logging Symposium, SPWLA, Paper O.

Hull, P. and Cooledge, J. E., 1960. Field Examples of Nuclear Magnetism Logging, J. Petrol. Techn. 12 (8): 14–22.

Itskovich, G., Mezzatesta, A., Strack, K. M. and Tabarovsky, L., 1998. High-Definition Lateral Log Resistivity Device–Basic Physics and Resolution, *SPWLA 39th Annual Well Logging Symposium*, Pap. V.

Juhasz, I, 1979. The Central Role of Q_v and Formation Water Salinity in the Evaluation of Shaly Formations, *SPWLA 20th 39th Annual Well Logging Symposium*, Pap. AA.

Juhasz, I., 1981. Normalized Q_v–The Key to Shaly Sand Evaluation using the Waxman-Smitts Equation in the Absence of Core Data, *SPWLA 22nd 39th Annual Well Logging Symposium*, Pap. Z.

Keller, G. V., 1949. An Improved Electrode System for use in Electric Logging, *Producers Monthly*, 13 (10): 12–15.

Keller, G. V., 1950. Modified Mono-Electrodes for Improved Resistivity Logging, Producers Monthly, 14 (9): 13–16.

Keller, G. V. and Licastro, P. H., 1959. Dielectric Constant and Electrical Resistivity of Natural-State Cores, USGS Bulletin 1052-H: 257–285.

Kenyon, W. E., Howard, J. J., Sezginer, A., Matteson, S. A., Horkowitz, K. and Ehrlich, R., 1989. Pore-Size Distribution and NMR in Microporous Cherty Sandstones, *SPWLA 30th Annual Logging Symposium*, Pap. LL.

Keys, W. S., 1989, *Borehole Geophysics Applied to Ground-Water Investigations*, National Water Well Association, Dublin, OH, 313 pp.

Kokesh, F. P., 1956. The Long Interval Method of Measuring Seismic Velocity, *Geophysics*, 21 (3): 724–738.

Little, J. W. and Youmans, A. H., 1964. The Neutron Lifetime Log–A Performance Report, *Oil and Gas J*. 62 (36): 165–168.

Loren, J. D., 1972. Permeability Estimates from NML Measurements, *Soc. Petrol. Engr.* Pap. No. 3334-PA.

Loren, J. D. and Robinson, J. D., 1970. Relations Between Pore Size, Fluid, and Matrix Properties, and NML Measurements, *Soc. Petrol. Engr.* Pap. No. 2529-PA.

Martin, D. W., Spencer, M. C. and Patel, H., 1984. The Digital Induction–A New Approach to Improving the Response of the Induction Measurement, *SPWLA 25th Annual Logging Symposium*, Pap. M.

Martin, P. M., 1956. Well Logging with an Atom Smasher, *AIME*, Pap. 722-G.

Moran, J. H., 1964. *Induction Method and Apparatus for Investigating Earth Formations Utilizing two Quadrature Phase Components of a Detected Signal*, US Patent No. 3,147,429.

Moran, J. and Finklea, E. E., 1962. Theoretical Analysis of Pressure Phenomena Associated with the Wireline Formation Tester, J. Petrol. Techn. 14 (8): 899–908.

Moran, J. H. and Kunz, K. S., 1962. Basic Theory of Induction Logging and Application Theory to the Study of Two-Coil Sondes, *Geophysics*, 27 (6): 829–838.

Nargolwalla, S. S., 1973. *Nuclear Technique for Borehole Logging of Geologic Materials*, Scintrex Limited, Concord, Ontario, 22 pp.

Nargolwalla, S. S., Robertshaw, P. and Hiscott, M., 1976. *Nuclear Metalog® Grade Logging in Lateritic Nickel and Porphyry Copper Deposits*, Application Brief 76-2, Scintrex Limited, Concord, Ontario, 27 pp.

Nargolwalla, S. S. and Seigel, H. O., 1977, In-Situ Mineral Deposit Evaluation with the Scintrex Metalog System, *Canadian Mining Journal*, 128 (3): 75–89.

Neuman, C. H., Sullivan, M. J. and Balanger, D. L., 1999. An Investigation of Density Derived from Pulsed Neutron Capture Measurements, *Soc. Petrol. Engr*. Pap. SPE-56647.

Owen, J. E. and Greer, W. J., 1951. The Guard Electrode Logging System, *Trans. AIME*, 192: 347–356.

Patnode, H. W. and Wyllie, M. R. J., 1950. The Presence of Conductive Solids in Reservoir Rocks as a Factor in Log Interpretation, *Trans. AIME*,189: 47–52.

Pickell, J. J. and Heacock, J. G., 1960. Density Logging, *Geophysics*, 25 (4): 891–904.

Pecor, C. J., 1979, *The Craft of Magic*, Prentice-Hall, Inc., Englewood Cliffs, 265 pp.

Poupon, A., Clavier, C., Dumanoir, J., Gaymard, R. and Misk, A., 1970. Log Analysis of Sand-Shale Sequences–A Systematic Approach, *J. Petrol. Technol.* 22 (7): 867–881.

Poupon, A. and Leveaux, J., 1971. Evaluation of Water Saturation in Shaly Formations, *SPWLA 12th Annual Meeting*, Pap. O.

Prensky, S. R., 1994a. A Survey of Recent Developments and Emerging Technology in Well Logging and Rock Characterization, *The Log Analyst*, 35 (2): 15–45.

Prensky, S. R., 1994b. A Supplement: A Survey of Recent Developments Logging Technology, *The Log Analyst*, 35 (5): 78–84.

Prensky, S. R., 2002. Recent Developments and Emerging Technology in Well Logging and Rock Characterization, *The Log Analyst*, 43 (3): 97–216.

Prensky, S. R., 2007. Recent Advances in Well Logging and Formation Evaluation, *World Oil*, 228 (3): 73–80.

Raymer, L. L., Hunt, E. R. and Gardner, J. S., 1980. An Improved Sonic Transit Time - to-Porosity Transform, *SPWLA 21st Annual Logging Symposium*, Pap. P.

Robinson, J. D., Loren, J. D., Vanjnar, E. A and Hartman, D. E., 1974. Determining Residual Oil With the Nuclear Magnetism Log, *Soc. Petrol. Engr.* Pap. No. 3797-PA.

Rodermund, C. G., Alger, R. P. and Tittman, J., 1961. Logging Empty Holes, *Oil and Gas J*, 59 (24): 119–124.

Ross, D., Alger, R., Bishop, D., Dumanoir, J., Millard, F., Wall, J. T. and Yarbough, D., ed. 1979. *The Art of Ancient Log Analysis, Society of Professional Well Log Analysts*, SPWLA, Houston.

Russell, W. L., 1944. The Total Gamma Ray Activity of Sedimentary Rocks as Indicated by Geiger Counter Determinations, *Geophysics*, 19, (2): 180–216.

Schlumberger, A. G., 1982. *The Schlumberger Adventure*, Arco Publishing, Inc., New York, 152 pp.

Schlumberger, C. and Schlumberger, M., 1929. Electrical Logs and Correlations in Drill Holes, Mining Metallurgy, 10, pp. 515–518.

Schlumberger, C., Schlumberger, M. and Leonardon, E. G., 1932. Electrical Coring - A Method of Determining Bottom Hole data by Electrical Measurements, T. P. 462, *Trans. AIME*, 101.

Schlumberger, C., Schlumberger, M. and Leonardon, E. G., 1933, A New Contribution to Subsurface Studies by Means of Electrical Measurements in Drill Holes, T. P. 503, *Trans. AIME*, 103: 1–18.

Schneider, L. E. and Watt, H. B., 1964. *The Densilog*, Lane-Wells Company, Houston, 24 pp.

Scott, J. H., Carroll, R. D. and Cunningham, R. D., 1967. Dielectric Constant and Electrical Conductivity Measurements of Moist Rock: A New Laboratory Method, *J. Geophy. Res.* 72 (20): 5101–5115.

Seevers, D. O., 1966. A Nuclear Magnetic Method for Determining the Permeability of Sandstones, *SPWLA 7th Annual Logging Symposium*, Pap. L.

Seevers, D. O., 1972. A Method of Determining Permeability by Means of a Pulsed NMR and an Application of the Method to the Study of a Gulf Coast Supernormally Pressured Well, *Soc. Petrol. Engr.* Pap. No. 3847-MS.

Seiler, D., Edmiston, C., Torres, D. and Goetz, J., 1990. Field Performance of a New Borehole Televiewer Tool and Associated Image Processing Techniques, *SPWLA 31st Annual Logging Symposium*, Pap. H.

Sentra, S. D. and Robinson, J. D., 1970. Nuclear Spin-Lattice Relaxation of Liquids Confine in Porous Solids, *Soc. Petrol. Engr.* Pap. No. 4689-PA.

Serra, O., 1989. *Formation MicroScanner Image Interpretation*, Schlumberger Educational Services, Houston, 117 pp.

Simandoux, P., 1963, Measures Dielectriques en Milieu Proeaux, Application a Measure des Saturations en Eau, Etude du Comportement des Massifs Argileaux, *Review de l'institut Français du Petrole*, Suppl. Issue, pp. 193–215.

Starcher, M., Mut, D., Fontanarosa, M., Davis, R. and Presmyk, C., 2002. Next Generation Waterflood Surveillance: Behind Casing Resistivity Measurement Successfully Applied in the 'A3–A6' Waterflood at Elk Hills Field, Kern County, Soc. Petrol. Engr. Pap. No. SPE-76730.

Stroud, S. G. and Schaller, H. E., 1958. New Radiation Log for the Determination of Reservoir Salinity, *Soc. Petrol. Engr.* Pap. No. 1118.G.
Summers, G. C. and Broding, R. A., 1952. Continuous Velocity Logging, *Geophysics*, 17 (3): 598–614.
Swulius, T. M., 1986. Porosity Calibration of Neutron Logs, SACROC Unit, *J. Petrol. Techn. 38* (4): 468–476.
Thomas, E. C., 2007. Personal communication from E. C. Thomas, Bayou Petrophysics, to D. G. Hill.
Tittle, C. W., Faul, H. and Goodman, C., 1951. Neutron Logging of Drill Holes: The Neutron-Neutron Method, *Geophysics*, 16 (4): 626–658.
Tittman, J., Sherman, H., Nagel, W. A. and Alger, R. P., 1965. The Sidewall Epithermal Neutron Porosity Log, *Soc. Petrol. Engr.* Pap. No. 1180.
Tittman, J. and Nelligan, W. B., 1959. Laboratory Studies of a Pulsed Neutron Source Technique in Well Logging, *Soc. Petrol. Engr.* Pap. No. 1227-G.
Tittman, J. and Wahl, J. S., 1965. The Physical Foundations of Formation Density Logging (Gamma-Gamma), *Geophysics*, 30 (2): 284–294.
Van Nostrand, R. G. and Cook, K. L., 1966. Interpretation of Resistivity Data, *US Geological Survey,* Prof. Pap. 499.
Vinegar, H. J. and Waxman, M. H., 1986. Induced Polarization of Shaly Sands, *Geophysics*, 49 (8): 1267–1287.
Vinegar, H. J., Waxman, M. H., Best, M. H. and Reddy, 1986. Induced Polarization Logging: Borehole Modeling, Tool Design and Field Tests, *The Log Analyst*, 27 (2): 25–61.
Vogel, C. B., 1952. A Seismic Velocity Logging Method, *Geophysics*, 17 (3): 586–597.
Von Hippel, A. R., 1954. *Dielectric Materials and Applications,* MIT Press, Cambridge, 438 pp.
Von Hippel, A. R., ed., 1954b. *Dielectrics and Waves*, MIT Press, Cambridge, 284 pp.
Wahl, J. S., Nelligan, W. B., Frentrop, A. H., Johnstone, C. W. and Schwartz, R. J., 1970. The Thermal Neutron Decay Time Log, *Soc. Petrol. Engr. J.* 10 (12): 365–379.
Waxman, M. H. and Smitts, L. J. M., 1968. Electrical Conductivities on Oil Bearing Shaly Sands, *Soc. Petrol. Engr. J.* 8 (6): 107–122.
Waxman, M. H. and Thomas, E. C., 1974. Electrical Conductivities in Oil Bearing Shaly Sands: I. The Relation Between Hydrocarbon Saturation and Resistivity Index; II. The Temperature Coefficient of Electrical Conductivity, *Soc. Petrol. Engr. J.* 14: 213–225.
Wichmann, P. A. and Webb, R. W., 1969. Neutron Activation Logging for Si to Al Ratios, *Soc. Petrol. Engr.* Pap. 2550.
Winchester, S., 2001. *The Map that Changed the World*, HarperCollins Publishers, New York, 329 pp.
Winsauer, W. O. and McCardell, W. M., 1952. Ionic Double-Layer Conductivity in Reservoir Rock, *AIME Trans.* 198: 129–134.
Worthington, A. E., 1973. An Automated Method for the Measurement of Cation Exchange Capacity of Rocks, *Geophysics*, 38 (2): 140–153.
Wylie, M. R. J., 1952, Procedures for the Direct Employment of Neutron Log Data in Electric Log Interpretation, *Geophysics*, 17 (5): 790–805.
Wyllie, M. R. J., Gregory, A. R. and Gardner, G. H. F., 1956. Elastic Wave Velocities in Heterogeneous and Porous Media, *Geophysics*, 21 (1): 41–70.

Wyllie, M. R. J., Gregory, A. R. and Gardner, G. H. F., 1958. An Experimental Investigation of Factors Affecting Elastic Wave Velocities in Porous Media, *Geophysics*, vol. 23, pp. 459–493.
Wyllie, M. R. J. and Southwick, P. K., 1954. An Experimental Investigation of the SP and Resistivity Phenomena in Dirty Sands, *J. Petrol. Techn.* 6 (2): 7–70.
Youmans, A. H., Hopkinson, E. Cl, Bergan, R. A. and Oshry, H. I., 1964. Neutron Lifetime, a New Nuclear Log, *J. Petrol. Techn.* 16 (3): 319–328.
Youmans, A. H., Hopkinson and Stewart, R. M., 1959. (D,T) Neutron Activation Logging, *Soc. Petrol. Techn.* Pap. No. 1304-G.
Zemanek, J., Glenn, E. E., Norton, L. J. and Caldwell, 1970. Formation Evaluation by Inspection with the Borehole Televiewer, *Geophysics*, 35 (2): 254–269.

Web Pages

Baker-Atlas Corporate History: http://www.bakerhughesdirect.com/cgi/atlas/resources/ExternalFileHandler.jsp?bookmarkable=Yes&channelId=4197399&programId=5876143&path=private/ATLAS/public/about/history_index.html
Halliburton Corporate History: http://www.halliburton.com/Default.aspx?navid=338&pageid=713
Schlumberger Corporate History Milestones: http://www.slb.com/content/about/history.asp?
Weatherford Corporate Background: http://www.weatherford.com/weatherford/groups/public/documents/aboutwft/ci_companyinformation.hcsp
Smithsonian National Air and Space Museum Sputnik-1:
http://www.nasm.si.edu/exhibitions/GAL100/sputnik.html
The Holy Grail:
http://en.wikipedia.org/wiki/Holy_Grail
http://www.britannia.com/history/arthur/grail.html
http://www.lib.rochester.edu/camelot/grlmenu.htm

Appendix B
Mechanics of Fluid Flow

B.1 Fundamental Equation of Fluid Statics

The fundamental equation of fluid statics states that pressure increases with depth, the increment per unit length being equal to the weight per unit volume (Binder, 1962, p. 13):

$$dp = -\rho g dz \tag{B.1}$$

where dp is the increment in pressure; dz is the increment in depth (z is a vertical distance measured positively in the direction of decreasing pressure); ρ is the density (mass per unit volume); and g is the gravitational acceleration. The minus sign indicates that pressure decreases with increasing z. The above relationship can be clearly understood on examining Figure B.1, which shows vertical forces on the infinitesimal element in the body of a static fluid. In this figure, dA represent an infinitesimal cross-sectional area, p is the pressure on the top surface of the element and (p + dp) is the pressure on the bottom surface.

Inasmuch as the pressure is due to the fluid weight, the weight of the element (ρg dz dA) is balanced by the force due to pressure difference (dp dA):

$$dp dA = -\rho g dz dA \tag{B.2}$$

or:

$$dp = -\rho g dz$$

In integral form, the above equation can be expressed as follows (see Figure B.1):

$$\int_1^2 \frac{dp}{\rho g} = -\int_1^2 dz = -(z_2 - z_1) \tag{B.3}$$

Mechanics of fluid flow is presented here because of its importance to geophysicists (FPS system of weights and measures was used) Basniev, K. S., Dmitriev, N. M., and Chilingar, G.V., 2011. *Mechanics of Fluid (Gas-Oil-Water) Flow*. Institute of Computer Science, Moscow, 566 pp.

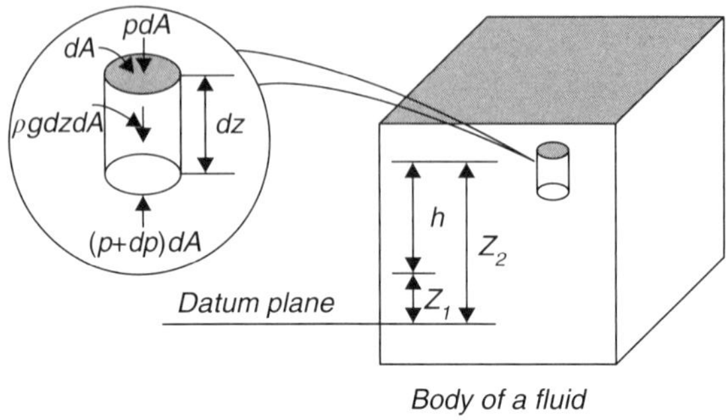

Figure B.1 Schematic diagram of vertical forces on an infinitesimal element in body of any fluid. (Modified after Binder, 1962, Figure 2.2, p. 13)

If ρ is assumed to be constant, eq. B.3 becomes:

$$p_2 - p_1 = -\rho\, g(z_2 - z_1) \tag{B.4}$$

or:

$$\Delta p = \gamma h \tag{B.5}$$

where h is the difference in depth between two points, which is commonly referred to as the "pressure head"; and $\gamma (= \rho g)$ is the specific weight. On expressing γ in lb/cu ft and h in ft, pressure difference Δp is found in lb/sq ft.

B.2 Buoyancy

When a body is completely or partly immersed in a static fluid, there is an upward vertical buoyant force on this body equal in magnitude to the weight of displaced fluid. This force is a resultant of all forces acting on the body by the fluid. The pressure is greater on the parts of the body more deeply immersed. The pressures at different points on the immersed body are independent of the body material, For example, if the same fluid is substituted for the immersed body, this fluid will remain at rest. This means that the buoyant, upward force on the substituted fluid is equal to its weight.

If the immersed body is in static equilibrium, the buoyant force and the weight of the body are equal in magnitude and opposite in direction, passing through the center of gravity of the body. For a comprehensive treatment of fluid statics, the reader is referred to an excellent book on fluid mechanics by Binder (1962).

B.3 General Energy Equation

The general energy equation can be summarized as follows:

$$\left\{\begin{array}{l}\text{The heat added}\\ \text{to unit weight of}\\ \text{the flowing fluid}\\ \text{between entrance}\\ \text{and exit.}\end{array}\right\} + \left\{\begin{array}{l}\text{The work transferred}\\ \text{to (done upon) unit}\\ \text{weight of the flowing}\\ \text{fluid between entrance}\\ \text{(1) and exit (2).}\end{array}\right\} = \left\{\begin{array}{l}\text{The total gain in}\\ \text{energy by unit}\\ \text{weight of the fluid}\\ \text{between entrance}\\ \text{and exit.}\end{array}\right\}$$

$$q + \frac{p_1 v_1}{778} - \frac{p_2 v_2}{778} + \frac{W}{778} = u_2 - u_1 + \frac{V_2^2 - V_1^2}{2g(778)} + \frac{Z_2 - Z_1}{778} \tag{B.6}$$

where P = pressure in psia; v = specific volume in ft^3/lb; V = velocity in ft/sec; Z = potential head in ft; q = heat transferred to fluid; $p_1 v_1/778$ = external work in pushing 1 lb of fluid across the entrance; W = work in ft-lb per lb fluid flowing; $u_2 - u_1$ = gain in internal energy; $= \dfrac{V_2^2 - V_1^2}{2g(778)}$ gain in kinetic energy; and $\dfrac{Z_2 - Z_1}{778}$ = gain in potential energy. Point 1 = entrance; point 2 = exit; 1 Btu = 778 ft-lb; $u_2 - u_1 = c_v (T_2 - T_1)$; c_v = specific heat at constant volume.

Inasmuch as enthalpy = $h = u + (p\,v)/778$, eq. B.6 becomes:

$$q + \frac{W}{778} = h_2 - h_1 + \frac{V_2^2 - V_1^2}{2g(778)} + \frac{Z_2 - Z_1}{778} \tag{B.7}$$

where $h_2 - h_1 = c_p (T_2 - T_1)$; c_p = specific heat at constant pressure.

For a number of cases, the process is adiabatic and change in internal energy is negligible. Thus:

$$\frac{p_1 v_1}{778} - \frac{p_2 v_2}{778} + \frac{W}{778} = \frac{V_2^2 - V_1^2}{2g(778)} + \frac{Z_2 - Z_1}{778} \tag{B.8}$$

and each term in the latter equation is in Btu/lb fluid flowing. On multiplying through by 778:

$$W = \left[\frac{p_2}{\gamma_2} + \frac{V_2^2}{2g} + Z_2\right] - \left[\frac{p_1}{\gamma_1} + \frac{V_1^2}{2g} + Z_1\right] \tag{B.9}$$

where γ = specific weight in lb/ft^3$(1/v)$; p/γ = pressure head in ft; $V^2/2g$ = velocity head in ft; and Z = potential head in ft.

For frictionless incompressible fluid with no work done:

$$\left[\frac{p_2}{\gamma_2}+\frac{V_2^2}{2g}+Z_2\right]=\left[\frac{p_1}{\gamma_1}+\frac{V_1^2}{2g}+Z_1\right]=\text{const.} \tag{B.10}$$

which is the well-known Bernoulli's equation.

B.4 Derivation of Formula for Flow Through Orifice Meter

A schematic diagram of incompressible fluid flow through an orifice meter is presented in Figure B.2. For an ideal flow with no friction losses the following relation will hold true:

$$\left[\frac{p_2}{\gamma_2}+\frac{V_2^2}{2g}+Z_2\right]=\left[\frac{p_1}{\gamma_1}+\frac{V_1^2}{2g}+Z_1\right] \tag{B.11}$$

where V = velocity in ft/sec; p = pressure in psia; r= specific weight in lb/cu ft; and Z = potential head above any datum plane in ft.

Inasmuch as volumetric rate of flow (in cu ft/sec) $Q = V_1A_1 = V_2A_2$:

$$V_1=\frac{V_2A_2}{A_1} \tag{B.12}$$

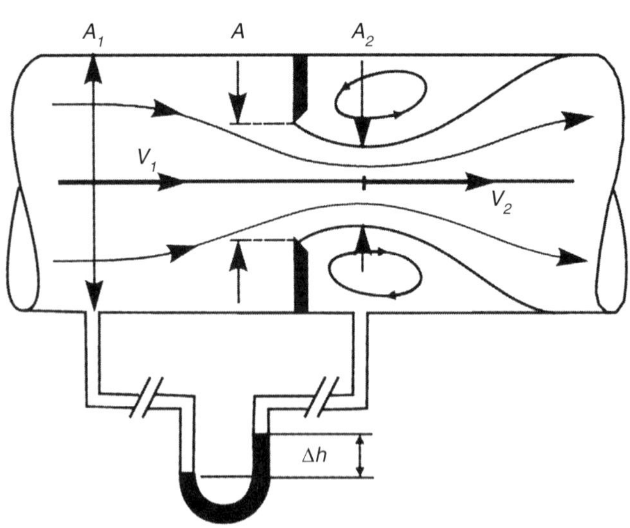

Figure B.2 Schematic diagram of an orifice meter.

Substituting eq. B.12 in eq. B.11 and solving for V_2:

$$V_2 = \left[\frac{2g(\frac{p_1}{\gamma} - \frac{p_2}{\gamma} + Z_1 - Z_2)}{1 - (\frac{A_2}{A_1})^2} \right]^{1/2} \tag{B.13}$$

For an actual flow one has to introduce correction factor for velocity (C_v) and correction factor for area (C_c), The latter is termed coefficient of contraction and is equal to A_2/A_1. Thus:

$$Q = C_v C_c V_2 A \tag{B.14}$$

The term discharge coefficient (C or C_d) often is substituted for $C_v C_c$. Another term flow coefficient (K) is defined as:

$$K = \frac{C}{[1 - (\frac{A_2}{A_1})^2]^{1/2}} \tag{B.15}$$

Thus:

$$\text{actual } Q = KA[2g(\frac{p_1}{\gamma} - \frac{p_2}{\gamma} + Z_1 - Z_2)]^{1/2} \tag{B.16}$$

If Δh = is manometer deflection in inches. of Hg, then;

$$\frac{p_1}{\gamma} + Z_1 - \frac{p_2}{\gamma} - Z_2 = \frac{\Delta h}{12} \frac{(sp\ gr_{Hg} - sp\ gr_f)}{sp\ gr_f} \tag{B.17}$$

where $sp\ gr_f$ = specific gravity of fluid flowing.

Flow equation for the Venturi meter (Figure B.3) can be derived similarly; however, $C_c = 1$ in this case.

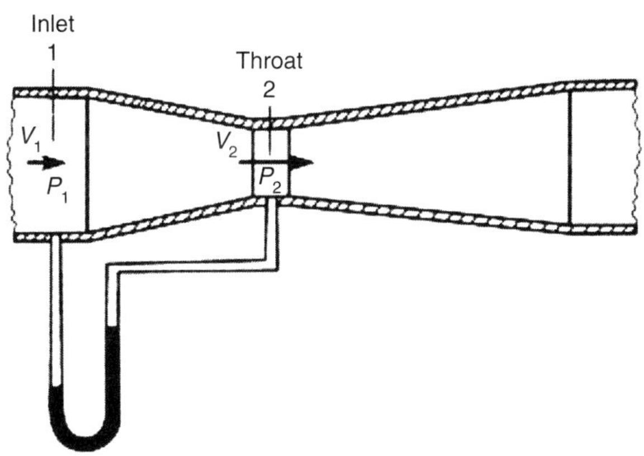

Figure B.3 Schematic diagram of a venturi meter.

B.5 Compressible Flow Formula

For a compressible flow, one can derive the following equation starting with the general energy equation (also see Binder, 1962):

$$G = CA_2\gamma_2 \left[\frac{\left[(2gk)(k-1)\right]\left(\frac{p_1}{\gamma_1}\right)\left[\left(1-\frac{p_2}{p_1}\right)^{\frac{k-1}{k}}\right]}{1-\left(\frac{A_2}{A_1}\right)^2\left(\frac{p_2}{p_1}\right)^{2/k}} \right]^{1/2} \tag{B.18}$$

where G = weight rate of flow in lb/sec, and k = (specific heat at constant pressure)/(specific heat at constant volume) = c_p/c_v. As shown in Nelson (1958, p. 211), k can be obtained for various hydrocarbons.

Example Problem B.1. Maximum reliable flow

Two reservoirs shown below are connected by a 4-in. 10,000-ft long pipe having friction factor of 0.02. Determine: (1) pump horsepower required to maintain a flow rate of 0.33 cu ft/sec of water (Y= 62.4 lb/cu ft); and (2) the maximum distance x for dependable (reliable) flow.

Solution:

1. One can use Bernoulli's equation between points 1 and 3:

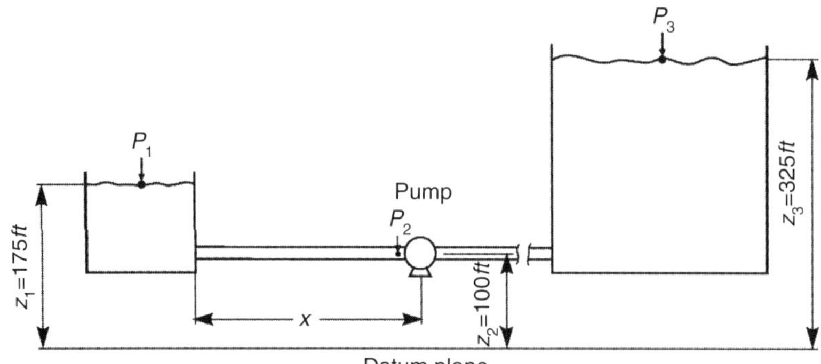

Figure B.4 Diagram for example problem B.1.

$$\left[\frac{p_1}{\gamma}+\frac{V_1^2}{2g}+Z_1+E_p\right]=\left[\frac{p_3}{\gamma}+\frac{V_3^2}{2g}+Z_3+h_{lf}+h_{le}+h_{lx}\right]$$

where E_p = energy output of the pump, ft-lb/lb of fluid flowing; h_{lf} = head loss due to friction = $f\ (l/d)(V_p{}^2/2g)$, ft-lb/lb: h_{le}= head loss due to entrance = $0.5V_p{}^2/2g$ in the case of sharp entrance. ft-lb/lb; h_{lx} = head loss due to the exit = dissipated kinetic energy ($=V_p{}^2/2g$); V_p = velocity in the pipe, ft/sec: d = inside diameter of the pipe, ft: l = length of the pipe, ft: γ = specific weight of the flowing fluid. lb/cu ft; g = gravitational acceleration, ft/sec^2: and; Z = elevation above some datum plane, ft.

Inasmuch as velocities at the surface of two reservoirs (V_1 and V_2) can be considered negligible and pressures p_1 and p_3 are atmospheric (0 gage), the above equation reduces to:

$$E_p=\left(Z_3-Z_1\right)+h_{lf}+h_{ie}+h_{lx}=\left(Z_3-Z_1\right)+\frac{V_p^2}{2g}(f\frac{l}{d}+0.5+1)$$

Inasmuch as:

$$V_p=\frac{Q}{A}=0.33\text{cu ft / sec/}\left(\pi\frac{\left(\frac{4}{12}\right)^2}{4}\right)=3.78\text{ft / sec,}$$

$$E_p = (325 - 175) + 3.78^{2/}64.4[(0.02)(10000)/(4/12) + 0.5 + 1.0)] = 285 \text{ ft-lb/lb.}$$

Thus, horsepower of the pump is equal to:

$$\text{HP} = Q\gamma E_p/550 = (0.33)(62.4)(285)/550 = 10.6, \text{ where } 550 \text{ ft-lb/sec} = 1 \text{ HP}$$

2. For maximum and yet reliable flow of water (i. e., no cavitation), the pressure at the inlet side of the pump (p_2) should be 2/3 of the barometric head of water. With safety factor incorporated, it is equal to –21 ft of water ($= p_2/\gamma$). Thus, using Bernoulli's equation between points 1 and 2:

$$\left[\frac{p_1}{\gamma}+\frac{V_1^2}{2g}+Z_1\right]=\left[\frac{p_2}{\gamma}+\frac{V_2^2}{2g}+Z_2+h_{lf}{}'+h_{le}\right],$$

one can solve for unknown distance x, inasmuch as terms p_1/γ and $V_1^2/2g$ can be neglected. Thus: $175 = -21 + (3.78)^2/64.4 + 100 + 0.02[(x)/(4/12)]\ (3.78^2/64.4) + 0.5(3.\ 78)^2/64.4$ and solving for x: $x = 7180$ ft.

Example Problem B.2. Compressible flow (nozzle)

A convergent-divergent nozzle is connected to a tank with air, having pressure of 100 psia and temperature of 100 °F. The tip diameter (point 3) is equal to two inches, and air discharges to atmosphere (p'_3 = 14.7 psi and T_3 = 60°F). Determine throat diameter (point 2) necessary to maintain maximum flow rate through this nozzle. Adiabatic constant k tor air is equal to 1.4. (See Figure B.5.)

Solution:

For maximum flow rate, the velocity in the throat must be sonic, because maximum velocity attainable in a convergent nozzle is sonic. Inasmuch as p_3/p_1 (=14.7/100= 0.147) is less than $(p_2/p_1)_{critical} = \{2/(k+1)\}^{k/k-1} = \{2/(1.4+1)\}^{1.4/1.4-1} = 0.528$), velocity V_3 in the divergent passage will be supersonic.

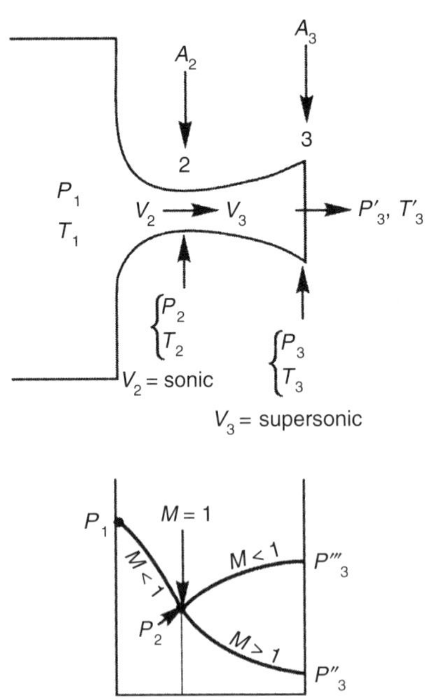

Figure B.5 Diagram for example problem B.2.

To attain sonic velocity in the throat (point 2), pressure p_2 must be critical:

$$p_2 = p_1\,(2/(k+1))^{\,k/k-1} = 100 * 0.528 = 52.8 \text{ psia.}$$

The specific weight of air in the tank, γ_1 is equal to:

$$\gamma_1 = \frac{p_1}{RT_1} = \frac{100 * 144}{53.3 * 560} = 0.483\,\text{lb}/\text{ft}^3,$$

where the gas constant for air, R, is equal to 53.3 and T_1 is the absolute temperature in °R (= °F + 460).

Inasmuch as $\frac{\gamma_1}{\gamma_2} = (p_2/p_1)^{1/k}$, $\gamma_2 = (0.528)^{1/1.4} * 0.483 = 0.3$ lb/cu ft.

In order to attain sonic velocity in the throat, temperature in the throat must be critical:

$$T_2 = T_1\,(2/k+1) = 560(2/2.4) = 466\ °\text{R}.$$

Thus, velocity V_2 is equal to V_c: $V_2 = V_c = c_2 = (kgRT_2)^{1/2} = (1.4 * 32.2 * 53.3 * 466)^{1/2} = 1060$ ft/sec.

Temperature at point 3 can be determined from the following equation[1]: $T_3 = T_1\,(p_3/p_1)^{k-1/k} = 560\,(0.147)^{0.4/1.4} = 320$ °R and γ_3 is equal to: $\gamma_3 = p_3/RT_3 = 14.7 * 144/53.3 * 320 = 0.12$ lb/cu. ft.

Velocity at point 3 can be determined on using the following equation:

$$V_3^2/2g = (p_1/\gamma_1)\,(k/k-1)\,\{1 - (p_2/p_1)^{(k-1/k)}\} \qquad \text{(B.19)}$$

$$V_3^2 = 64.4 * (100 * 144/0.483)\,(1.4/0.4)\{1 - (0.147)^{0.4/1.4}\}$$

Solving for V_3:

$$V_3 = 1700 \text{ ft/sec, } \textit{i.e.}, \text{ supersonic speed.}$$

Inasmuch as for adiabatic flow the weight rate of flow in the throat (W_2) is equal to the weight rate of flow at the exit (W_3):

$$W_2 = A_2V_2\gamma_1 = W_3 = A_3V_3\gamma_3 \qquad \text{(B.20)}$$

[1] $p_1v_1 = RT_1$; $p_3v_3 = RT_3$; $v_1 = 1/\gamma_1$; $v_3 = 1/\gamma_3$. Thus: $T_1 = p_1/\gamma_1R$ and $T_3 = p_3/\gamma_3R$. Dividing T_3 by T_1: $T_3/T_1 = (p_3/p_1)(\gamma_1/\gamma_3)$, and inasmuch as $\gamma_3/\gamma_1 = (p_3/p_1)^{1/k}$, $T_3/T_1 = (p_3/p_1)(p_3/p_1)^{-1/k} = (p_3/p_1)^{k-1/k}$

$\left(\pi d_2^2 / 4 * 144\right)(1060)(0.3) = \left(\pi 2^2 / 4 * 144\right)(1700)(0.12)$, one can solve for throat diameter d_2: $d_2 = 1.161$ inches.

Example problem B.3: Compressor problem

Air at standard conditions is handled at a rate of 1000 lb/hr by a compressor. Cross-sectional area of inlet is 0.6 ft^2 and that of outlet is 0.11 ft^2. Air is compressed to 100 psia and 180 °F, and the heat taken from air is 50,000 Btu/hr; $c_p = 0.239$. If the change in elevation is negligible, what is the work done on the air?

Solution:

$$q = \frac{-50{,}000\,\text{Btu/hr}}{1000\,\text{lb/hr}} = -500\ \text{Btu/hr}$$

Weight rate of flow:

$$G = AV\gamma\ \text{lb/sec}$$

where A = cross-sectional; area in ft^2; V = velocity in ft/sec; γ = specific weight in lb/cu ft.

$$G = A_1 V_1 \gamma_1 = A_2 V_2 \gamma_2 \tag{B.21}$$

$$V_1 = \frac{G}{A_1\gamma_1} = \frac{1000/3600}{(0.60)*(0.07651)} = 6.06\ \text{ft/s}$$

where 0.07651 is the specific weight of standard sea-level air (59 °F and 14.7 psia).

$$\gamma_2 = \frac{p_2}{RT_2} = \frac{100(144)}{(53.3)(640)} = 0.421\ \text{lb/ft}^3$$

$$V_2 = \frac{G}{A_2\gamma_2} = \frac{1000/3600}{(0.11)*(0.421)} = 5.97\ \text{ft/s}$$

$$q + \frac{W}{778} = h_2 - h_1 + \frac{V_2^2 - V_1^2}{2gJ}$$

$$h_2 - h_1 = 0.239(180 - 59) = 29 \text{ Btu / lb}$$

$$\frac{W}{778} = 50 + 29 + \frac{(5.97)^2 - (6.06)^2}{(64,4) * (778)}$$

$$W = 61{,}600 \text{ ft lb/lb}$$

If the answer is desired in HP then one has to use the following equation:

$$\text{HP} = \frac{(W \text{ ft} - \text{lb / lb}) * (G \text{ lb / sec})}{(550 \text{ ft} - \text{lb / sec}) / \text{HP}} \tag{B.22}$$

Example Problem B.4. Friction Losses in Circular Pipes

Calculate the pressure drop in benzene-flowing 200 feet commercial steel pipe 6 inches in diameter. Given: temperature = 50 °F; sp. gr. = 0.90; velocity =11.0 ft/sec; absolute roughness of pipe = 0.00015 ft; dynamic viscosity of benzene = 1.6 * 10^{-5} slugs/ft-sec; and g = gravitational acceleration = 32.2 ft/sec/sec.

Solution:

The Reynolds Number is equal to:

$$\text{Re} = \frac{Vd\rho}{\mu} = 11.0 * \left(\frac{6}{12}\right) * \left(\frac{0.90 * 62.4}{32.2}\right) / 1.6 * 10^{-5} = 6 * 10^{-5}$$

Using Figure B.6,

$$\frac{e}{D} = 0.0003 - 0.0004.$$

Thus, from Figure B.7, the friction factor $f = 0.016$.
The head loss, h, is equal to:

$$h = f \frac{l}{d} \frac{V^2}{2g} = 0.016 \left(\frac{200}{0.5}\right) \left(\frac{11.0^2}{2 * 32.2}\right) = 12.02 \text{ ft or ft lb / lb}$$

Pressure drop $\Delta p = \gamma h = 0.90(62.4) \dfrac{12.02}{144} = 4.69 \text{ psi}$.

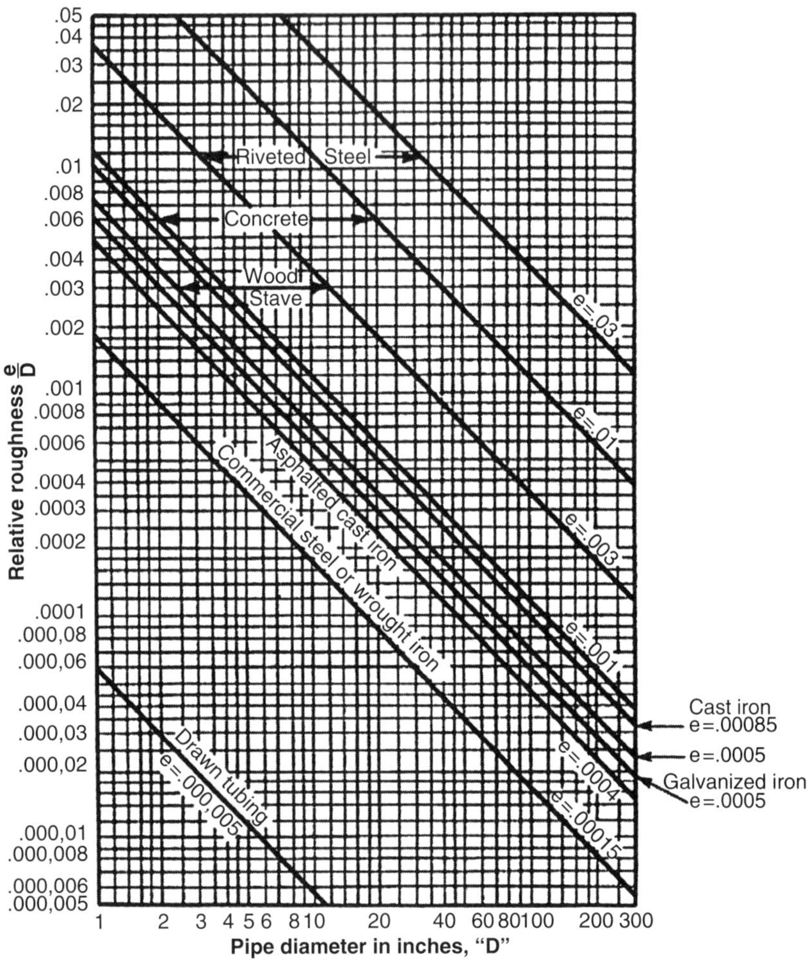

Figure B.6 Chart for determining relative roughness of pipe. (After Moody, 1944.)

B.6 Farshad's Surface Roughness Values and Relative-Roughness Equations

Surface roughness for various recently-developed pipes is presented in Table B.1. The surface roughness of internally plastic-coated pipes is the lowest compared to the other pipes in this group. The bare Cr 13 pipe exhibits the highest average surface roughness value.

A relative-roughness chart (Figure B.8) was developed for (1) internally plastic- coated, (2) honed-bare carbon steel, (3) electro-polished bare Cr 13, (4) cement- lining, (5) bare carbon steel,

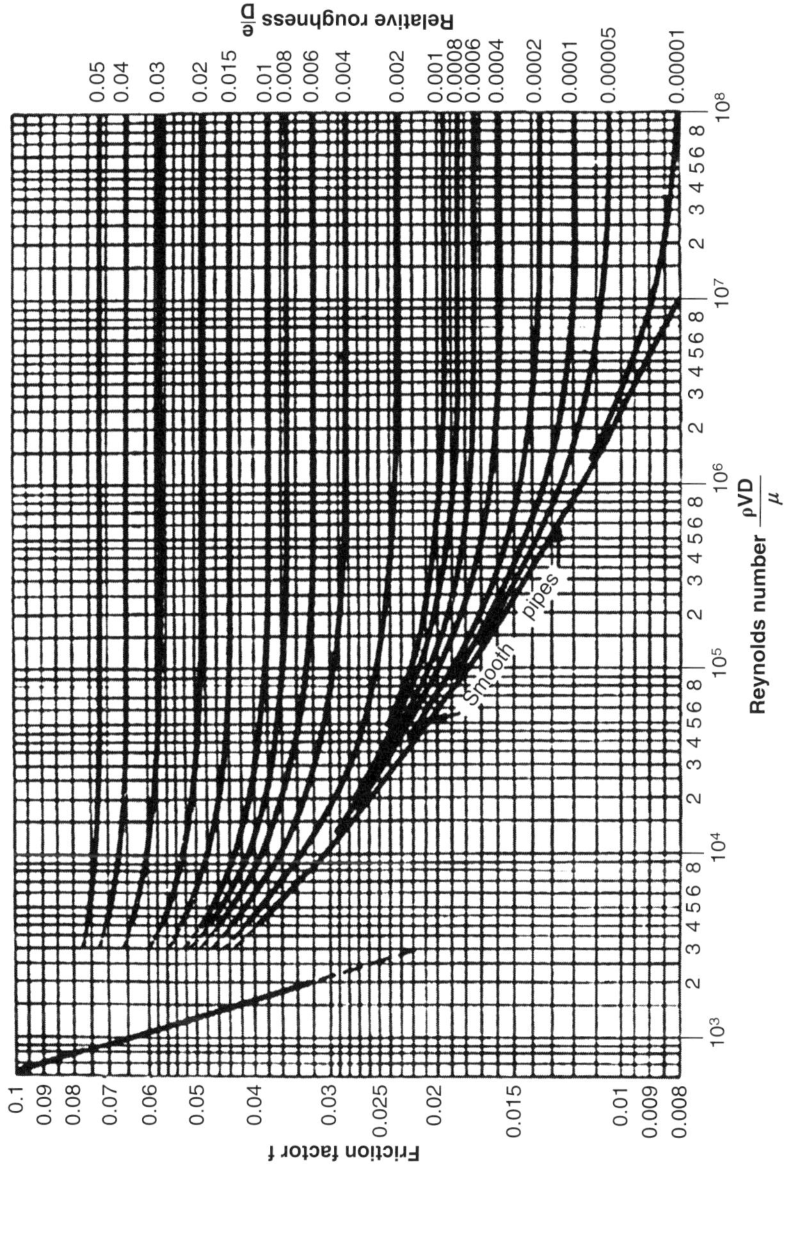

Figure B.7. Friction factors for flow in circular pipes. (After Moody, 1944).

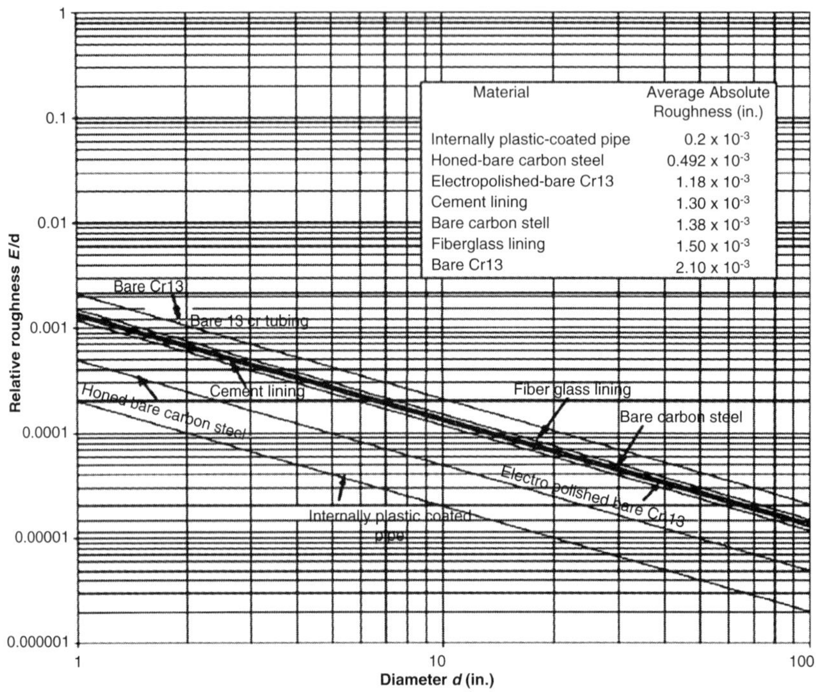

Figure B.8 Farshad's surface roughness chart for modem pipes.

Table B.1 Farshad's surface roughness values for modern pipes.

Material	**Average Measured Absolute Roughness (inches * 10^{-3})**	**Average Measured Absolute Roughness (micrometers)**
Internally plastic coated	0.2	5
Honed bare carbon steel	0.492	12.5
Elcetro-polished bare Cr 13	1.18	30
Cement lining	1.3	33
Bare carbon steel	1.38	36
Fiber glass lining	1.5	38
Bare Cr 13	2.1	55

(6) fiberglass lining, and (7) bare Cr 13 pipes. The relative roughness e/d (dimensionless) is related to the absolute roughness e (in inches) and pipe diameter d (in inches).

Table B.2 Farshad's relative roughness *(eld)* equations for modem pipes.

Material	Equation (Diameter, in Inches)
Internally plastic coated	$e/d = 0.0002\ d^{-1.0098}$
Honed bare carbon steel	$e/d = 0.0005\ d^{-1.0101}$
Elcetro-polished bare Cr 13	$e/d = 0.0012\ d^{-1.0089}$
Cement lining	$e/d = 0.0014\ d^{-1.0105}$
Bare carbon steel	$e/d = 0.0014\ d^{-1.0112}$
Fiber glass lining	$e/d = 0.0016\ d^{-1.0086}$
Bare Cr 13	$e/d = 0.0021\ d^{-1.0055}$

A set of nonlinear mathematical models is offered to describe the log/log relationship between the average relative roughness and pipe diameter for various modem pipes (Table B.2).

B.7 Flow Through Fractures

The writers have placed strong emphasis on the importance of fractures in carbonate reservoirs. It has been shown in the geological and engineering literature that fractures can constitute the most important heterogeneity affecting production. Craze (1950) cited carbonate reservoirs in Texas, U.S.A., which have low matrix permeabilities, that produce moveable oil from fractures and vugs. Also, Daniel (1954) discussed the influence of fractures on oil production from carbonate reservoirs of low matrix permeability in the Middle East. Reservoirs are not mechanically continuous owing to the presence of fractures. In this sense, the reservoir rock is a discontinuum rather than a continuum. The nature and spatial relationship of discontinuities, such as fractures, dissolution channels, and conductive stylolites that affect fluid flow in carbonate rocks are best evaluated using large-core analysis.

Geological conditions which create fractures and control fracture spacing in rocks include: (1) variations in lithology; (2) physical and mechanical properties of the rocks and fluids in the pores; (3) thickness of beds; (4) depth of burial; (5) orientation of the earth's stress field: (6) amount of differential stress (tectonic forces); (7) temperature at depth; (8) existing mechanical discontinuities; (9) rate of overburden loading or unloading; (10) gravitational compaction (rock or sediment volume reduction as a result of water loss during compaction); (11) anisotropy; and (12) continuum state at depth (competent versus incompetent character of the rocks).

B.8 Permeability of a Fracture-Matrix System

One is interested in the total permeability of the fracture-matrix system rather than the permeability contributions of its various parts. The studies of Huitt (1956) and Parsons (1966) provided the following two equations for determining permeability values in a horizontal direction (k_H) through an idealized fracture-matrix system (using English units):

$$k_H = k_m + 5.446 * 10^{10}\, w^3 \cos^2(a/L) \tag{B.23}$$

where k_m is the matrix permeability (mD); w is the fracture width (in.); L is the length of the fracture; and α is the angle of deviation of the fracture from the horizontal plane in degrees. If w and L are expressed in mm, then Eq. B.23 becomes:

$$k_H = k_m + 8.44 * 10^{7}\, w^3 \cos^2(a/L) \tag{B.24}$$

Various mathematical models have been proposed to describe the velocity of a fluid in a fracture, to estimate tank oil-in-place in fractured reservoirs, to determine the fracture porosity, and to calculate average "height" of fractures (Chilingarian *et al.*, 1992).

B.9 Fluid Flow in Deformable Rock Fractures

Witherspoon *et al.*, (1980) proposed a model analyzing fluid flow in deformable rock fractures. This study has ramifications with respect to the migration and production of subsurface fluids. The withdrawal of fluids from carbonate rocks can cause a fracture to close due to induced compaction of the reservoir.

The above proposed model consists of a single-phase fluid flowing between smooth parallel plates. The pressure drop is proportional to the cube of the distance between plates (w = width or aperture of a fracture). For laminar flow (Witherspoon *et al.*, 1980):

$$q = 5.11 * 10^6 \left[w^3 \Delta p \left[\frac{a}{L\mu} \right] \right] \tag{B.25}$$

where q is the volumetric rate of flow (bbl/D); w is the width (or aperture) of a fracture (in.); Δp is the pressure drop (psi); a is the width of the fracture face (ft); L is the length of the fracture (ft); and μ is the viscosity of the fluid (cP).

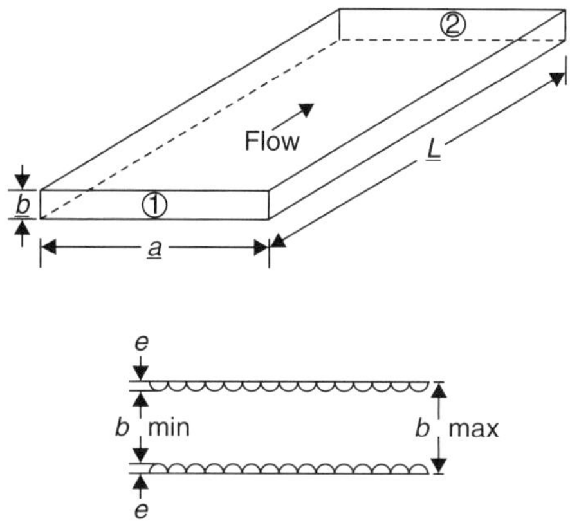

Figure B.9 Simple fracture-fluid flow model showing the length of the fracture, *L*; width, *a*; thickness, *b*; and the absolute roughness, *e*.

But natural fractures are rarely smooth and, therefore, head loss owing to friction, h_{Lf}; is equal to:

$$h_{Lf} = f\left[\frac{LV^2}{d_e 2g}\right] \tag{B.26}$$

where f is the friction factor, which is a function of the Reynolds Number, N_{Re}, and relative roughness that is equal to the absolute roughness, e, divided by the width (height or aperture) of the fracture, w (or b) (Figure B.9). The Reynolds Number is equal to $Vd_e\rho/\mu$, where V is the velocity of flowing fluid (ft/sec); d_e is the equivalent diameter (ft); ρ is the mass per unit volume, *i. e.*, specific weight, γ in lb/ft^3 divided by the gravitational acceleration, g, in ft/sec/sec (= 32.2). Effective diameter, d_e, is equal to hydraulic radius, R_h times four (R_h = area of flow/wetted perimeter).

Lomize (1951) and Louis (1969) studied the effect of absolute and relative roughness on flow through induced fractures, sawed surfaces and fabricated surfaces (e. g., by gluing quartz sand onto smooth plates). They found that results deviate from the classical cubic law at small fracture widths. Jones *et al.*, (1988) studied single-phase flow through open-rough natural fractures. They found that N_{Rec} (critical Reynolds Number where laminar flow ends) decreases with decreasing fracture width (b or w) for such fractures.

Jones *et al.* (1988) suggested the following equations for open, rough fractures with single-phase flow:

$$q = 5.06 * 10^4 a[\Delta\, pw^3 / f\, L\, \rho]^{0.5} \tag{B.27}$$

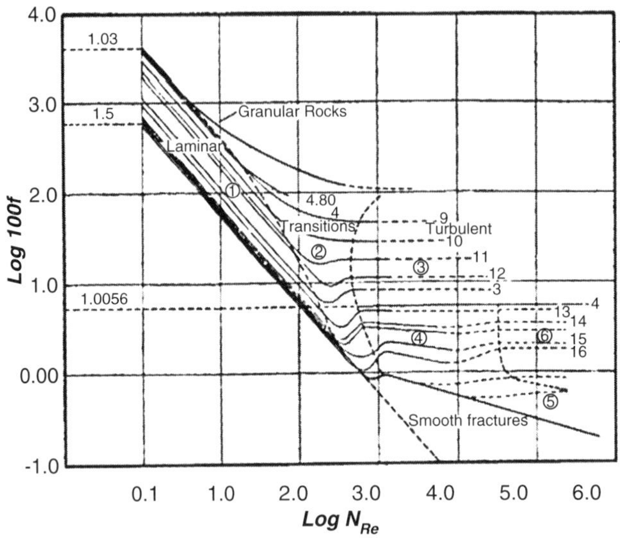

No.	e, cm	e/b
3	0.055	0.327
4	0.055	0.205
9	0.175	0.854
10	0.175	0.687
11	0.175	0.574
12	0.175	0.432
13	0.055	0.150
14	0.055	0.120
15	0.055	0.069
16	0.055	0.054

Figure B.10 Chart showing the relation between friction factor, *f*. and Reynolds number, N_{Re}, for laminar, transitional and turbulent fluid flow in granular rocks and smooth fractures. (Modified after Lomize, 1951).

and

$$k = 5.39 * 10^5 \mu \, [wL / f \Delta \, p \, \rho]^{0.5} \tag{B.28}$$

where *k* is the permeability in darcys; ρ is the density of the fluid (lb/ft^3); and *f* is the friction factor, which is dimensionless.

Based on experimental data, Lomize (1951) developed many equations relating friction factor (*f*) and Reynolds Number (N_{Re}) for both laminar and turbulent flows. He also prepared elaborate graphs relating friction factor, Reynolds Number, and relative roughness of fractures (*e*/*b* or *e*/*w*) (Figure B.10).

Lomize (1951) found that at the relative roughness (*e*/*b*) of less than 0.065, fractures behave as smooth ones (*e*/*b* = 0) and friction factor (*f*) is equal to:

$$f = 6/N_{Re} \tag{B.29}$$

In the turbulent zone, with *e*/*b* varying from 0.04 to 0.24 and $N_{Re} <$ 4000–5000, friction factor is equal to:

$$f = B/(N_{Re})^a \tag{B.30}$$

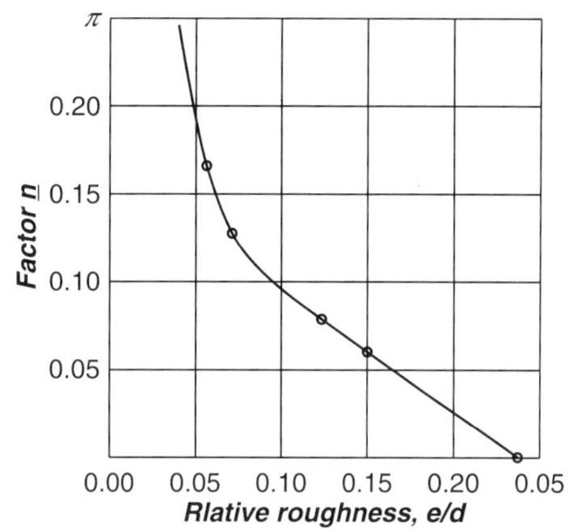

Figure B.11 Chart showing the relation between coefficient, n, and the relative roughness, e/b. where the coefficient $B = 0.056$. (Modified after Lornize, 1951).

Coefficient B is equal to 0.0056 and n can be found from Figure B.11 or by using the following equation:

$$n = 0.163 - [0.684\ (e/b)] + [2.71/e^{76.5\ (e/b)}] \tag{B.31}$$

The following example illustrates how to use the discussed equations and graphs. and the significance of the results.

Example problem B.5: Effect of fractures on total permeability

If $w = 0.005$ in., $L = \mathrm{I}$ in., $a = 0°$, and $k_m = 1$ mD, then using eq. B.23. $k_H = 6{,}800$ mD.

This example shows the overwhelming contribution which relatively small fracture can exert on total permeability.

Example problem B.6: Pressure drop in a vertical fracture

Determine the pressure drop in psi in a vertical fracture (flow is in upward direction) given the following information: absolute roughness, $e = 0.065$ mm; fracture width (w) or height (b) = 0.68 mm; width of fracture face, $a = 5$ mm ($a > b$); length of fracture, $L = 5$ cm; volumetric rate of flow, $q = 1\ \mathrm{cm^3/sec}$; specific gravity of flowing oil (sp. gr.) = 0.8; and Reynolds Number ($N_{Re} = 4000$) (see Figure B.9).

Using Bernoulli's Equation for flow from point 1 to point 2:

$$\frac{p_1}{\gamma} + \frac{V_1^2}{2g} + z_1 = \frac{p_2}{\gamma} + \frac{V_2^2}{2g} + z_2 + h_{lf} \quad \text{(B.32)}$$

and

$$\frac{p_1}{\gamma} - \frac{p_2}{\gamma} = \frac{p}{\gamma} = (z_2 - z_1) + h_{lf} = L + h_{lf} \quad \text{(B.33)}$$

where p_1 and p_2 are pressures at points 1 and 2, respectively, in lb/ft^2 absolute; V = velocity of flowing fluid in ft/sec; z_1 and z_2 = potential heads at points 1 and 2 in ft; g = gravitational acceleration, ft/sec/sec (=32,2): h_{lf}= head loss due to friction in ft. All terms in the above equation are in ft-lb per lb of fluid flowing or in ft.

$$\begin{aligned} q = 1\ \text{cm}^3/\text{sec} &= 1\ \text{cm}^3/\text{sec} * 3.531 * 10^{-5}\ (\text{ft}^3/\text{cm}^3) \\ &= 3.531 * 10^{-5}\ \text{ft}^3/\text{sec} \end{aligned}$$

A (cross-sectional area of flow) = $a * b$ = 5 * 0.68 mm * (1.07639 * 10^{-5} ft^2/mm^2) = 3.6597 * 10^{-5} ft^2.

Thus:

$V = q/A = 3.531 * 10^{-5}/3.6597 * 10^{-5} = 0.965$ ft/sec

Hydraulic radius R = (flow area)/(wetted perimeter) = $(a * b)/(2a + 2b) = 9.814 * 10^{-4}$ ft

Equivalent diameter = d_e= 4R = $2ab/(a + b) = 3.9277 * 10^{-3}$ ft

Inasmuch as N_{Re} is 4000 and relative roughness, $e/b = 0.065/0.68 = 0.095$. one can use eq. B.31 (and Figure B.11 to determine n):

$$f = B/(N_{Re})^n = 0.056/(4000)^{0.12} = 0.0207$$

Thus:

$$\begin{aligned} h_{lf} = f\,(L/d_e)\,(V^2/2g) &= 0.0207(0.164/3.93 * 10^{-3}) \\ &[(0.965)^2/(2 * 32.2)] = 0.0197\ \text{ft} \end{aligned}$$

and

$$\Delta p = \gamma(L + h_{lf}) = [(0.8 * 62.4)(0.164 + 0.0197)] : 144 = 0.062\ \text{psi}$$

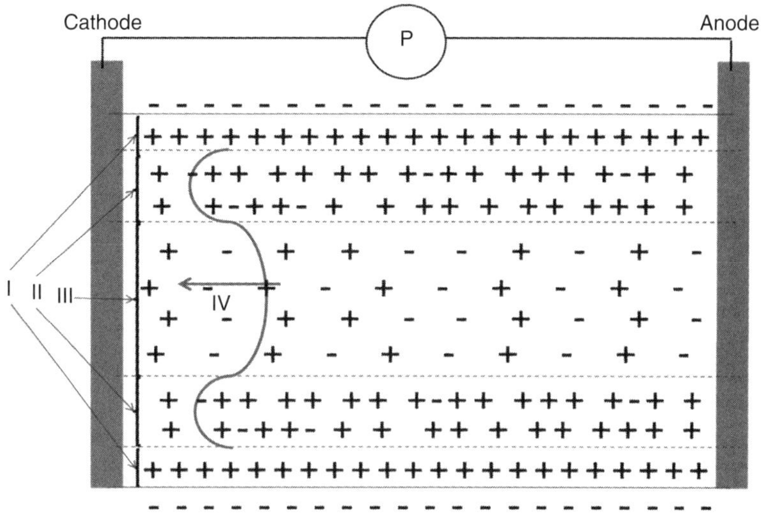

Figure B.12 Schematic diagram of electrokinetic double layer (I: Immobile Double Layer, II: Mobile Double layer, III: Free water, IV: Velocity Profile) as envisioned by Dr. George V. Chilingar (personal communication). Solid curved line – velocity profile in a capillary. P = D.C. current power supply.

B.10 Electrokinetic Flow

Inasmuch as EEOR (Electric Enhanced Oil Recovery) technology is gaining popularity (Wittle *et al.*, 2008, 2010), it is important to discuss here the electrokinetic flow mechanism. This technology is based on double-layer theory as illustrated in Figure B.12, where the negatively-charged surface of the clays attracts the positive ions of aqueous medium, forming the immobile double layer. This immobile double layer is followed by a thick mobile layer with a predominance of positively-charged ions (cations), with a few negatively-charged ions (anions).

Upon application of direct electric current, the mobile double-layer moves towards the cathode, dragging free water with it. A schematic diagram of EEOR field set-up, as used by Wittle *et al.*, (2008, 2011) in Canada and California oilfields, is presented in Figure B.13.

In case where the imposed electrical potential gradient, E, is in the same direction as the pressure drop (Δp).

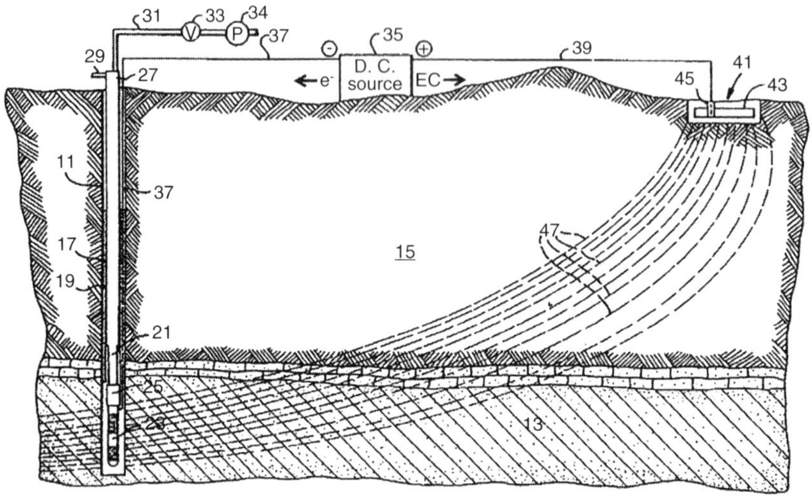

Figure B.13 Schematic diagram of EEOR electrode arrangement in oilfields as used by Wittle *et al.* (2008, 2011). (After Ambah *et al.*,1965, and Titus *et al.*, 1985).

The total flow rate (q_t) upon application of D.C. current is as follows:

$$q_t = \frac{A k \Delta p}{\mu L} + \frac{A k_e E}{\mu L} \tag{B.34}$$

where k = hydrodynamic permeability; $k_e = \left(\frac{D\zeta}{4\pi F}\right)$ = electrokinetic permeability; Δp = pressure drop; μ = viscosity; L = length of porous media; A = cross sectional area of porous media; ζ = zeta potential; F = formation resistivity factor (Archie's); and D = dielectric constant.

Equation B.34 can be presented in a dimensionless form by normalizing the flow rates and, thus, eliminating the μ, A, and L terms (Chilingar *et al.*, 1970):

$$\frac{q_t}{q_i} = 1 + \left(\frac{k_e E}{k\ \Delta p}\right) \tag{B.35}$$

and

$$\frac{q_t - q_i}{q_i} = \frac{k_e E}{k\ \Delta p} \tag{B.36}$$

where q_i = initial hydrodynamic stabilized flow rate.

Equation B.35 shows that an increase in flow rate is dependent upon the zeta potential, dielectric constant, brine concentration,

Darcy permeability, and pressure drop. Equation B.35 indicates that as the hydrodynamic permeability decreases, the percent increase in the flow rate becomes more significant. Thus, k_e/k ratio determined in the laboratory can be used as an index for predicting the success of EEOR application.

References

Ambah, S. A., Chilingar, G. V. and Beeson, C. M., 1965. Application of electrokinetic phenomena in civil and petroleum engineering. *NewYork Acad. Sci.*, 118 (14):585–602

Binder, R. C., 1962. *Fluid Mechanics*. Prentice-Hall, Englewood Cliffs, New Jersery, 4th ed., 453 pp.

Chilingar, G. V., El-Nassir, A. and Stevens, R. G., 1970. Effect of direct electrical current on permeability of sandstone cores. *J. Petrol. Tech.*, 22 (7): 830–836.

Chilingarian, G. V., Mazzullo, S. V., and Rieke, H. H., 1993, *Carbonate Reservoir Characterization: A Geologic-Engineering Analysis*, Part 1. Elsevier, Amsterdam, The Netherlands, 639 pp.

Craze, R. C., 1950. Performance of limestones reservoirs. *Trans. AIME*, 189:287–294

Daniel, E. J., 1954. Fractured reservoirs of Middle East. *Bull. Am. Assoc. Petrol. Geologists*, 38:774–815.

Huitt, J. L., 1956. Fluid flow in simulated fractures. *AICHE J.*, 2(2):259–264

Jones, T. A, Wooten, S. O., and Kaluza, T. J., 1988. Single-phase flow through natural fractures. *Soc Petrol. Engrs (SPE)*, 63d Ann. Tech Conf. and Exhibition, Houston, Tx, 7 pp.

Lomize, G. M., 1951, *Flow in Fractured Rocks*. Gos. Energ. Lzd., Moscow-Leningrad, 127 pp.

Louis, C. 1969. A study of ground water in jointed rocks and its influence on the stability of rock masses. *Rock Mech. Res.*, Imp. Coll., London, Report No. 10, 90 pp.

Nelson, W. L., 1958, *Petroleum Refinery Engineering*. McGraw-Hill, New York, N.Y., 4th ed., 960 pp.

Parsons, R. W., 1996. Permeability of idealized fractured rocks. *Trans. AIME*, 237:126–136.

Titus, C. H., Wittle, J. K., and Bell, C. W., 1985. Apparatus for passing electrical current through an underground formation. U. S. Patent No. 4,495,990.

Wilson, J. T., 1985. Biotransformation of soil. *Applied Environmental Microbiology*, 1:242–243.

Witherspoon, P. A, Wang, J. S. Y., Iwai, K., and Gale, J. E., 1980. Validity of the cubic law for fluid flow in a deformable rock fracture. *Water Resources Res.*, 16(6):1016–1024.

Wittle, J. K., Hill, D. G., and Chilingar, G. V., 2008a. Direct current stimulation for heavy oil production. Paper 2008–374. *Second World Heavy Oil Congress*, Edmonton, March 10–12.

Wittle, J. K., Hill, D. G. and Chilingar, G. V., 2011. Direct electric current oil recovery (EEOR)—A new approach to enhancing oil production. *Energy Sources, Part A: Recovery, Utilization, and Environmental Effects*, 33:805–822.

Appendix C
Glossary

A

Abnormal pressure	Pressure exceeding or falling below the normal hydrostatic pressure to be expected at a given depth.
Acoustic basement	The deepest more-or-less continuous seismic reflector; often an unconformity below which seismic energy returns are poor or absent.
Absolute permeability	A measure of the ability of a single-phase fluid (such as water, gas, or oil) to flow through a rock formation when the formation is totally filled (saturated) with this fluid.
Acoustic (sonic) log	Measurement of porosity, lithology and cement bonding by the use of acoustic (sonic) waves.
Acoustic (sonic) wave	An elastic wave that contains and transmits sound energy. There are three types of acoustic waves: compression-dilation waves (P-waves, Primary or longitudinal waves), shear waves (S-waves, Secondary or transverse waves), and surface waves (L-waves). In the solid Earth, acoustic waves are the P and L types of waves.
Alkaline (soft) water	Water containing not more than 60 ppm (60 mg/l by volume or mg/kg by weight) of hardness-forming constituents (calcium, magnesium and iron) expressed as $CaCO_3$ equivalent. Cf: Hard water.
Allochthonous	Sediments or rocks formed elsewhere than where they are ultimately deposited; of foreign or introduced origin. Syn: Allogenous.
Alluvial fan	An outspread, gently slopping mass of alluvium deposited by a stream, especially in an arid or semiarid region where a stream issues from a narrow canyon onto a plain or valley floor.

Alluvium	A general term for clastic deposits made by streams on river beds, flood plains, and alluvial fans; especially a deposit of silt or silty clay laid down during the time of flood.
Amplitude	The maximum departure of a wave from the average value.
Anisotropic medium	A medium physical properties of which vary in different directions. Cf: Isotropic medium.
API Gravity	The standard method of expressing the specific weight of oils: $°API = 141.5/SG - 131.5$, where SG is the specific gravity of oil at 60°F compared to water at 60°F.
Aquiclude	A body of relatively impermeable rock functioning as an upper or lower boundary of an aquifer and does not transmit water. Cf: Aquifer.
Aquifer	A permeable reservoir or portion of reservoir containing formation water or groundwater. Cf: Aquiclude.
Argillaceous	Containing clay minerals as impurities in carbonate, siliciclastic, or evaporate sediments.
Aromatics	Group of cyclic hydrocarbons found in oils. Contain a benzene ring nucleus in their structure, with a general formula C_nH_{2n-6}.
Asphalt	Black to dark-brown solid or semisolid materials, which gradually liquify when heated. Composed principally of carbon and hydrogen, but contains appreciable quantities of nitrogen, sulfur and oxygen; largely soluble in carbon disulfide.
Asphaltenes	Any of the solid, amorphous, black to dark-brown dissolved or dispersed constituents of crude oils and bitumens, which are soluble in carbon disulfide. They consist of carbon, hydrogen, and some nitrogen and oxygen.
Atomic number	The number of protons in the nucleus and the number of electrons outside the nucleus.

Atomic weight — The weight of any atom as measured on an arbitrary scale based on the weight of an oxygen atom, which is chosen to be 16.

Authigenic — Formed or generated in place; specifically refers to minerals that have precipitated in place or which have replaced other minerals or particles in various diagenetic environments.

Autochthonous — Refers to sediments or rocks that have accumulated in place. Cf: Allochthonous, Allogenous. Syn: Authigen1c.

B

Bank — A buildup consisting of skeletal matter formed by in-place organisms or sediments deposited generally in shallow water; surrounded by deeper water.

Barite — A mineral often used as an addition to drilling mud (or fluid) to add weight: barium sulfate ($BaSO_4$), SG ≈ 4.3.

Barrier reef — Long, linear reef oriented parallel to shoreline and separated from it at some distance by a lagoon of considerable depth and width; generally occur along the margins of shallow-water platforms, and pass seaward into deeper-water environments.

Basin — (*a*) A depressed area.
(*b*) A low area in the Earth's crust, of tectonic origin, in which sediments have been accumulated (See: Sedimentary basin).

Benzene — An aromatic hydrocarbon (C_6H_6) found in petroleum. Used as solvent for petroleum products. Used as synonym for gasoline in many European countries.

Bindstone — Reef rock that has accumulated as a result of the presence of tabular or lamellar fossils that encrusted or otherwise bound sediments during deposition. Partial syn: Boundstone, Biolithite.

Biolithite — General term for reef rocks that have accumulated as a result of the activity of organisms. Partial syn: Boundstone.

Bitumen — A generic term for natural, inflammable substances that are composed of a mixture of hydrocarbons which are substantially free of oxygenated bodies.

Blowout — Blowing out of gas and fluids when excessive well pressure exceeds the pressure of the drilling fluid head. Can occur during drilling through overpressured formations, *i.e.,* when formation pressure exceeds the hydrostatic pressure in the well.

Boundstone — General term for reef rock that has accumulated as a result of the activity of organisms; or a nonreef rock that has accumulated as a result of extensive syndepositional marine lithification.

Bound water — The thin layer of water, which adheres to the surface of clay minerals and is not producible.

Bright spot —
1. A local increase of amplitude on a seismic section for any reason.
2. An increase of amplitude of seismic wave assumed to be caused by hydrocarbon accumulation. Syn: Amplitude anomaly. Cf: Dim spot.

Bulk modulus (K) — The ratio of applied uniform triaxial stress (e.g., hydrostatic pressure) to volumetric strain in a body. Symbol: K, where K = (force/surface area) ÷ (change in volume/original volume).

$[\text{lb/in}^2] \div [\text{in}^3/\text{in}^3] = \text{lb/in}^2$.

$K = G\,(2(1 + \nu)/3(1 - 2\nu))$

where G is the shear modulus and ν is the Poisson's ratio. Reciprocal of matrix compressibility. c_r.

C

Caliper log — Determination of the inside diameter of casings using the spring-activated caliper arms that measure the varying hold widths as the device is drawn upward.

Capillarity	The action by which a fluid, such as water, is drawn up (or depressed) in small-diameter interstices (or tubes) as a result of surface tension. Syn: Capillary action.
Capillary	Small-diameter canals and pores throats in rocks through which fluids can move by capillary forces.
Capillary action	The upward and outward movement of fluids through the porous rock as a direct result of surface properties of rock.
Caprock	An impervious body of a rock that forms a vertical seal to upward hydrocarbon migration. In a salt dome, an impervious body of anhydrite and gypsum, with minor calcite and sulfur, which overlies the top of the salt body, or plug. It probably results from accumulation of the less soluble minerals of the salt body during leaching of its top in the course of its ascent.
Capture	Any process in which a neutron, on colliding with an atomic nucleus, sticks to it or is absorbed into it, or from which fission results.
Carbonate rock	A rock consisting chiefly of carbonate minerals, such as limestone, and dolomite. A sedimentary rock composed of more that 50% by weight of carbonate minerals (calcite, dolomite and magnesite).
Carbonates	Group of minerals found mostly in limestones and dolomites. Calcite ($CaCO_3$) is the most abundant and most important. Aragonite has the same formula as calcite but is less stable; dolomite contains magnesium: $CaMg(CO_3)_2$. Often used as a synonym for carbonate rocks.
Casing	Pipe used to keep the borehole walls from collapsing and to seal the borehole to prevent fluids outside the well from moving from one portion of the well to another (*i.e.,* crossflow).
Catagenesis	Term applied to changes in existing sediments after lithification (in rocks) during deep burial at elevated temperatures and pressures short of metamorphism. Adj: Catagenetic. Syn: Mesogenesis, Epigenesis.

Cation-exchange capacity (CEC)	The total amount of exchangeable cations that a particular clay can adsorb at a given pH. Exchangeable cations are held mainly on the surface of clay minerals, and are measured in milligram-equivalents per 100 g of the clay.
Caustobiolith	A combustible organic rock, usually of plant origin.
Cement	Naturally occurring (biogenic or abiogenic) precipitate of minerals, usually calcite, aragonite, or dolomite in sedimentary rocks, that binds particles together into a lithified framework.
Cementation	Process by which a sedimentary rock particles or fragments are cemented together after deposition. Partial syn: Consolidation.
Cementing	Pumping of the cement slurry down the casing and then back up the annular space between the casing and the borehole.
Cement slurry	A mixture of cement and water in a liquid form which is pumped behind the casing. The slurry is allowed to set until it hardens.
Chalk	Carbonate rock of low-magnesium carbonate composition consisting mainly of the remains of coccoliths and coccospheres.
Check shots	Shots into a borehole seismometer to check the results of integrating a continuous velocity or sonic log. Syn: Well shooting.
Chlorite	A group of green, soft, and platy minerals that are related in structure and composition to the micas. They are hydrated silicates and have general formula $(MgFe)_5Al(AlSi_3)O_{10}$.
Circumgranular	Cement which completely lines the pores in a rock.
Clarke	The average abundance of a chemical element in the Earth's crust. It is named in honor of Frank W. Clarke (1847-1931), a Chief Chemist of the US Geological Survey from 1884 to 1925.

Clastic	Term used in reference to particles (carbonate, siliciclastic, or other mineral composition) that commonly are transported by fluids.
Claystone	A compacted, non-fissile, fine-grained, sedimentary rock composed predominantly of clay-sized particles. Syn: Mudstone.
Coated grain	Carbonate particle consisting of fragments surrounded by a cortex of chemically precipitated carbonate (e.g., ooids, pisoliths) or cortex composed of organic encrustation (e.g., oncolites, rhodolites).
Compaction	Reduction in bulk volume and/or thickness of a sedimentary deposit resulting from either physical processes of grain readjustment (close packing) in response to an increased weight of overburden (mechanical compaction), or chemical processes such as dissolution, grain interpenetration, and stylolitization (chemical compaction). Partial syn: Consolidation.
Compressibility	A material property to be compressed (reduction in volume on application of pressure). Many different formulas are used to determine the compressibility. In general, it is equal to the change in volume upon application of pressure divided by the initial pore or bulk volume.
Compton scattering	Inelastic scattering in which the quantum, *i.e.*, the discrete portion of energy, disappears and its energy is divided between the electron and the fresh quantum of lower energy and larger wavelength.
Condensate	Liquid hydrocarbons, generally clear or pale straw-colored and of high °API gravity (above 60°), that are produced with wet gas. Syn: Distillate, Natural gasoline.
Conduction log	Measurement of the formation electric conductivity. Syn: Induction log.

Conductivity	The ability of a material to conduct electrical current. Syn: Specific conductance. Reciprocal of resistivity.
Connate water	Water ultimately of marine origin that has been entrapped in sediment pores after its burial, and which has been out of contact with the atmosphere for an appreciable period of geologic time.
Consolidation	Any process (including compaction and cementation) that changes loose sediments to a coherent rock. Syn: Lithification.
Contact angle	The angle which the oil – water interface makes with the solid (rock). Usually, it is measured from the solid through the water phase (if oil and water are both present) to the oil – water interface.
Continental slope	The part of continental margin that is located between the continental shelf and continental rise (or oceanic trench). It is characterized by relatively steep slope of 3–6°.
Contour map	A map with continuous lines connecting data points of equal value, such as elevation, formation thickness, rock porosity, etc.
Core	A sample of the rock (usually cylindrical) taken from the well during drilling operations.
Core analysis	Cores from borehole are analyzed for porosity, permeability, fluid content (water and/or oil saturation), lithology, and structure (fractures, cross bedding, etc.).
Core sampling	Taking a sample of geological strata for examination. Syn: Coring.
Correlation	1. Determination of equivalence in stratigraphic positions; for example, in different wells based upon similarities in the well-log characteristics. 2. The matching of different well logs and other well data either in the same well or in different wells.

Critical velocity	A velocity at the transitional point between laminar and turbulent types of fluid flow. This point occurs in the transitional range of Reynolds numbers of approximately 2000 to 3000.
Cross-section	A geologic diagram showing the formations and structures cut by vertical plane.
Cross-stratification	Layers or laminae of sedimentary rock deposited at angles to the horizontal (not exceeding the angle of repose in air or water) as a normal consequence of transport by air or water. Syn: Cross-bedding.
Crude oil	Petroleum as it is produced from the formation. The two categories of crude oils are distinguished: (*a*) heavy crude – thick (sticky) oil with an °API gravity of less than 17°, and (*b*) light crude – thin (light) oil with an °API gravity greater than 25°.
Cryptocrystalline	Term used in reference to crystal components (e.g., cement or architectural elements of shells) having a very fine size, generally not resolvable without the use of at least a petrographic microscope; also refers to a rock with such texture. Syn: Microcrystalline.
Cuttings	Rock chips or fragments produced by drilling and brought to the surface. Syn: Drill cuttings.
Cyclic sedimentation	Sedimentation involving a vertical repetition of rock types representative of distinct depositional environments. Syn: Rhythmic sedimentation.

D

Darcy	A unit of permeability; the permeability, which will allow a fluid flow of 1 mm/sec with viscosity of 1 cP through a cross-sectional area of 1 cm^2 under a pressure gradient of 1 atm/cm. The commonly used unit is the millidarcy = 10^{-3} Darcy.
Darcy's law	A relationship for the fluid flow rate $q = (kA/\mu)\,\Delta p/\Delta x$, where k = permeability, A = cross-sectional area, μ = fluid viscosity, and Δp = pressure gradient across the Δx distance.

Decay	The process of spontaneous transformation of radioactive nuclei.
Decay law	An expression describing a physical decay of radioactive nuclei. If the time of decrease of a quantity of radioactive nuclei is proportional to the quantity at that time, then the decay law is exponential, *i.e.*, $N(t) = N_0e^{-\lambda t}$ where $N(t)$ is the quantity at time t, λ is the decay constant and N_0 is the quantity at time $t = 0$.
Deformation	A general term used to describe the structural processes that may affect rocks after their formation. Includes folding and faulting.
Delta	The nearly flat alluvial tract of land at the mouth of a river, commonly forming a triangular or fan-shaped plain resembling the Greek letter "delta", Δ, in plan view. It is crossed by many distributaries, and resulted from the accumulation of sediment supplied by the river. Most deltas are partly subaerial and partly below water.
Deltaic	Pertaining to or characterized by a delta, e.g., deltaic sedimentation or deltaic coast.
Density	The mass or weight of a substance per unit volume (in ppg, lb/ft^3, g/cm^3, or kg/m^3). The authors of this book like to distinguish between the density (ρ) and specific weight (γ): $\gamma = g(\rho)$, where g is the gravitational acceleration.
Density logging	Measurement of formation porosity as a function of bulk density of the rocks. Involves bombarding the formation with gamma rays, with detectors measuring the number of gamma rays that are reflected from the formation.
Desiccation	Loss of interstitial water from sediments as a result of drying.
Detrital	Term generally restricted to sediments derived from the erosion of preexisting rocks. Syn: Terrigenous, Siliciclastic.

Diagenesis	All chemical, physical, and biologic changes in sediments or rocks that have altered their original textures and mineralogies, operative from the time of their deposition, and through lithification (conversion of sediments to rocks).
Diagenetic facies	An assemblage of rocks with similar diagenetic attributes or which have been affected by similar diagenetic processes.
Differential pressure	The difference between fluid pressures at two different points; e.g., the difference between the pressure in a reservoir and in a borehole drilled through the reservoir.
Diffusion	The spontaneous movement and scattering of particles of liquids, gases, or solids.
Dim spot	A local decrease of the amplitude of a seismic event. Cf: Bright spot.
Directional drilling	Intentional deviation of a borehole from the vertical.
Directional survey	A logging method that records drift angle, or deflection from the vertical, and direction (azimuth) of the drift.
Dispersion	A suspension of fine particles in a liquid (e.g., colloids in water).
Dissolution	Process of dissolving substances. Syn: Leaching.
Dissolution – enlargement/ enhancement	In porosity studies, the process of enlarging or otherwise enhancing the size of preexisting pores by solution.
Dissolution – reprecipitation	Common process of carbonate dissolution and void formation at the microscale or macroscale followed by the precipitation of another mineral phase.
Distally-fining	A sequence of rocks wherein sediment size decreases either away from shore, toward deeper water (marine), or from the point of sediment input (marine or terrestial).

Dolomite "dikes"	Tabular bodies of dolomite generally at high angles to the bedding in the Precambrian Wyman Formation, east-central California. The word "dike" is descriptive only, with no genetic implication intended.
Dolomite front	A relatively sharp contact between laterally adjacent limestone and dolomite.
Dolomitic	Where used in a rock name, 'DOLOMITIC' refers to those rocks that contain 5–50% of the mineral dolomite, as cement and/or grains or crystals. Dolomitic can be applied to the large spectrum of sedimentary rocks that are dolomite-bearing, and also to limestones which have been dolomitized to a degree but not completely.
Dolomitic mottling	Incipient or arrested dolomitization, or arrested (or incomplete) dedolomitization. Common to limestones that have large particulate skeletal or non-skeletal material embedded in fine-textured matrix. Under the effects of dolomitization there is a preferential replacement or alteration of the matrix but not of the large particles. Also common to more or less homogenous textured limestones that have been incompletely dolomitized, leaving patches, blotches, lamiae, and other structures unaffected.
Dolomitized	Refers to rocks or portions of rocks in which limestone textures are discernible, but which have been changed to dolomite.
Dolomitization	Replacement of pre-existing carbonate sediment or rock by dolomite.
Dorag dolomitization	A model proposed by Badiozamani (1973) for dolomitization by a mixture of meteoric ground water and seawater. ("Dorag" is a Farsi word for "mixed blood".)
Drilling mud	Fluid composed of water or oil, clays, chemicals, and weighting materials used to lubricate the bit and to move cuttings out of the hole. Syn: Drilling fluid.

Drillpipe	Heavy seamless tubing used to rotate the bit and circulate the drilling mud.
Drillstem test (DST)	Drillstem test employs equipment, which allows a well to flow for a short period of time, gathering information on reservoir fluids, pressure, and the ability of the reservoir to produce fluids.
Drillstring	A long continuous string of tubular goods of tubing, drill collars, bit, and subsurface tool.
Drive mechanism	The natural force present in a reservoir which causes the fluid to move toward the borehole, the action of one fluid pushing another. Drive mechanisms include water drive, gas-cap expansion, solution -gas drive, gravity drainage, compaction drive, and combination drive.
Dry gas	Natural gas consisting principally of methane and devoid of readily condensable constituents such as gasoline. Cf: Wet gas.

E

Economic geology	The study and analysis of geologic bodies and materials (systems) that can be utilized profitably by man, including fuels, metals, nonmetallic minerals, and water. Tthe application of geologic knowledge and theory to the search for and the understanding of mineral deposits.
Edgewater	Water around the edge of a reservoir – water presses inward.
Effective permeability	A measure of the ability of the porous medium to transmit a particular fluid at the existing saturation, which is normally less than 100%.
Effective porosity	The percentage of the bulk volume of a rock that is composed of the interconnected pores that allows the passage of fluids through the rock. (As used in the U.S.A.) See: Porosity.
Effective pressure	An effective stress, p_e, which is equal to the total overburden pressure, p_t, minus the pore (fluid) pressure, p_p: $p_e = p_t - p_p$. Syn: Net overburden pressure, NOBP.

Elastic constants	Characteristics of elastic property of rocks: each of elastic constants is the ratio of stress to strain for different modes of elastic deformation. In linear, isotropic elastic substances, only four constants are needed to characterize fully the elastic behavior of these substances: Young's modulus, E; shear (or rigidity) modulus, G; bulk modulus, K; and Poisson's ratio, ν. In anisotropic media, up to 36 elastic constants are required.
Elastic deformation	Any change in shape in response to an applied force where the initial shape is recoverable within a reasonable time delay after the applied force is removed.
Elasticity	A property a material may have of recovering its original shape and size "immediately" after a deforming force has been removed.
Electrical resistivity	The electrical resistance per unit length and per unit reciprocal cross-sectional area of a medium.
Electric log	Measures the electric characteristics of a formation; the tool transmits signals to the surface. Downhole tools include either normal or lateral device. In both devices (sondes) a constant electric current is passed between electrodes A and B; the resultant potential difference is measured between electrodes M and N. In the normal device, electrodes A and M are on the sonde, and electrodes B and N are, theoretically, located an infinite distance away. In the lateral device, electrodes A, M and N are on the sonde, and electrode B is located an infinite distance away.
Electric logging	The process of recording the electrical resistivity, spontaneous potential, and induced potential versus depth within the uncased borehole.
Electron	The smallest atomic particle having a negative charge.
Electron-volt	A unit of energy: the amount of energy acquired by an electron when it passes through a potential difference of 1 volt in vacuo. An electron volt equals to the energy of a single hydrogen atom travelling at about 550 miles a minute. Symbol: eV.

Enhanced recovery — Techniques that supplement the natural primary recovery mechanism to increase the flow of the reservoir fluids to the borehole.

Eogenetic stage — The time interval between final deposition and burial of the newly deposited sediment or rock below the depth of significant influences by processes that either operate from the surface or depend for their effectiveness on proximity to the surface (Choquette and Pray, 1970, p. 269).

Eolian — Referring to processes and products of sediment transport, erosion, or deposition by wind.

Epigenesis — As used here, it includes all processes at low temperature and pressure that affect sedimentary rocks after diagenesis and up to metamorphism. Epigenesis haven subdivided into juxta- and apo-epigenesis (Wolf, 1965). It is possible that under unusual conditions syngenesis and diagenesis grade directly into metamorphism. For example, unconsolidated sediments may be exposed to volcanic high temperatures and metasomatic material and undergo metamorphism before diagenesis. Also sediments partly undergoing cementation may be metamorphosed by shallow intrusions causing an increase of temperature and possibly pressure before epigenesis could occur:

$$\left.\begin{array}{l}\text{syngenesis}\\ \text{diagenesis}\\ \text{epigenesis}\end{array}\right\} \rightarrow \text{metamorphism}$$

$$\downarrow$$

$$\text{metamorphism}$$

It appears, however, that the term has been used by different workers in different ways. It was used by many Soviet workers to embrace the modifications and transformations occurring within rocks prior to weathering and metamorphism.

Erosion — The physical and/or chemical processes whereby the earthy and rocky materials of the Earth's crust are loosened, dissolved, or worn away, and simultaneously removed from one place to another by running water (including rainfall), waves and currents, moving ice, or wind.

Ethane — A colorless, odorless, water-insoluble, gaseous paraffin hydrocarbon occurring in natural gas; formula: C_2H_6.

Evaporate — Sediments deposited from aqueous solution as a result of extensive or total evaporation. Examples include anhydrite, rock salt, and various nitrates and borates.

Extraclasts — Particles (fragments) produced by erosion of older rocks derived from outside the basin of carbonate deposition. Cf: Intraclasts.

F

Facies — Sum of all lithologic, biologic, and diagenetic attributes in a rock from which the origin and environment of deposition can be inferred. The term can be restricted to lithologic facies (lithofacies), depositional facies, biotic facies (biofacies), diagenetic facies, and seismic facies. Nicolaus Steno introduced the term in geologic literature in 1669.

Fan — See: Alluvial fan.

Fault — A break and displacement in subsurface strata.

Faulting — The process of fracturing and displacement of subsurface strata that produces faults and/or fractures.

Field — An area consisting of a single reservoir or multiple reservoirs all grouped on, or related to, the same individual structural feature and/or stratigraphic condition.

Filter cake — The layer of concentrated solids from the drilling mud or cement slurry that forms on the walls of the borehole opposite the porous and permeable formation. Syn: Mud cake.

Filtrate	The liquid portion of drilling mud that is forced into porous and permeable formations next to the borehole.
Flat spot	A horizontal seismic reflection not conformable with other reflections and attributed to an interface between two fluids such as gas and water or gas and oil.
Floatstone	Reef rock composed of matrix-supported organic particles, the particles being of allochthonous (transported) rather than in-place origin.
Flow	Movement of fluids through the reservoir.
Flow test	Preliminary test to confirm flow rate through a testing tool prior to going downhole. Syn: Formation testing.
Fluidity	The measure of rate with which a fluid is continuously deformed by a shearing stress; ease of flowing. Cf: Viscosity.
Fluorescent	Instantaneous re-emission of light of a greater wavelength than that originally absorbed.
Flushed zone	The area near the borehole, from which almost all of the original formation water and/or movable hydrocarbons have been replaced by the invading mud filtrate. Cf: Uninvaded zone.
Fluvial	Of or pertaining to rivers; growing or living in a stream or river; produced by the action of a stream or river.
Folding	Folding (or bending) of strata is usually the result of compression that causes the formation of the geological structures known as anticlines, synclines, monoclines, isoclines, etc. The amplitude of the structure is the vertical distance from crest to trough of a fold.
Formation [stratigraphy]	A bed or deposit composed throughout of substantially the same kind of rock. The fundamental unit in lithostratigraphy. Specific features distinguish one rock formation from another.

Formation [drilling]	A general term applied by drillers without stratigraphic connotation to a sedimentary rock that can be described by certain drilling or reservoir characteristics: e.g., hard formation, porous formation, productive formation, etc.
Formation evaluation	The analysis and interpretation of well-log data, drill-stem tests, cores, drill cuttings, etc., in terms of the nature of the formations and their fluid content. The objectives of formation evaluation are to ascertain if commercially productive hydrocarbons are present and, if so, the best means for their recovery.
Formation resistivity factor	$F = R_0/R_w$, where R_0 is the electrical resistivity of a formation 100% saturated with formation water and R_w is the formation water resistivity. $F = a\phi^{-m}$, where ϕ is the porosity, a is the empirical constant, and m is the cementation factor (varies from 1.3 for unconsolidated sands and oolitic limestones to 2.2 for dense sandstones and dolomites). F embodies the effects of grain size, shape, distribution, sorting and packing.
Formation resistivity index	$I_R = R_t/R_0$, where R_t is the true resistivity of a formation saturated with formation fluids (oil, gas, water) in different proportions and R_0 is the resistivity of a formation 100% saturated with formation water. $I_R = a(S_w)^{-n}$, where S_w is the water saturation, a is the empirical constant, and n is the saturation exponent (varies from 1.6 for clayey sands to 2 for clean, non-clayey sands and limestones).
Formation testing	The gathering of pressure data and fluid samples from a formation to determine its production potential before choosing a completion method. Syn: Flow test.
Formation volume factor (*FVF* or *B*)	The volume of oil (and the solution gas dissolved in it) at reservoir pressure, p, and temperature, T, per unit volume of stock-tank oil (at surface, $T = 60$ °F and $p = 1$ atm).

Fossil	Any remains, trace, or imprints of a plant or animal that has been preserved, by natural processes, in the Earth's crust since some past geologic time; any evidence of past life.
Fracturing	Artificial opening up (fracturing) of a formation, by pumping fluids under high pressure, to increase permeability and flow of oil to the well. See also: Hydraulic fracturing.
Funicular water	Capillary water contained in a cluster of grains, the interstices of the cluster being completely filled with water by a single closed capillary meniscus.
Funnel viscosity	Viscosity as measured by the Marsh funnel, based on the number of seconds it takes for 1000 cm^3 of drilling fluid to flow through the funnel.

G

Gamma-ray logging	Measurement of the natural formation radioactivity to determine lithology.
Gas-oil ratio (GOR)	A measure of the volume of gas produced with oil, expressed in ft^3/barrel or m^3/tonne.
Geiger counter	An instrument for measuring ionizing radiation, with a tube carrying a high-voltage wire in an atmosphere containing argon plus halogen or organic vapor at low pressure, and an electronic circuit, which quenches the discharge and passes on an impulse to record the event. Also: Geiger-Muller Counter.
Gel strength	A measure of the ability of colloidal dispersion to develop and retain a gel form, based on its resistance to shear.
Grain-packing geometry	The way by which grains are distributed in the rock matrix. The more important geometrical parameters are (*a*) proximity of grains, (*b*) density of grains, and (*c*) density of cement determined from the thin-section investigation.

(*a*) Proximity of grains is the number of grain contacts crossed by micron-scale ruler divided by the total number of grains crossed by the ruler (expressed in percent).

(*b*) Density of grains is the total number of micron-scale ruler points crossing a single grain at all positions of the ruler divided by the total length of the ruler and multiplied by the magnification of optical microscope (expressed in percent).

(*c*) Density of cement is the difference between 100% and the total number of micron-scale ruler points covering *i*-th site of cement divided by the number of observed micron-scale ruler positions.

Grainstone — Grain-supported carbonate rock textural type, generally mud-free. Syn: Sparite.

Gypsum — A naturally occurring crystalline form of calcium sulfate ($CaSO_4 \cdot 2H_2O$).

H

Half-life — The length of time it takes a sample of radioactive isotope to decrease to half of its original amount by radioactive decay (disintegration). The number is a physical constant characteristic of the isotope, and independent of the particular amount originally present, and of external conditions such as temperature and pressure. The half-life $T_{1/2}$ is related to the decay constant λ: $T_{1/2} = 0.693/\lambda$.

Hard water — Water containing more than 60 mg/l (60 ppm) of hardness-forming constituents (calcium, magnesium and iron) expressed as $CaCO_3$ equivalent. Cf: Alkaline (soft) water.

Hydraulic fracturing — Fracturing the formation by pumping in a specifically blended liquid under pressure high enough to cause the formation to crack open.

Hydrocarbon	Any organic compound, gaseous, liquid, or solid, consisting solely of carbon and hydrogen. They are divided into four specific groups: paraffin, cycloparaffin or naphthene, olefin, and aromatic. Crude oil is essentially a complex mixture of hydrocarbons.
Hydrophilic	Having strong affinity for water; refers to colloids that swell in water and are not easily coagulated.
Hydrophobic	Lacking strong affinity for water; refers to colloids whose particles are not highly hydrated and coagulate easily.
Hydrostatic head	Pressure (p) exerted at the bottom of a column of liquid, $p = \gamma D$, where γ is the specific weight of liquid (e.g., in lb/ft^3 or kg/m^3), and D is the depth (e.g., in ft or m). Syn: Pressure head; Hydrostatic pressure.
Hysteresis	A phenomenon exhibited by a system or material in which response depends nonlinearly on past responses.

I

Igneous rock	A rock mass formed by the solidification of material poured (when molten) into the Earth's crust or onto its surface.
Illite	Very common and widely occurring group of clay minerals; crystals form tiny flakes; formed by the weathering decomposition or hydrothermal alteration of muscovite or feldspar. Syn: Hydrous mica, Hydromica.
Induction log	Measurement of the formation resistivity response to the induced current.
Inflow performance relationship (IPR)	Plot of the flowing bottomhole pressure versus the flow rate, greatly influenced by the reservoir drive mechanism.
Injection	Placing fluids into the reservoir under pressure, e.g., carbon dioxide (EOR), caustic flood, and water injection (secondary recovery).

Interfacial tension	The surface tension occurring at the interface between two liquids that do not mix, such as oil and water.
Interstitial	Interparticle (either pore space, cement, or fluids).
Interval transit time	The time required for compressional acoustic wave to traverse a unit distance (foot or meter) of subsurface formation. It is the reciprocal of the velocity of the compression-dilation acoustic wave. Symbol: Δt.
Invaded zone	The portion about a borehole into which drilling fluid has penetrated, displacing some of the formation fluids. Invasion takes place in porous, permeable zones because the pressure of the drilling fluid is greater than that of the formation fluids.
Irreducible fluid saturation	Equilibrium saturation of the wetting phase (oil or water), which cannot be lowered by flowing indefinitely a non-wetting phase through a porous medium, providing evaporation does not take place.
Irreducible water saturation	The equilibrium situation when the fraction of the pore volume of a reservoir is occupied by water at maximum hydrocarbon saturation. It represents water that has not been displaced by hydrocarbons because it has been trapped by adhering to rock surfaces, trapped in small pores and narrow interstices, etc.
Isotope	One of a set of chemically identical species of atom, which have the same atomic number but different mass numbers.
Isotropic medium	A medium physical properties of which are the same in all directions. Cf: Anisotropic medium.

K

Kaolinite	A common, white to grayish, or yellowish clay mineral of a kaolin group; occurs as a secondary mineral produced by alteration of aluminosilicates.

Karst — Topography (surficial and subsurface) formed as a result of the dissolution of soluble rocks, such as limestones, dolomites, and evaporates, and characterized by closed depressions, caves, and underground drainage.

Kick — An entry of water, gas, oil, or other formation fluid into the borehole during drilling. If prompt action is not taken to control the kick or kill the well, a blowout may occur.

Kill a well — To overcome pressure in a well by use of mud or water so that surface connections may be removed.

L

Lacustrine — Pertaining to lakes or deposits of lakes.

Laterolog — A resistivity log made with sonde that is focused by use of guard or electrodes which force the electric current to flow nearly at right angles to the logging sonde. Syn: Focused log.

Leaching — Pertaining to dissolution of soluble minerals or rocks. Partial syn: Dissolution.

Lift mechanisms — Various methods of lifting oil from the borehole to the surface, including: electric submersible lift, gas lift, hydraulic lift, and rod lift.

Lithification — Process of converting unconsolidated sediments to rocks by the addition of mineral cement or by compaction. Syn: Consolidation. Partial syn: Cementation.

Lithoclast — Mechanically or biogenically formed and deposited fragment (larger than 2 mm) of a weakly lithified sediment or rock, formed within the basin of deposition. Cf: Extraclast.

Lithofacies — See: Facies.

Lithology — Descriptive characteristics of rocks expressed in terms of their mineral composition, structure, texture, color, grain-size, and arrangement of their component parts.

Lithostatic pressure	The pressure due to the weight of the overlying rock (including interstitial fluids), which generally is different from the pressure of fluids in the rock's pore space. Syn: Geostatic pressure.
Log	Continuous record of certain data obtained from a logging tool lowered into the borehole.
Logging	Recording of data (various physical, chemical, and mechanical properties of reservoir) versus depth obtained by lowering of various types of measuring tools into a borehole. Well-logging includes: acoustic (sonic) log, caliper log, conduction log, density log, driller's log, electric log, gamma-ray log, induction log, nuclear log, neutron log, pressure log, production log, resistivity log, spontaneous potential log, temperature log, etc.

M

Matrix	The continuous material (sediment, cement) composing rocks; the continuous material enclosing interstices in rocks.
Maturation [sediments]	Term pertaining to the mineralogic composition of siliciclastic or carbonate rocks as they approach a pure quartz or calcite end-member composition, respectively.
Meniscus [lithology]	The hourglass shape of interparticle cements precipitated from fluids held by capillarity at grain-to-grain contacts; commonly indicative of cement precipitation in the vadose environment.
Meniscus [petroleum]	The curved upper surface of a liquid column, concave when the containing walls are wet by the liquid (negative meniscus) and convex in the case of non-wetting liquid (positive meniscus).
Mesh	A measure of fineness of a screen or sieve; e.g., a 200-mesh sieve has 200 openings per linear inch through which a particle of 74 microns can pass.
Mesogenetic	Postdiagenetic physicochemical changes in rocks occurring in the deep burial environment prior to metamorphism. Syn: Catagenetic, Epigenetic.

Metamorphic rock	A rock derived from preexisting rocks by mineralogical, chemical, and structural alterations caused by processes within the Earth's crust at high temperatures and pressures.
Methane	A colorless odorless inflammable gas; the simplest paraffin hydrocarbon, formula CH_4. It is the principal constituent of natural gas and is often associated with crude oil.
Mica	A silicate mineral characterized by sheet cleavage, *i.e.*, it separates in thin sheets. There are two major types of micas: biotite – a ferromagnesian black mica, and muscovite – a potassic white mica.
Micrite	Particulate, fine-grained matrix of carbonate rocks; by various definitions, the particles being less than 0.01 mm in size; a carbonate rock textural type composed dominantly of mud. Syn: Carbonate mudstone.
Micritization	Organic or inorganic process of converting preexisting carbonate cement or grains into micrite.
Microcrystalline	See cryptocrystalline; refers to a rock with such a texture.
Microfacies	Term for the features, composition, and appearance of rocks, or of specific diagenetic features, as identified in thin sections.
Microspar	Fabric of carbonate crystals resulting from recrystallization of micrite-size crystals or grains that range in size from 5 to about 30 microns.
Migration	The movement of liquid and gaseous hydrocarbons from their source or generating beds, through permeable formations into reservoir rocks.
Mobile offshore drilling unit (MODU)	A drilling rig that is used to drill offshore exploration and development wells and that floats upon the surface of the water when being moved from one drillsite to another. Two basic types of MODU's are used: bottom-supported drilling rig and floating drilling rig.

Mobility — The ability of a fluid to move through a reservoir.

Mole — The fundamental unit of mass of a substance expressed as the number of grams or pounds indicated by its molecular weight (lb/mole or g/mole).

Molecular weight — The sum of atomic weights in a molecule. For example, the molecular weight of water (H_2O) is 18 because the atomic weight of hydrogen is 1 and that of oxygen is 16.

Monominerallic — Composed of one mineral species.

Montmorillonite — An expanding-lattice clay mineral, white to gray with tints of blue, pink, and green; usually occurs in massive, microcrystalline aggregates of very fine, scale-like crystals. It is characterized by a three-layer crystal lattice (one sheet of aluminum and hydroxyl between two sheets of silicon and oxygen); subject to swelling on wetting (and shrinking on drying) due to introduction of considerable interlayer water. Syn: Smectite.

Mud [sed] — Fine-grained particles, by various definitions less than 20 microns or 4 microns in size. Syn: Particulate micrite.

Mud [drill] — See: Drilling mud.

Mud cake — A clay layer ("fish scale") formed where the drilling mud loses water upon filtration into a porous and permeable formation during rotary drilling; adheres to the borehole walls. Syn: Filter cake.

Mud logging — A continuous analysis of the drilling mud and cuttings to determine the presence or absence of oil, gas, or water in the formations penetrated by borehole; ascertain the depths of any oil- and gas-bearing formations using chemical analysis, chromatograph interpretation, and sample description.

Mudstone — (*a*) Argillaceous or clay-bearing sedimentary rock which is non-plastic and has a massive or non-foliated structure. See: Claystone.

(*b*) Carbonate rock textural type composed dominantly of mud (micrite) with less than 10% grains. See: Lime mud.

N

Naphthene	Group of saturated homocyclic hydrocarbons having the formula C_nH_{2n}. Syn: Cycloparaffin.
Natural gas	A naturally occurring mixture of hydrocarbon and non-hydrocarbon gases found in porous media at depth. It is often associated with crude oil. Composed mainly (≈70 to 90%) of methane gas.
Net overburden pressure (NOBP)	Grain-to-grain stress, p_e, which is equal to the total overburden pressure, p_t, minus the pore (fluid) pressure, p_p: $p_e = p_t - p_p$. Syn: Effective pressure.
Neutron	An atomic particle with no electric charge, but with mass approximately equal to that of the proton, about 1.67×10^{-24} gram.
Neutron-gamma log	A neutron log that records the varying intensity of gamma rays resulting from synthetic neutron bombardment. The induced gamma radiation is related to the hydrogen content (and hence the fluid content) of the rocks penetrated by borehole.
Neutron logging	Measurement of porosity; also valuable information concerning rock composition and fluid content. The logging tool bombards the formation with neutrons.
Neutron-neutron log	A neutron log that detects neutrons produced artificially by neutron bombardments. It is sensitive to hydrogen content and used for porosity determination.
Newtonian fluid	A fluid, the viscosity of which remains constant for all rates of shear if constant conditions of temperature and pressure are maintained. Syn: Newtonian flow, Viscous flow. Cf: Non-Newtonian fluids.

Non-Newtonian fluid A fluid in which the relationship of the shear stress to the rate of shear is nonlinear, *i.e.*, a fluid the viscosity of which is not constant. Most drilling fluids behave as non-Newtonian fluids. Syn: Non-Newtonian flow, Viscous-plastic flow. Cf: Newtonian fluid.

Nuclear logging Measurement of porosity, type of fluids and/or gas, lithology, etc., by recording the nuclear properties of the formation.

Nucleus The central portion of the atom, which makes up most of the weight of the atom and is charged positively. Atomic nuclei are made up of two kinds of fundamental particles: protons and neutrons.

O

Occlusion Refers to porosity reduction as a typical consequence of cementation or compaction.

Offlap Progressive offshore migration of the up-dip terminations of sedimentary beds within a conformable sequence of rocks. Cf: Onlap.

Oil-base mud A drilling or workover fluid in which oil is the continuous phase and which contains about 2 to 5% of water.

Oilfield The surface area overlaying an oil reservoir(s). The term usually includes not only the surface area, but also the reservoir, the wells, and the production equipment.

Olefins Group of unsaturated hydrocarbons (one double bond), such as ethylene ($C_{2n}H_{4n}$).

Onlap Progressive onshore migration of the up-dip terminations of sedimentary beds within a conformable sequence of rocks. Cf: Offlap.

Overbalance The extent to which the hydrostatic pressure of the mud column exceeds formation pressure.

Overburden Section of rocks overlying a given stratum or strata.

Overburden-pressure	Total pressure, p_t, exerted on a reservoir by the weight of the overlying rocks and fluids. It is balanced by the pore pressure, p_p, plus the grain-to-grain stress, p_e (or effective pressure, p_e), $p_t = p_p + p_e$.
Overpressured reservoirs	Porous rocks characterized by greater than normal hydrostatic pressure resulting, for example, from undercompaction due to rapid sedimentation.
Oxidation	Process in which a given substance loses electrons or a share of its electrons.

P

Packer	Mechanical device set in the casing (attached to the tubing) to prevent communication between the tubing and annulus.
Packing	Three-dimensional arrangement of particles in a rock.
Packstone	Muddy, but grain-supported carbonate rock textural type.
Paleoenvironment	Ancient depositional (or diagenetic) environment.
Paleogeomorphic	Term generally used in reference to a buried landscape; in reference to hydrocarbon reservoir traps in or along certain buried landscape features.
Paraffin	Denotes a solid, waxy material.
Paraffins	Group of saturated aliphatic hydrocarbons with general formula C_nH_{2n+2}.
Pay zone	The zone of production where commercially recoverable oil and/or gas are present, *i.e.*, productive or producing formation. Syn: Pay sand.
Pelagic	Pertaining to open ocean water as an environment; deep-sea sediments without terrigenous material (either inorganic red clays or organic oozes).
Pellet	A particle composed of fecal material. Partial syn: Pelletoid, Peloid.

Pellicular water	Water in layers more than one or two molecules thick that adhere to the surfaces of rock grains.
Peloid	A cryptocrystalline carbonate particle of unrecognizable origin, most likely a completely micritized grain, less likely a fecal pellet. Partial syn: Pellet.
Pendular water	Capillary water ringing the contact points of adjacent grains or particles of rock.
Perforating	Making holes in the casing (or liner), cement and formation so that formation fluids can enter the borehole. There are three basic types of perforation: mechanical, gun and jet.
Permeability	The ability of a porous medium to transmit fluids. Permeability is equal to one Darcy if 1 cm^3 of fluid flows through 1 cm^2 of cross-section of rock per second under a pressure gradient of 1 atm/cm, the fluid viscosity being 1 cP. See also: Absolute permeability, Effective permeability, Relative permeability.
Petroleum	A naturally-occurring complex of liquid hydrocarbons that may contain varying amounts of impurities (sulfur, nitrogen), which after distillation yields a range of combustible fuels, petrochemicals, and lubricants. Syn: Crude oil.
Petrophysics	The physical properties of reservoir rocks and the methods of their study based on core analysis and well logs.
pH	The negative logarithm of the hydrogen ion concentration in aqueous solution, $pH = -\log_{10}[H^+]$. The pH is measured on a scale of 0 to 14; a neutral medium (such as pure water) has pH of 7; numbers above 7 indicate relative alkalinity; numbers below 7 indicate relative acidity. Most pH values in natural systems lie in the range of 4 to 9.
Piezoconductivity	A strain-dependent pressure conductivity exhibited by fluid-saturated anisotropic medium (e.g., an aquifer or oil-saturated reservoir).
Pinnacle reef	An isolated, long (thick), spire or column-shaped reef.

Pisolite	An accretionary carbonate particle, usually larger than sand-size, composed of a particulate nucleus surrounded by a cortex, generally laminated, of precipitated calcium carbonate; term commonly used for rocks containing pisoids or pisoliths. Syn: Pisolith, Pisoid.
Plastic deformation	The permanent deformation of a rock or mineral upon application of stress.
Plastic fluid	A complex, non-Newtonian fluid in which the shear force is not proportional to the shear rate. Most drilling muds are plastic fluids. Syn: Non-Newtonian fluid. Cf: Newtonian fluid.
Plasticity	A property of certain materials by which the deformation due to a stress is largely retained after removal of the stress. The ability of a substance to be deformed without rupturing.
Platform [drill]	Working surface or deck of a drilling rig, e.g., an immobile offshore structure from which development wells are drilled and produced.
Platform [sed]	A linear region of variable width of shallow-water calcium carbonate deposition.
Play	The extent of a petroleum-bearing formation, e.g., a productive structure.
Plug and abandon (P&A)	To place cement plugs into a borehole and abandon it.
Poisson's ratio (v)	The ratio of the lateral contraction per unit breadth to the longitudinal extension per unit length when a material is stretched. For most substances its value lies between 0.2 and 0.4. Symbol: v = lateral strain/axial strain = (in/in)/(in/in); measured at 50% of ultimate strength. $v = (3K - 2G)/2(3K + G)$, where K is the bulk modulus and G is the shear modulus.
Polymer	Compound having many repeated linked units. In oilfield operations, various types of polymers are used to thicken drilling mud, fracturing fluids, acid, water, and other liquids.
Polymineralic	Composed of more than one mineral.

Pore	A hole, opening, or passageway in a rock. Syn: Interstice.
Porosity	Absolute or total porosity is the total of all voids spaces present within a rock, but not all these spaces are interconnected and, thus, able to contain and transmit fluids. The effective porosity (open porosity) as used in the U.S.A. is defined as the proportion of the rock containing interconnected pores. Porosity is expressed as a percentage or decimals of the bulk volume of the rock. The writers of this book define effective porosity as open (interconnecting porosity) minus the irreducible fluid saturation.
Porosity terms	(*a*) Fabric-selective porosity: pores that occur with respect to specific elements in the rock. Cf: Non-fabric-selective. (*b*) Framework porosity: porosity in the matrix of rocks, exclusive of fractures. Syn: Matrix porosity. (*c*) Porosity-specific: porosity occurrence within a given rock type or paleodepositional facies. (*d*) Pore system: the total petrophysical attributes of a porous unit. (*e*) Primary porosity: porosity inherited from the depositional environment. Cf: Secondary porosity. (*f*) Secondary porosity: developed after deposition as a result of dissolution, for example. Cf: Primary porosity.
Postdepositional	Physical and/or chemical changes in sediments or rocks after deposition and burial.
Pressure gradient	The difference in pressure at two given points, divided by the distance between these two points.
Pressure head	The height (h) of a column of liquid giving rise to a pressure at a certain point in a liquid: $h = p/\gamma$, where h is the head of liquid (e.g., in ft or m), p is the pressure (e.g., in lb/ft^2 or kg/m^2), and γ is the specific weight of liquid (e.g., in lb/ft^3 or kg/m^3). Hydrostatic head in the case of water.
Pressure log	Measurement of the formation pressure at various depths.

Production log	Measurement of production status of a completed well. Yields information on the nature and movement of fluids within the well.
Production test	A test of the well's producing potential usually done during the initial completion phase.
Productivity Index (PI)	Is equal to $[q/(p_r - p_{wf})]$ value, where q is the flow rate (e.g., in bbl/d or t/d); pr is the average reservoir pressure (e.g., in psia or kg/cm^2); and p_{wf} is the flowing bottomhole pressure.
Progradation	The accretion and migration of sedimentary bodies seaward. General term that refers to delta growth seaward. Cf: Regression [sed].
Proton	An atomic particle with a positive charge equal numerically to the negative charge of electron: 1.60×10^{-10} Coulomb.
Provenance	The source or place of origin of detrital sediments.
Pseudospar	Fabric of carbonate crystals, resulting from recrystallization of micrite-size crystals or grains, which are larger than 30 microns in size.
PVT analysis	An examination of reservoir fluid in a laboratory under various pressures, volumes, and temperatures to determine the characteristics and behavior of the fluid at reservoir conditions.

R

Radial flow	Two-dimensional flow from all points around a 360° circle within a formation to a centered well.
Radioactive isotope	Naturally-occurring or artificially produced isotope exhibiting radioactivity.
Radioactivity	Spontaneous decay or disintegration of certain natural heavy elements (e.g., radium, actinium, uranium, thorium) accompanied by the emission of (1) alpha rays, which are positively charged helium nuclei; (2) beta rays, which are fast electrons; and (3) gamma rays, which are short-wavelength electromagnetic waves. Due to this disintegration, the original elements gradually transmuting into others of different chemical properties, the ultimate end-product being lead.

Radioactivity logging	The recording of the natural or induced radioactive characteristics of subsurface formations. A radioactivity log normally consists of two recorded curves: a gamma-ray curve and a neutron curve. The measurements can be made in both open and cased holes filled with any fluid (including air).
Rate of penetration (ROP)	Speed with which the drilling bit cuts through the formation.
Rate of shear	Rate (in rpm) at which an action resulting from applied forces causes or tends to cause two adjacent parts of a body to slide relative to each other in a direction parallel to their plane of contact.
Recovery	The petroleum produced from the reservoir in % (or fraction) of the total oil-in-place reserves; it is subdivided into primary, secondary, and tertiary recovery.
Recrystallization	Term that refers to an increase in the size of existing crystals without a change in mineralogy.
Redox potential	A scale of values, measured as electric potential in volts, indicating the ability of a substance or solution to cause reduction or oxidation reactions under non-standard conditions. The higher the value of redox potential, the more oxidizing the environment. Symbol: Eh.
Reduced pressure	The absolute pressure (p) at which the gas exists divided by the absolute critical pressure (p_{cr}): $p_r = p/p_{cr}$.
Reduced temperature	The absolute temperature (T) divided by the absolute critical temperature (T_{cr}): $T_r = T/T_{cr}$.
Reduction	Process in which a given substance gains electrons.
Reef	An organic buildup made of the remains of sedentary calcareous organisms, especially corals.
Regression [sedimentation]	Retreat of the sea from land areas; also, any change that converts the offshore, deep-water environment to the near-shore, shallow-water one. Cf: Transgression.

Relative permeability	Ratio of the effective permeability at a given saturation of that fluid to the absolute (single-phase fluid) permeability at 100% saturation.
Replacement	Replacement of one mineral for another mineral, e.g., dolomitization, silicification.
Resendimentation	Refers to sediments, which were originally formed and deposited in one environment and subsequently transported to a completely different environment.
Reservoir	1. A subsurface body (layered or massive) made-up of porous and permeable rock which may accumulate various formation fluids (water, oil and/or natural gas). 2. A subsurface accumulation of crude oil or natural gas under adequate trap conditions.
Reservoir pressure	The average pressure within the reservoir at any given time.
Reservoir rock	A porous and permeable rock holding an accumulation of crude oil and/or natural gas. Sandstones and carbonates (limestones and dolomites) are the usually encountered reservoir rocks.
Resin	Semisolid or solid complex, amorphous mixture of organic compounds having no definite melting point or tendency to crystallize.
Resistivity	The property of a material which resists the electrical current flow. Syn: Specific resistance. Reciprocal of conductivity.
Resistivity-contrast factor	The ratio $(\rho_2 - \rho_1)/(\rho_2 + \rho_1)$, which appears in the analysis of resistivity relationships between materials of resistivity ρ_1 and ρ_2.
Resistivity log	Defines the reservoir contents. Electric current flows in the formation between two electrodes on a logging tool and measures resistivity between those two points.
Reynolds number	The dimensionless number defined as $R_e = \rho v d/\mu$, where ρ is the density of a fluid with viscosity μ traveling at a velocity v in a pipe having diameter d. Above $R_e = 2000 - 3000$, flow is turbulent, whereas below, it is laminar.

Rheology — The study of fluid flow especially of fluid elasticity, viscosity and plasticity.

Rhodolite — An accretionary carbonate particle, larger than sand-size, with or without a nucleus surrounded by a laminated to massive cortex constructed by red (*rhodophyte*) algae. Syn: Rhodolith, Rhodoid.

Rift — A long graben/trough associated with a pull-apart zone.

Roentgen (Röntgen) — Quantity of radiation which produces one electrostatic unit of electricity of either sign per cubic centimeter of air at standard pressure and temperature. Symbol: *r*.

Rotary drilling — A drilling method that imparts a turning or rotary motion to the drill- string to drill the hole.

Rudstone — Reef rock composed of grain-supported texture of allochtonous (transported) rather than in-place organic particles.

S

Salt dome — A dome that is caused by an intrusion or piercing of rock salt into overlying deposits, which are usually arched so that they dip in all directions away from the center of the dome.

Sample — A representative unit of a rock, cuttings, fluid, ore, fossil population, or other entity for analysis or display.

Saturation — Percentage of a particular fluid in a porous medium, expressed as the percent of the pore volume.

Screen liner — Perforated pipe or wire mesh screen placed at the bottom of the well to prevent larger formation particles from entering the wellbore.

Seal — An impermeable bed that acts as a barrier to the vertical or lateral migration of hydrocarbons.

Sea-marginal — Environments close to the sea, such as lagoons, tidal flats, beaches, or deposits in these environments. Syn: Marine-transitional.

Sediment	Solid fragmental material or a mass of such material, either inorganic or organic that originates from weathering of rocks and is suspended in, transported by, and deposited by air, water, or ice. Sediment can also form as a result of coagulation of clays, chemical precipitation from solution, or secretion by organisms. It forms in layers on the Earth's surface at ordinary temperatures in a loose, unconsolidated form; e.g., gravel, sand, silt, mud, loess, alluvium.
Sedimentary basin	A low area in the Earth's crust, of tectonic origin, in which sediments have been accumulated.
Sedimentary rock	A layered rock formed as a result of compaction and/or consolidation of sediments, including a clastic rock such as sandstone or siltstone, a chemical rock such as rock salt or gypsum.
Seismic wave	A general term for all elastic waves produced by earthquakes or generated artificially by explosion or vibration.
Sequence	A succession of geologic events, processes, or rocks, arranged in chronological order.
Shadow (blind) zone	1. An area in which there is little penetration of acoustic waves, usually because of their velocity distribution. 2. A portion of the subsurface from which reflections are unobservable because their ray-paths do not reach the surface.
Shear	Action or stress that results from applied forces and that causes or tends to cause two adjoining portions of a substance or body to slide relative to each other in a direction parallel to their plane of contact.
Shear modulus (G)	The ratio of the shear stress to the deformation angle, radians. Symbol: $G = 3KE/(9K - E)$. where E = Young's modulus and K = bulk modulus.

Shelf	A gently sloping or near-horizontal, shallow, marine platform. A stable cratonic area that was periodically flooded by shallow marine waters and received a thin, well-winnowed cover of sediments.
Sidetracking	Drilling a new section of borehole parallel to a previously drilled hole but blocked with junk.
Sidewall core	Core generally one inch in diameter taken from the side of the borehole, often by wireline.
Siliciclastic	In reference to terrigenous detrital sediment composed of silicate mineral grains.
Smectite	A highly-swelling family of clay minerals that includes montmorillonite and bentonite.
Solution gas	Gas dissolved so thoroughly in the crude oil that the solution behaves as one phase. Gas bubbles dissolved in the oil push the latter towards the wellbore.
Sonic logging	Measurement of velocity (or interval transit time) of sound (acoustic) waves as they travel over short distances in the cement or formation rocks. Sonic logs reflect lithologic variation and are used in correlation and formation evaluation.
Sorting	A measure of the spread or range of particle size distribution about the mean in a sediment population.
Sour gas (or crude oil)	An acid gas (or crude oil) containing a significant amount of hydrogen sulfide (H_2S) and carbon dioxide (CO_2). Cf: Sweet gas (or crude oil).
Source rock (petroleum)	Sedimentary rock deposited together with the organic material, which under pressure, heat, and time was transformed to liquid or gaseous hydrocarbons. Source rock is usually shale and limestone.
Source rock (sediment)	The rock from which fragments and other detached pieces have been derived to form later a sedimentary rock. Syn: Parent rock, Mother rock.

Sparite	Grain-supported, mud-free carbonate rock textural type. Syn: Grainstone.
Specific surface	Surface of pores and pore channels per unit of pore volume (commonly), per unit of bulk volume, or per unit of grain volume.
Spontaneous potential (SP)	The difference in electrical potential between the subsurface formations as a result of chemical reactions and differences in solution concentration (electrochemical component of SP), or the movement of fluids through porous media (electrokinetic component of SP).
Spontaneous potential logging	Measurement of spontaneous potential curve showing spontaneous potential distribution versus depth and representing small electromotive forces caused by infiltration (by the drilling mud) of the reservoir rocks or possibly by an electrochemical reaction between mud and reservoir fluid. It is used to indicate the lithology variation (especially sand--shale alternation) and to identify porous and permeable intervals penetrated in the wellbore.
Strain	The dimensional change in the shape or volume of a body as a result of an applied stress.
Stress	A measure of the intensity of force (F) acting upon an area A of a body. Stress = F/A can be expressed in units of N/m^2. Stress can be resolved into two components: (1) compressive (tensile) stress acts normal to the surface and changes the volume of the body, and (2) shear stress acts parallel to the surface and changes the shape of the body.
Stylolite	A pressure-solution feature, generally formed in moderately to deeply buried rocks, characterized by a thin seam or suture of irregular, interlocking, saw-toothed appearance.
Subaerial	Referring to exposure on land, to meteoric fluids.

Subsidence	Local or regional downwarping of a land surface due to tectonic or sediment loading. Subsidence of land surface can occur as a result of fluid (oil and/or water) withdrawal from the oil/gas- and water-producing formations.
Sucrosic	General, non-genetic term for coarse crystalline texture, used mostly in reference to dolomites; a porosity term referring to intercrystalline pores within coarse crystalline dolomites.
Surface tension	The tendency of liquids to maintain as small a surface as possible. It is caused by the cohesive attraction between the molecules of liquid.
Surfactant	A soluble compound that concentrates on the surface boundary between two substances such as oil and water and reduces the surface tension between them.
Sweet gas (or crude oil)	Gas or crude oil devoid of hydrogen sulfide and carbon dioxide.
Syndepositional	Physical, biologic, or diagenetic processes occurring during sediment deposition. Syn: Synsedimentary.

T

Tar	A thick black or dark brown viscous liquid obtained by the destructive distillation of coal, wood or peat.
Temperature log	Measurement of the formation temperature at various depths. It is also used to determine the height of cement behind the casing and to locate the source of water influx into the borehole. Syn: Temperature survey, Thermal log.
Terrigenous	Sediments, typically siliciclastic, derived from the erosion on land of preexisting rocks. Syn: Detrital.
Texture	General physical appearance or characteristics of a rock, including parameters such as size, shape, sorting, and packing of constituent particles.

Tidal flat	Environment and deposits therein, formed in the intertidal zone (including neighboring supratidal and upper subtidal environments and deposits). Syn: Peritidal flat.
Tight formation	A petroleum- or water-bearing formation of relatively low porosity and permeability. Syn: Hard rock.
Tortuosity	The inverse ratio between the linear dimension (L_1) of rock sample and the curvilinear dimension of the equivalent path (L_2) of fluid flow within the sample.
Tract	A region or area of land that may be precisely or indefinitely defined.
Transgression	The spread of the sea over land areas; also, any change that brings offshore, deep-water environments to areas previously occupied by nearshore, shallow-water environments. The term transgressive is used in reference to sediments deposited during a transgression. Cf: Regression. Syn: Sea invasion.
Transmissibility	The rate of water flow through each vertical strip of the aquifer having a height equal to the thickness of the aquifer and being under a unit hydraulic gradient and at a prevailing water temperature. Syn: Hydroconductivity.
Trap	The subsurface structure in which a petroleum accumulation in a reservoir rock existed under such sealing conditions that petroleum migration and escape is prevented. Sealing can be achieved by: (*a*) the abutment of impermeable formations against the reservoir, e.g., stratigraphic trap, unconformity trap, (*b*) the lateral variation within the bed to reduce permeability, e.g., pinch-out trap, shale-out trap, (*c*) the presence of water preventing downward migration, e.g., structural trap, or (*d*) the presence of impermeable fault preventing lateral migration, e.g., fault trap.

Tubing	Relatively small-diameter pipe that is run into the well to serve as a conduit for the passage of oil and gas to the surface.
Turbidite	Sediment deposited from a turbulent current. It is characterized by graded bedding, moderate sorting, and well-developed primary structures, especially lamination.
Turbodrilling	A rotary drilling method in which fluid pumped down the tubing turns the drill bit. The downhole motor consists of multistage vane-type rotor and stator section, bearing section, drive shaft, and bit-rotating sub.
Turbulent flow	The erratic, nonlinear flow of a fluid caused by high velocity. Cf: Laminar flow.

U

Ultraviolet light	Light waves shorter than the visible blue violet waves of the spectrum. Crude oil, colored distillates, residuum, a few drilling fluid additives, and certain minerals and chemicals fluoresce in the presence of ultraviolet light.
Unconformity	A substantial break or gap in the geologic record where a rock unit is overlain by another that is not next in the stratigraphic succession.
Underbalanced	Of or relating to a condition in which pressure in the borehole is less than the pressure in the formation.
Uninvaded zone	The area far enough away from the borehole so as not to have been invaded by the filtrate of drilling mud. Cf: Flushed zone.
Upward-shoaling	A vertical section of deposits that records continually decreasing paleowater depths.

V

Viscoelasticity	Property of a solid or liquid which when deformed exhibits both viscous and elastic behavior through the simultaneous dissipation and storage of mechanical energy.

Viscosity	Measure of the internal resistance of a fluid to flow. Viscosity is equal to the ratio of shearing stress to the rate of shearing strain. There are two kinds of viscosity: (1) absolute or dynamic viscosity and (2) kinematic viscosity. Cf: Fluidity.
Vitrinite	A coal maceral group that is the dominant organic constituent of humic coals. Vitrinite forms the familiar brilliant black bands of coal. Vitrinite particles are found in about 80% of the clays and sands.
Vitrinite reflectance	One of the petrographic components of coals used for the determination of paleotemperatures based on the thermogenic changes in optical properties of the organic matter. The reflectance of vitrinite is determined by microscopic examination of polished sections, measuring and comparing the electric currents generated in a photomultiplier under the influence of light reflected from the polished surface of the rock sample and from the standard with a known reflection. The reflectance of vitrinite is determined in immersion oil (R°) and also in air (R^{a}).
Vug	A small cavern in a rock, larger than a pore.

W

Wackestone	A mud-supported carbonate rock with greater than 10% particles.
Water-base mud	A drilling mud in which the continuous phase is water.
Water cut	Percentage (by volume) of water associated with the produced oil.
Weathering	The destructive physical and/or chemical processes constituting that part of erosion whereby earthy and rocky material on exposure to atmospheric agents at or near the Earth's surface are changed in character (color, texture, composition, firmness, or form), with little or no transport of the loosened or altered material.

Well completion	Finishing a well, *i.e.*, placing the casing, cement and perforations opposite the productive zone. The activities and methods of preparing a well for testing productive intervals and production of oil and gas or for other purposes, such as injection of water.
Well shooting	A method of determining the average velocity as a function of depth by lowering a geophone into a borehole and recording energy from shots fired from surface shotholes. Often run in addition to a sonic log to supply a reference time at the base of the casing and to check the integrated time. Syn: Check shots.
Wet gas	A natural gas containing liquid hydrocarbons. Cf: Dry gas.
Wettability	The ability of a liquid to form a coherent film on a solid rock surface, due to the dominance of molecular attraction between the liquid and the solid surface over the cohesive force of the liquid itself.
Wireline	A rope made from steel wire.
Wireline formation tester	A formation fluid sampling device, actually run on conductor line rather than wireline, that also logs flow and shut-in pressure in formation near the borehole.
Workover	Remedial work on a well, *i.e.*, cleaning, repairing, servicing, stimulating, etc., after commencement of production.

X

X-rays	Non-nuclear electromagnetic radiation of very short wavelength, in the interval of 0.1-100 angstroms, *i.e.*, between that of gamma rays and ultraviolet radiation. Syn: Roentgen rays.

Y

Yield point	The minimum unit stress in the material at which the material deforms appreciably without an increase of load.

Young's modulus (E)	The ratio of the tensile stress (σ) to the tensile strain (ε) in a linear elastic material at loads less than proportional elastic limit. Syn: Elastic modulus. Symbol: E. $E = 9(KG)/(3K + G)$

Z

Zone of lost circulation	Openings in the formation (fractures, etc.) into which the drilling mud is lost without returning to the surface during the drilling operations.

References

Badiozamani, K., 1973. The dorag dolomitization model – application to the Middle Ordovician of Wisconsin. *J. Sed. Petrol.*, 43:965–984.

Choquette P. W. and Pray L. C., 1970. Geologic nomenclature and classification of porosity in sedimentary carbonates. *Bull. Am. Assoc. Petrol. Geol.*, 54:207–250.

Wolf, K. H., 1965. Littoral environment indicated by open-space structures in algal reefs. *Palaeogeogr., Palaeoclimatol., Palaeoecol.*, 1:183–223.

Bibliography

Abasov, M.T., Buryakovsky, L.A., Azimov E.Kh., Badalov, T.A., Palatnik, G.G. and Putkaradzhe, L.A., 1989. *Geophysical and Hydrodynamic Methods of Prediction of Water Encroachment in Oil Pools*. Baku, Azerneshr Press, 262 pp.

Abasov, M.T., Dzhafarov, I.S. and Dzhafarova, N.M., 1987. *Basics of Computerized Geological Interpretation and Analysis*. Baku, Elm Press, 228 pp.

Abasov, M.T., Buryakovsky, L.A., Kondrushkin, Yu.M., Dzhevanshir, R.D., Bagarov, T. Yu. and Chilingar, G.V., 1997. Some methods of oil and gas reserve estimation in Azerbaijan. *Energy Sources*, 19 (7):731–759.

Adler, Yu.P., 1969. *Introduction to Experimental Design*. Moscow, Metallurgy Press, 157 pp.

Aleksandrov, B.L., 1987. *Anomalously High Formation Pressures in Oil- and Gas-bearing Basins*. Nedra, Moscow, 216 pp.

Alford, R.M., 1986. Shear data in the presence of azimuthal anisotropy. 56th *SEG Meeting*. Expanded Abstracts, pp. 476–479.

Ammot, E., 1959. Observations relating to the wettability of porous rock. *Trans. AIME*, 216:156–162.

Aoyagi, K., Kazama, T., Sekiguchi, K. and Chilingarian, G.V., 1985. Experimental compaction of Na-montmorillonite clay mixed with crude oil and seawater. *Chem. Geol.*, 49:385–392.

Archie, G.E., 1942. The electrical resistivity log as an aid in determining some reservoir characteristics. *Trans. AIME*, 146:54–62.

Archie, G.E., 1950. Introduction to petrophysics of reservoir rocks. *Am. Assoc. Petrol. Geologists Bull.*, 34(5):943–961.

Archie, G.E., 1952. Classification of carbonate reservoir rocks and petrophysical considerations. *Am. Assoc. Petrol. Geologists Bull.*, 36(6):278–298.

Arns, C.H., Knackstedt, M.A., Pinczewski, W.V., and Garboczi, E.J., 2002. Computation of linear elastic properties from microtomographic images: methodology and agreement between theory and experiment, *Geophysics*, 67:1396–1405.

Arps, D.D., 1984. Engineering concepts useful in oil finding. *Am. Assoc. Petrol. Geologists Bull.*, 157–165.

Athy, L.F., 1930. Density, porosity and compaction of sedimentary rocks. *Am. Assoc. of Petrol. Geologists*, 14(1):1–24.

Avchan, G.M., 1972. *Physical Properties of Sedimentary Rocks at High Pressure and Temperature*. Moscow, Nedra Press, 145 pp.

Avchan, G.M., Matveyenko, A.A. and Stefankevich, Z.B., 1979. *Petrophysical Parameters of Sedimentary Rocks under Reservoir Conditions*. Moscow, Nedra Press, 224 pp.

Avchan, G.M., Volarovich, M.P., Dortman, N.B. and Khramov, A.M., 1966. Physical properties of rocks. *Soviet Geology*, 2:78–94.

Avdusin, P.P. and Tsvetkova, M.A., 1938. On the structure of porous space of oil-saturated sands and sandstones. *Oil Economy*, 6:24–27.

Avseth, P., Mukerji, T. and Mavko, G., 2005. *Quantitative Seismic Interpretation: Applying Rock Physics Tools to Reduce Interpretation Risk.* Cambridge University Press, 376 pp.

Baechle, G., Weger, R., Eberli, G. and Colpaert, 2006. A. Pore size and pore type effects on velocity – implication for carbonate rock physics models. Workshop Sounds of Geology, Norway, Bergen. Available from World Wide Web. <http:/ mgg.rsmas.miami.edu/students/greg/cipr_baechle-final.pdf> [Cited 2007-10-20]

Bagrintseva, K.I., 1972. About porosity of carbonate rocks of various composition and genesis. *Lithology and Mineral Resources*, 4:99–106.

Bagrintseva, K.I., 1977. *Carbonate Oil and Gas Reservoir Rocks.* Moscow, Nedra Press, 231 pp.

Bagrintseva, K.I., 1982. *Fracturing of Sedimentary Rocks.* Moscow, Nedra Press, 256 pp.

Bagrintseva, K.I., and Per'kova, Ya.N., 1971. About influence of genetic characteristics on formation of reservoir properties of carbonate rocks. In: *Materials of IV All-Union Conference on Reservoirs.* Izd. Nedra, Moscow, pp. 297–301.

Bakhtin, V.V., and Martirosova, A.O., 1972. Variation in density and porosity of clayey rocks as related to their depth of burial. *Geol. Neft Gaza*, 10:57–60.

Baldwin, B. and Butler, C.O., 1985. Compaction curves. *Am. Assoc. Petrol. Geologist*, 69:622–626.

Batzle, M. and Wang, Z., 1992. Seismic properties of pore fluids. *Geophysics*, 57 (11):1396–1408.

Beeson, C.M., 1950. The Kobe porosimeter and the oil well research porosimeter. *Trans. AIME*, 189 (313).

Bernabe, Y., Brace, W.F. and Evans, B., 1982, Permeability, porosity and pore geometry of hot-pressed calcite. *Mechanics of Materials*, 1:73–83.

Berner, R.A., 1971. *Principles of Chemical Sedimentology.* McGraw-Hill, New York, NY, 240 pp.

Berryman, J.G., 1995. Mixture theories for rock properties. In: *Handbook of Physical Constant*, T.J. Ahrens, Ed., *Am. Geophys. Union*, pp. 205–228.

Binder, R.C., 1962. *Fluid Mechanics*, 4th ed., Prentice-Hall, Englewood Cliffs, NJ, 453 pp.

Birch, F., 1942. Thermal conductivity and diffusivity. In: F. Birch, Ed., *Handbook of Physical Constants. Geol. Soc. Am., Spec. Pap.*, 36:243–266.

Bissell, H.J. and Chilingar, G.V., 1967. Classification of sedimentary carbonate rocks. In: G.V. Chilingar, H.J. Bissell and R.W. Fairbridge, Eds., Carbonate Rocks, A. Elsevier, Amsterdam, pp. 87–168.

Bolt, G.H., 1956. Physico-chemical analysis of compressibility of pure clays. *Geotechnique*, 6:86–93.

Bone, M.R., Giles, B.F. and Tegland, E.R., 1976. 3-D high resolution data collection, processing and display. *Proc. 46th Annual SEG Meeting*, Houston, TX.

Bourbié, T., Coussy, O., and Zinszner, B., 1987. *Acoustics of Porous Media.* Butterworth-Heinemann, 352 pp.

Box, G.E. and Wilson, K.B., 1951. On the experimental attainment of optimal conditions. *J. Roy. Stat. Soc.*, Series B (1).

Bridgman, P.W., 1918. Stress—strain relations in crystalline cylinders. *Am. J. Sci.*, 45:269–280.

Bridgman, P.W., 1936. Shearing phenomena at high pressure of possible importance for geology. *J. Geol.*, 44(6):653–669.

Bridgman, P.W., 1958. *The Physics of High Pressure.* Bell Lab., 2nd ed., 445 pp.

Bridgman, P., Lowson, A., and Lazarus, L., 1966. *Solids Under High Pressure.* Moscow, Mir Press, 524 pp.

Brooks, C.S. and Purcell, W.R., 1952. Surface area measurements on sedimentary rocks. *Trans. AIME* 195, TP 3458, pp. 89–296.

Brown, A.R., 1986. *Interpretation of Three Dimensional Seismic Data.* Am. Assoc. Petrol. Geologists Memoir 42, Tulsa, Oklahoma. 341 pp.

Brown, H.W., 1951. Capillary pressure investigations. *J. Petrol. Techn.*, 3(3):67–74.

Brown, P.R., 1969. Compaction of fine-grained terrigenous and carbonate sediment – a review. *Bull. Can. Petrol. Geologists*, 17:486–495.

Bryant, W.R., 1973. Consolidation of marine clays. *Paper presented at ONR Symposium and Workshop on the Physical and Engineering Properties of Deep-Sea Sediments*, April 24–27, 23 pp.

Buckley, S.E. and Leverett, M.C., 1941. Mechanism of fluid displacement in sands. *Trans. AIME*, 146:107–116.

Burjakovskij, L.A., 1960. Permeabilitätsbestimmung mit Hilfe der Widerstandskarottage. Berlin, *Zeitschrift für Angewandte Geologie*, Band 6, Heft 3:112–114.

Buryakovsky, L.A., 1968. Entropy as criterion of heterogeneity of rocks. *International Geology Review*, 10(7):761–764.

Buryakovsky, L.A., 1970. Classification of sandy-silty-shaly rocks based on grain-size distribution and reservoir-rock properties. Moscow, *Soviet Geology*, 2:60–69.

Buryakovsky, L.A., 1974b. Anisotropy of stratified heterogeneous rocks and its numerical evaluation. Moscow, *Soviet Geology*, 11:148–153.

Buryakovsky, L.A., 1974c. Distribution patterns of oil and gas deposits within the Absheron Archipelago. *Internat. Geol. Review*, 16(7):749–758.

Buryakovsky, L.A., 1977. *Reservoir Rock Parameters of Offshore Oil and Gas Fields in Azerbaijan as Determined from the Well Logs.* Baku, Elm Press, 152 pp.

Buryakovsky, L.A., 1982. Passive experiment in petrophysics (Azeri and English summary). *Proc. Azerbaijan Academy of Sciences, Earth's Science Series*, 3:107–116.

Buryakovsky, L.A., 1985a. Experimental design in petrophysics (Azeri and English summary). *Proc. Azerbaijan Academy of Sciences, Earth's Science Series*, 2:101–108.

Buryakovsky, L.A., 1985b. *Petrophysics of Oil and Gas Reservoirs of the Productive Series, Azerbaijan*: Baku, Elm Press, 196 pp.

Buryakovsky, L.A., 1992. Scientific bases, methods and results of mathematical simulation and prediction of structure and behavior of petroleum geology systems. *Energy Sources*, 14:273–295.

Buryakovsky, L.A., 1993a. Outline of general and petroleum geology in Azerbaijan and the South Caspian Basin. *Houston Geol. Soc. Bull.*, 35(6):16–23, 43–47.

Buryakovsky, L.A., 1993b. Offshore oil and gas fields in Azerbaijan: History and description, *Houston Geol. Soc. Bull.*, Part I, 35(10):2–17, 41–47; Part II, 36(3):23–27.

Buryakovsky, L.A. and Agamaliyev, R.A., 1989b. *Natural Electric and Radioactivity Fields in Oil and Gas Wellbores.* Baku, Azineftekhim Publication. 128 pp.

Buryakovsky, L.A., Arkharova, I.M., and Agamaliyev, R.A., 1986b. Surface-active properties of rocks in the Productive Strata of Azerbaijan. Viegrad, Hungary, *Proc. II Symposium in Mining Chemistry.* 139–149.

Buryakovsky, L.A. and Chilingar, G.V., 2005. Experimental design in petrophysical studies: Fundamentals. *Energy Sources*, 27:1503–1510.

Buryakovsky, L.A., Chilingar, G.V. and Aminzadeh, F., 2001. *Petroleum Geology of the South Caspian Basin*: Gulf Professional Publishing (an imprint of Butterworth-Heinemann), Boston-Oxford-Johannesburg-Melbourne-New Delhi-Singapore, 442 pp.

Buryakovsky, L.A. and Dzhafarov, I.S., 1980. Permeability anisotropy in stratified oil- and water-saturated rocks. *Internat. Geol. Review*, 22(9):1067–1069.

Buryakovsky, L.A. and Dzhafarov, I.S., 1984a. Stratigraphic modeling of barren strata using well-log data (English summary). Tallinn, Estonia, *Eesti NSV Teaduste Akademia Toimetised, Geology Series*, 33(1):8–13.

Buryakovsky, L.A., Dzhafarov, I.S. and Dzhevanshir, R.D., 1982. *Prediction of Petrophysical Properties of Oil and Gas Reservoirs and Seals*. Moscow, Nedra Press, 200 pp.

Buryakovsky, L.A., Dzhafarov, I.S., and Dzhevanshir, R.D., 1990a. *Mathematical Modeling of Petroleum Geology Systems*. Moscow, Nedra Press, 295 pp.

Buryakovsky, L.A., Dzhafarov, I.S., Dzhevanshir, R.D. and Kuzmina-Gerasimova, V.L., 1981. Simulation of physical properties of reservoir and sealing rocks in deposits of oil and natural gas. *Proc. International Symposium on the Computerizing Geological Data-COGEODATA*, pp. 26–27.

Buryakovsky, L.A., Dzhafarov, I.S. and Khalaf, Kh.A., 1985d. Classification of carbonate oil-and gas-bearing rocks of Shilou Formation based on well-logging data. *Oil and Gas*, 12:9–13.

Buryakovsky, L.A. and Dzhevanshir, R.D., 1986a. Interaction of clay mineral transformation with thermobaric conditions at depth. *Geochemistry International*, 23(8):99–106.

Buryakovsky, L.A. and Dzhevanshir, R.D., 1989a. Consolidation of sedimentary rocks (mathematical modeling). Paris, France, Ecole des Mines, *Proc. Thematic Meeting on IAS: Compaction and Decompaction of Sediments*, pp. 10–11.

Buryakovskiy, L.A., Dzevanshir, R.D. and Aliyarov, P.Yu., 1986. *Geophysical Methods of Studying Geofluid Pressures*. Akad. Nauk Azerb. SSR, Elm Publ., Baku, 147 pp.

Buryakovsky, L.A., Dzhevanshir, R.D., and Chilingarian, G.V., 1991. Mathematical simulation of sediment compaction. *J. Petrol. Sci. Eng.*, 5:151–161.

Buryakovsky, L.A. and Khalaf, Kh.A., 1984b. On the justification of geometry structure model of carbonate rocks of Shilou Formation, Syria, *Oil and Gas*, 11:13–18.

Buryakovsky, L.A. and Khalaf, Kh.A., 1985c. Well-logging characteristics, reservoir properties and petrophysical parameters of carbonate rocks of Shilou Formation, Syria, *Oil and Gas*, 3:11–17.

Buryakovsky, L.A. and Kukhmazov, M.S., 1975. Classification of sandy-silty-shaly rocks according to petrography and reservoir indexes. *Petroleum Geology*, 13(3):25–129.

Buryakovsky, L.A. and Kuliev, R.D., 1990c. *Experimental Design in Petrophysical Simulation at High Pressure and Temperature*. Moscow, VINITI Publication No. 3550-B90, 30 pp.

Buryakovsky, L.A., Kuliev, R.D. and Stefankevich, Z.B., 1990b. Experimental study on rock permeability increase under repeated influence of filtrated fluids. Siófok, Hungary, *Proc. III Symposium in Mining Chemistry*, pp. 67–74.

Buryakovsky, L.A., Listengarten, B.M. and Bairamalibeili, N.I., 1974a. Correlation of conjugated petrophysical and geochemical parameters of oil and gas reservoirs. *Proc. Azerbaijan Academy of Sciences, Earth's Science Series*, 5:80–88.

Buryakovsky, L.A. and Samedov, F.I., 1961. *Well-logging Techniques for Studying the Reservoir Rocks of Absheron Archipelago*. Baku, Azerneshr Press, 28 pp.

Buryakovsky, L.A., *et al.*, 1985e. *Integrated Experimental Studies of Petrophysical Parameters of Carbonate Rocks of Shilou Formation, Syria*. Proc. Institute of Deep Oil and Gas Deposits, Academy of Sciences of Azerbaijan, 29 pp.

Buryakovsky, L.A. and Dzhevanshir, R.D., 1986a. Interaction of clay mineral transformation with thermobaric conditions at depth. *Geochemistry International*, 23(8):99–106.

Buch, R., 1990. *Laboratory Manual in Physical Geology*. New York, AGI/NAGT, Macmillan Publishing Co., 216 pp.

Carman, P.C., 1937. Fluid flow through granular beds. London, *Trans. Institute Chem. Eng.*, 15:150–156.

Carman, P.C., 1938. The determination of specific surface of powders. London, *J. Soc. Chem. Industry*, 57(7):225–234.

Carman, P.C., 1939. Permeability of saturated sands, soils and clays. *J. Soc. Chem.*, also: *J. Agr. Sci.*, 29:57–58.

Carman, P.C., 1955. The determination of the specific surface of powders, In: *The Colloid Chemistry of Silica and Silicates,* R.K. Iler, Ed., Cornell University Press Ithaca, NY, 324 pp.

Carman, P.C., 1956. *Flow of Gases Through Porous Media*, London, Butterworths Scientific, 182 pp.

Carpenter, C.B. and Spencer, G.B., 1940. Measurement of compressibility of consolidated oil-bearing sandstones. *U.S. Bur. Min. Rept. Invest*, 37(10):3540–3551.

Chalkley, H.W., Cornfield, J. and Park, H., 1949. A method for estimating volume-surface ratios, *Science*, 9(110):295

Chamberlin, T.C., 1897. The method of multiple working hypotheses. *J. Geol.*, 5(8):837–848, reprinted in *J. Geol.*, 1931, (39):155–165.

Chapman, R.E., 1995. *Physics for Geologists*. London, UCL Press, 143 pp.

Chayes, F., 1950. Measurement of intercept distances in thin-sections. *Trans. American Geophysical Union*, 31:870–872.

Chernikov, O.A. and Kurenkov, A.I., 1977. *Lithologic Studies of Productive Sands*. Moscow, Nauka Press, 112 pp.

Chierici, G.L., Ciucci, G.M., Eva, F., and Long, G., 1967. Effect of the overburden pressure on some petrophysical parameters of reservoir rocks. *Proc. 7th World Petroleum Conference*, Mexico City, pp. 309–330.

Chilingar, G.V., 1957. Khanin's classification of reservoir rocks. *Compass*, 34(4):335–339.

Chilingar, G.V., 1964. Relationship between porosity, permeability, and grain-size distribution of sands and sandstones. In: *Deltaic and Shallow Marine Deposits, Developments in Sedimentology*, Vol. 1, L.M.J.U. van Straaten, Ed., Vol.1 Elsevier, Amsterdam, pp. 71–74.

Chilingar, G.V., Bissel, H.J., and Wolf, K.H., 1979. Diagenesis of carbonate sediments and epigenesis (or catagenesis) of limestones. In: *Diagenesis in Sediments and Sedimentary Rocks, Development in Sedimentology*, 25A. G. Larsen and G.V. Chilingar, Eds., Elsevier Science Publishers, Amsterdam, pp. 247–422.

Chilingar, G.V., Khilyuk, L.F. and Rieke, H.H., 2005. *Probability in Petroleum and Environmental Engineering*. Gulf Publishing Company, Houston, Texas, 275 pp.

Chilingar. G.V., Mannon, R.W., and Rieke, H.H., 1972. *Oil and Gas Production From Carbonate Rocks*. New York, American Elsevier, 408 pp.

Chilingar, G.V., Main, R. and Sinnokrot, A., 1963. Relationship between porosity, permeability, and surface areas of sediments, *J. Sediment. Petrology,* 33(3):759–765.

Chilingar, G.V., Rieke, H.H., III and Robertson, J.O. Jr., 1963. Relationship between high overburden pressures and moisture content of halloysite and dickite clays, *Bull. Am. Geol. Soc.,* 74(8):1041–1048.

Chilingar, G.V., Rieke, H.H., III and Sawabini, C.T., 1969. Compressibilities of clays and some means of predicting and preventing subsidence, In: *Symposium on Land Subsidence*, Internat. Assoc. Sci. Hydrology and UNESCO, Tokyo, Japan, 89(II):377–393.

Chilingar, G.V., Zenger, D.H., Bissell, H.J., and Wolf, K.H., 1979. Dolomites and dolomitization. In *Diagenesis in Sediments and Sedimentary Rocks,* G. Larsen and G.V. Chilingar Eds., Amsterdam, Elsevier, pp. 423–536.

Chilingarian, G.V, Chang, J., and Bagrintseva, K.I., 1990. Empirical expression of permeability in terms of porosity, specific surface area, and residual water saturation. *J. Petrol. Sci. Eng.,* 4:317–322.

Chilingarian, G.V, Haroun, M., and Bagrintseva, K.I., 2008. Example of Fractured Carbonate Reservoir Rocks of Russia, and Relationship between Porosity and Permeability in Carbonate Rocks. *SPE* 114177, CIPC/SPE Joint Conf., Calgary, Alberta, Canada, 10 pp.

Chilingarian, G.V., Mazzullo, S.V. and Rieke, H.H., 1992. *Carbonate Reservoir Characterization: A Geologic Engineering Analysis, Part I.* Elsevier Science Publishers, Amsterdam, 639 pp.

Chilingarian, G.V. and Rieke, H.H., 1976. Compaction of argillaceous sediments. In: *Abnormal Formation Pressures.* In: W.H. Fertl Ed., Elsevier Science Publishers, Amsterdam, pp. 49–100.

Chilingarian, G.V. and Rieke, H.H., III, 1968. Data on consolidation of fine-grained sediments. *J. Sediment. Petrology.* 38(3):811–816.

Chilingarian, G.V. and Wolf, K.H., Eds, 1975. *Compaction of Coarse-Grained Sediments.* Elsevier Science Publishers, Amsterdam, 552 pp.

Chilingarian, G.V., Wolf, K.H. and Allen, D.R., 1975. Introduction. In: G.V. Chilingarian and K.H. Wolf, Eds., *Compaction of Coarse-Grained Sediments,* Vol. I. Elsevier, Amsterdam, pp. 1–42.

Chilingarian, G.V. and Yen, T.F., 1983. Some notes on wettability and relative permeability of carbonate rocks. *Energy Sources,* 7(1):67–75.

Chilingarian, G.V., Yen, T.F. Rieke, H.H. III, and Fertl, W.H., 1983. Compressibilities of sands and clays. In: E.C. Donaldson and H. Van Domselaar, Eds., *Proc., US DOE/Venezuela Forum on Subsidence Due to Fluid Withdrawal,* CONF-821199, NTIS, Springfield, Va., pp. 25–32.

Chrisoph, H., Arns, M.A., Knackstedt, W., Pinzcewski, V. and Garboczi, E.J., 2002. Computation of linear elastic properties from microtomographic images: Methodology and match to theory and experiment. *Geophysics,* 67(5):1396–1405.

Coberly, C.J., 1937. Selection of screen openings for unconsolidated sands. *API Drill. and Prod. Pract.,* 189.

Cochran, W.G. and Cox, G.M., 1957. *Experimental Designs*: New York, John Wiley and Sons, Inc., 611 pp.

Cornell, D. and Katz, D.L., 1953. Flow of gases through consolidated porous media, *Ind. Eng. Chem.,* 45:2145–2152.

Cohen, A.J., 2008. *Rock Properties Basis for Seismic Attributes: Interpretation. Class Notes*. Self Publ., University of Louisiana at Lafayette, Lafayette, LA.

Craze, R.C. 1950. Performance of limestone reservoirs. *Trans. AIME*, 189:287–294.

Craft, B.C., Holden, W.R. and Graves, E.D., Jr,. 1962. *Well Design (Drilling and Production)*, Prentice-Hall, Englewood Cliffs, New Jersey, 539 pp.

Degens, E.T. and Chilingar, G.V., 1967. Diagenesis of subsurface waters. In: G. Larsen and G.V. Chilingar, Eds., *Diagenesis and Sedimentary Rocks*. Elsevier, Amsterdam, pp. 447–502.

Dakhkilgov, T.D., 1967. Study of changes in porosity, permeability, and electric conductivity values under pressure influence. *Proc. Grozny Petroleum Institute*, 30:68–69.

Dakhnov, V.N. and Dolina, L.P., 1959. *Geophysical Techniques of Investigation of Oil and Gas Saturated Reservoir Rocks*. Moscow, Gostoptekhizdat, 268 pp.

Dakhnov, V.N., 1982. *Well Logging Data Interpretation*, 2nd edition: Moscow, Nedra Press, 448 pp.

Dakhnov, V.N., 1985. *Well Logging Techniques for Determining Reservoir Rock Properties and Oil and Gas Saturation of Subsurface Formations*, 2nd edition, Moscow, Nedra Press, 310 pp.

Darcy, Henry, 1856. *Les Fontaines Publiques de la Ville de Dijon*, Victor Dalmomt, Paris.

Dickinson, G., 1953. Geological aspects of abnormal reservoir pressures in Gulf Coast Louisiana. *Am. Assoc. Petrol. Geologists, Bull., 37*:410–432.

Dobrynin, V.M., 1962. Effect of overburden pressure on some properties of sandstones. *J. Soc. Pet. Eng.*, pp. 360–366.

Dobrynin, V.M., 1970. *Deformation and Changes of Physical Properties of Oil and Gas Reservoir Rocks*. Nedra, Moscow, 231 pp.

Dobrynin, V.M., 1970. *Deformation and Alteration of Physical Properties of Oil and Gas Bearing Rocks*, Moscow, Nedra Press, 240 pp.

Dobrynin, V.M. and Serebryakov, V.A. 1978. *Methods of Diagnosing Anomalously High Formation Pressures*. Nedra, Moscow, 231 pp.

Doll, H.G., 1948. The SP Log: Theoretical analysis and principles of interpretation. *Trans. AIME*, 179:146–180.

Doll, H.G., 1949. Introduction to Induction Logging. *J. Petrol. Techn.*, 1(6):148–162.

Doll, H.G., 1950. The SP Log in shaly sands. *Trans. AIME*, 189:205–214.

Doll, H.G., 1951. The Laterolog. *J. Petrol. Techn.*, 3(11):305–316.

Donaldson, E.C., Thomas, R.D., and Lorenz, P.B., 1969. Wettability determination and its effect on recovery efficiency. *J. Soc. Petrol. Eng.*, 9(1):13–20.

Donaldson, E.C., Ewall, N. and Singh, B., 1991. Characteristics of capillary pressure curves. *J. Petrol. Sci. Eng.*, 6:249–261.

Donaldson, E.C. and Alam, W., 2008. *Wettability*. Gulf Publ. Co., Houston, TX, 336 pp.

Donaldson, E.C., Kendall, R.F., Baker, B.A. and Manning, F.S., 1974. Surface area measurement of geologic materials.*J. Soc. Pet. Eng.*, 15:111–116.

Dortman, N.B., Ed., 1985. *Physical Properties of Rocks and Commercial Minerals (Petrophysics)*, Moscow, Nedra Press, 455 pp.

Dresser-Atlas, Inc., 1982. *Well logging and Interpretation Techniques – Course for Home Study. Dresser Atlas*, Dresser Industries, Houston, TX, 207 pp.

Dunham, R.J., 1962. Classification of carbonate rocks according to depositional texture. In: *Classification of Carbonate Rocks*, W.E. Ham, Ed., Am Assoc. Petrol. Geologists Memoir. No.1, pp. 108–121.

Durmish'yan, A.G., 1973. Towards a question of consolidation of clays. *Izv. Akad. Nauk SSSR, Ser. Geol.*, 8:85–89.

Dzhafarov, I.S. and Buryakovsky, L.A., 1979. Development of mathematical model relating natural parameters to electrical resistivity of sedimentary rocks. *Computers & Geosciences*, 5(2):269–271.

Dzevanshir, R.D., Buryakovskiy, L.A. and Chilingarian, G.V., 1986. Simple quantitative evaluation of porosity of argillaceous sediments at various depths of burial. *Sediment. Geol.*, 46:169–175.

Efros, D.A., 1963. *Flow of Heterogeneous Fluid Systems*. Moscow, Gostoptekhizdat.

Embree, P., 1985. Resolution and rules of thumb. *Proc. SEG Seismic Field Techniques Workshop*, Monterey, CA.

Elenbaas, J.R. and Katz, D.L., 1948. A radial turbulent formula. *AIME Petrol. Techn.*, T.P. 2304, 174:25–40.

Ellanskiy, M.M., 1972. Warscheinlichkeitsmodelle für Erdöl- und Erdgas-speichergesteine in der Erdöl-Erdgas-Geophysik und Geologie. *Berlin, Deutsche Ges. Geol. Wiss. A Geol. Paläont.* 17(3):311–324.

Ellanskiy, M.M., 1978. *Petrophysical Relationships and Integral Interpretation of Well Logging Data*: Moscow, Nedra Press, 215 pp.

Ellis, D.V., 1987. *Well Logging for Earth's Scientists*. Elsevier, Amsterdam, 532 pp.

Engelhardt, W. von, 1960. *Der Porenraum der Sedimente*. Springer-Verlag, Berlin-Gottingen-Heidelberg, 232 pp.

Eremenko, N.A. 1960. *Geology of Petroleum (Handbook)*, Tom I, GosTopTekhIzdat, Moscow, 592 pp.

Faris, S.R., Gournay, L.B., Lipson, L.B. and Webb, T.S., 1954. Verification of tortuosity Equations. *Am. Assoc. Petrol. Geologists. Bull.*, 38:2226–2232.

Fatt, I., 1953. The effect of overburden pressure on relative permeability. *J. Petrol. Techn.*, 5(10):15–16.

Fatt, I., 1956. The network model of porous media. *Trans AIME*, 207:160–181 also: *J. Petrol. Techn.*, 8(7):144–177.

Fatt, I., 1957a. Compressibility of a sphere pack. Comparison of theory and experiment. *J. Applied Mech.*, 24(1):148–149.

Fatt, I., 1957b. Effect of overburden and reservoir pressure on electric logging formation factor: *Am Assoc. Petrol. Geologists Bull.*, 41(11):2456–2466.

Fatt, I, 1958a. Pore volume compressibility of sandstone reservoir rocks. *J. Pet. Techn.*, 10(3):64–66.

Fatt, I., 1958. Pore structure in sandstones by compressive sphere-pack models: *Am Assoc. Petrol. Geologists Bull.*, 42(8):1914–1923.

Fatt, I., 1958b. Compressibility of sandstone at low to moderate pressure. *Am. Assoc. Petrol. Geologists. Bull.*, 42(8):1924–1957.

Fatt, I. and Dykstra, H., 1951. Relative permeability studies. *J. Petrol. Techn.*, 3(9):249–255.

Fertl, W.H., 1976. *Abnormal Formation Pressures*. Elsevier Scientific Publishing Co., Amsterdam, 382 pp.

Fertl, W.H. and Chilingarian, G.V., 1989. Prediction of tectonically-caused overpressures by using resistivity and density measurements of associated shales. *J. Pet. Sci. Eng.*, 3:203–208.

Fertl, W.H. and Rieke, H.H. III, 1981. In-situ clay mineral identification using gamma-ray spectral logging. *Abstr. Int. Clay Conf.* 1981, Bologna/Pavia, p.92.

Fisher, R.A., 1922. On the interpretation of chi-square from contingency tables and the calculation of P.J., *Royal Statistic Soc.*, 85:87–94.

Fisher, R.A., 1942. *The Design of Experiments*, 3rd edition, Oliver and Boyd Ltd., Edinburg and London: 263 pp.

Fisher, R.A., 1948. *Statistical Methods for Research Workers*, 10th edition: Oliver and Boyd Ltd., Edinburg and London: 354 pp.

Fisher, R.A., 1951. *The Design of Experiments*, 6th edition, Oliver and Boyd Ltd., Edinburg and London: 263 pp.

Fisher, R.A. and Yates, F., 1963. *Statistical Tables for Biological, Agricultural and Medical Research*, 6th edition, Hafner, New York: 145 pp.

Folk, R.L., 1951a. A comparison chart for visual percentage estimation. *J. Sediment. Petrol.*, 21:32–33.

Folk, R.L., 1951b. Stage of textural maturity in sedimentary rocks. *J. Sediment. Petrol.*, 21:127–130.

Folk, R.L., 1968. *Petrology of Sedimentary Rocks.* Hemphill's Book Store, Austin, TX, 170 pp.

Foster, J.B. and Whalen, H.E., 1966. Estimation of formation pressure from electrical surveys-offshore Louisiana. *J. Pet. Techn.*, 18:165–171.

Fraser, H.J. and Graton, L.C., 1935. Systematic packing of spheres with particular relation to porosity and permeability, *J. Geol.*, 11(12):785–909.

Friedman, G.M., 1962. On sorting coefficients and the lognormality of the grain-size distribution of sandstones. *J. Geol.*, 70:737–753.

Frolova, E.K., 1962. Lithologic characteristics of productive carbonate formations of Kuleshov Oil Field. In: *Geology, Geochemistry, and Geophysics Symposium*, Kuybyshev Science Research Institute of Petroleum Industry, USSR, Vl(11), pp. 62–84.

Frolova, E.K., 1963. Lithologic Characteristics of productive horizon of Bashkir suite (A4 Stratum). Kuybyshev Sci. Research Inst. Petroleum Industry USSR, 20:22–63.

Füchtbauer, H., 1961. Zur Quarzneubildung in Erdöllagerstätten. Erdöl Kohle, 14:169–173.

Füchtbauer, H., 1963. Zum Einfluss des Ablagerungsmilieus auf die Garbe von Biotiten und Turmalinen. *Fortschr. Geol. Rheinl. Westf.*, 10:331–336.

Füchtbauer, H., 1967a. Influence on different types of diagnosis on sandstone porosity. Proc. *7th World Pet. Congr.*, 2:353–369.

Füchtbauer, H., 1967b. Der Einfluss des Ablagerungsmilieus auf die Sandstein-Diagenese im mittleren Buntsandstein. *Sediment. Geol.*, 1:159–179.

Füchtbauer, H. and Müller, G., 1970. *Sedimente und Sedimengesteine.* Schweizerbart, Stuttgart, 726 pp.

Füchtbauer, H. and Reineck, H.E., 1963, Porosität und Verdichtung Rezenter, Mariner Sedimente. *Sedimentology*, 2:294–306.

Gaither, A., 1953. A study of porosity and grain relationships in experimental sands. *J. Sediment. Petrology.*, 23:180–191.

Gassmann, F., 1951, Über die Elastizität poroser Medien. *Vierteljahrsschrift der Naturforschenden Gesellschaft in Zürich*, 96:1–23.

Gassman, F., 1951. Elastic waves through a packing of spheres. *Geophysics*, 16(4):673–685.

Georgescu-Roegen, N. 1971. *The Entropy Law and the Economic Process.* Harvard University Press, Cambridge, MA, 457 pp.

Gipson, M., Jr., 1966. A study of the relations of depth, porosity and clay mineral orientation in Pennsylvanian shales. *J. Sediment. Petrology*, 36:888–903.

Glumov, I.F. and Dobrynin, V.M., 1962. Variation in electrical resistivity of water-saturated rocks under the influence of overburden and reservoir pressure. *Applied Geophysics*, Issue 33:190–206.

Gomaa, E.M., 1970. Compressibility of rocks and factors affecting them. *Am. Inst. Min. Metall. Pet. Eng., Annual Stud. Contest Meet., April, Univ. Calif., Berkeley, Presented Paper.*

Grachev, Y.P., 1979. *Mathematical Methods Used for Experimental Design*, Moscow, Food Industry Press: 200 pp.

Graton, L.C. and Frazer, H.J., 1935. Systematic packing of spheres–with particular relation to porosity and permeability. *J. Geol.*, 43:785–909.

Greer, F.C., Main, R., Beeson, C.M. and Chilingar, G.V., 1962. Determination of surface areas of sediments. *J. Sediment Petrology*, 32:140–145.

Gregory, A.R., 1976. Fluid saturation effects on dynamic elastic properties of sedimentary rocks. *Geophysics*, 41:895–921.

Griffiths, J.C., 1952. Grain-size distribution and reservoir-rock characteristics. *Am. Assoc. Petrol.* Geologists *Bull.*, 36:205–229.

Griffiths, J.C., 1967. *Scientific Methods in Analysis of Sediments*. McGraw-Hill Book Co. New York 508 pp.

Griffiths, J.C., 1981. Systems behavior and geoscience problem-solving. In: Merriam, D.F., Ed., *Computer Applications in the Earth Sciences: An Update of the 70s*: Plenum Press, New York, pp. 1–21.

Grunberg, L. and Nissan, A.H., 1943. The permeability of porous solids to gases and liquids. *J. Inst. Petrol. Techn.*, 29 (236)

Gudok, N.S., 1970. *Studies of Physical Properties of Porous Media*, Moscow, Nedra Press, 205 pp.

Gullikson, D.M., Caraway, W.H. and Gates, G.L., 1961. Chemical analysis and electrical resistivity of selected California oilfield waters. *U.S. Bur. Mines, Rept. Invest.*, No. 5736, 21 pp.

Guéguen, Y. and Palciauskas, V., 1994. *Introduction to the Physics of Rocks*. Princeton University Press, Princeton, NJ, 294 pp.

Guyod, H. and Shane, L.E., 1969. *Geophysical Well Logging, I.* Hubert Guyod, Houson, TX, 256 pp.

Hall, H.N., 1953. Compressibility of reservoir rocks. *Trans. AIME*, 198:309–311.

Ham, H.H., 1966. New charts help estimate formation pressure. *Oil Gas J.*, 64:58–63.

Hamilton, E.L., 1956. Low sound velocities in high porosity sediments. *J. Acoustical Soc. Am.*, 28:16–19.

Hamilton, E.L., 1959. Consolidation and lithification of deep-sea sediments. *Geol. Soc. Am. Bull.*, 70(11):1399–1424.

Han, D., 1986. *Effects of Porosity and Clay Content on Acoustic Properties of Sandstones and Unconsolidated Sediments*, PhD dissertation, Stanford University, CA, 219 pp.

Harbaugh, J.W. and Bonham-Carter, G., 1970. *Computer Simulation in Geology*. Willey Interscience, 318 pp.

Harbaugh, J.W., Doveton, J.H., and Davis, J.C., 1977. *Probability Method in Oil Exploration*. John Wiley & Sons, New York, 269 pp.

Hashin, Z. and Shtrikman, S., 1963, A variational approach to the elastic behavior of multiphase materials. *J. Mech. Phys. Solids*, 11:127–140.

Hawkins, M.E., Dietzman, W.D. and Pearson, C., 1964. Chemical analysis and electrical resistivities of oilfield brines from field in East Texas. *U.S. Bur. Rept. Ivest.*, No. 6422, 20 pp.

Hedberg, H.D., 1926. The effect of gravitational compaction on the structure of sedimentary rocks. *Am. Assoc. Petrol. Geologists Bull.*, 10:1035–1072.

Hedberg, H.D., 1936. Gravitational compaction of clays and shales. *Am. Jour. Sci.*, 31:241–287.

Hicks, Ch., 1967. *Basic Principles of Experimental Design* (translated from English). Moscow, Mir Press, 407 pp.

Hottman, C.E. and Johnson, R.K., 1965. Estimation of formation pressure from log-derived shale properties. *J. Petrol. Techn.*, 16:717–722.

Hoshino, K., Koide, H., Inami, K., Iwamura, S. and Mitsui, S., 1972. Mechanical properties of Japanese Tertiary sedimentary rocks under high confining pressure. *Geol. Surv. Japan. Rep.*, 224: 200 pp. (in Japanese).

Ivankin, V.N., Karus, Ye.V., and Kuznetsov, O.L., 1978. *Acoustic Logging*: Moscow, Nedra Press.

Jodry, R.L., 1972. Pore Geometry of Carbonate Rocks (Basic Geologic concepts), pp. 35–83. In: G.V. Chilingar, R.W. Mannon and H.H. Rieke, *Oil and Gas Production from Carbonate Rocks*, Am. Elsevier Publ. Co., Inc., New York, 408 pp.

Johnston, N. and Beeson, C.M., 1945. Water permeability of reservoir sands. *Trans. AIME*, 160:43–55.

Johnson, E.F., Bossler, D.P., and Naumann, V.O., 1959. Calculation of relative permeability from displacement experiments. *Trans. AIME*, 216:370–372.

Jones, T., 1983. *Wave Propagation in Porous Rocks and Models for Crystal Structure.* Ph.D. Thesis, Stanford University, CA.

Kahnn, J.S., 1956. The analysis and distribution of the properties of packing in sand-size sediments. *J. Geol.*, 64 (4, 6):385–395, 578–606.

Kempthorne, O., 1952. *The Design and Analysis of Experiments.* New York, John Wiley and Sons, Inc., 631 pp.

Khanin, A.A., 1956. About classification of petroleum and natural gas reservoir rocks. *Razved. Okhr. Nedr*, 1:7–16.

Khanin, A.A., 1958. Toward the equation of determining reservoir properties of non-cemented sandstones. *Tr. Vses. Nauchn. Issled. Inst. Khim. Pererabotki Gazov*, 4(12):1–8.

Khanin, A.A., 1966. *Fundamentals of Studies of Oil and Gas Reservoirs.* Moscow, Gostoptekhizdat, 208 pp.

Khanin, A.A., 1969. *Oil and Gas Reservoir Rocks and Their Study.* Moscow, Nedra Press, 366 pp.

Khanin, A.A., 1976. *Petrophysics of Oil and Gas Reservoirs.* Moscow, Nedra Press, 295 pp.

Kiefer, J., 1959. Optimal experimental design. *J. Roy. Stat. Soc.* Series B, (21):279–319.

Klinkenberg, L.J., 1941. The permeability of porous media to liquids and gases. *API Drilling and Production Practices*, 7(3):200–208.

Klinkenberg, L.J., 1951. Analogy between diffusion and electrical conductivity in porous rocks. *Bull. Geol. Soc. Amer.*, 62 (6):559–564.

Knutson, C.F. and Bohor, B.F., 1963. Reservoir rock behavior under moderate confining pressure. In: C. Fairhurst, Ed., *Rock Mechanics.* Pergamon. New York, N.Y., pp. 627–658.

Kobranova, V.N., 1962. *Physical Properties of Rocks*. Moscow, Gostoptekhizdat, 490 pp.

Kobranova, V.N., 1986. *Textbook on Petrophysics*, 2nd edition, Moscow, Nauka Press, 392 pp.

Kondrushkin, Yu.M. and Buryakovsky, L.A., 1987. Natural reservoirs in volcanic rocks and methods of estimating oil reserves in reservoirs with complex porous space: Moscow, *Geol. Oil and Gas*, (1):33–36.

Kotyakhov, F.I., 1949. Relationship between major physical parameters of sandstones. Moscow, *Oil Economy*, 12:29–32.

Kotyakhov, F.I., 1949. Interrelationship between major physical parameters of sandstones. *Neft. Khoz.*, 12:29–32.

Kotyakhov, F.I., 1956. *Basics of Physical Properties of Oil Reservoirs*. Moscow, Gostoptekhizdat: 363 pp.

Kotyakhov, F.I., 1977. *Physical Properties of Oil and Gas Reservoirs*. Moscow, Nedra Press, 287 pp.

Kozel'skiy, L.T. and Matveev, A.K., 1973. Relationship between lithologic characteristics of sedimentary rocks of coal-bearing deposits and their geophysical parameters. In: V.E. Khanin *et al.* (Eds.), *Present-Day Problems of Geology and Geochemistry of Combustible Deposits*. Akad. Nauk SSSR, Izd. Nauka, Moscow: pp. 184–191.

Kozeny, J., 1927. Über Kapillare Leitung des Wassers im Boden. *Sitzungsber. Akad. Wiss.* Wien, 136:271–306.

Kozeny, J., 1927. Über Kapillare Leitung des Wassers im Boden (Aufstieg, Versikerung und Anwendung auf die Bemasserung). *Sitzungberichte das Akademie der Wissenschaften*, Wien, 136(11a):271–306.

Krumbein, W.C., 1955. Statistical analysis of facies maps. *J. Geol.*, 63:452–470.

Krumbein, W.C. and Graybill, K.A., 1965. *An Introduction to Statistical Models in Geology*, New York, McGraw-Hill Book Co., 398 pp.

Krumbein, W.C., Kauffman, M.E., and McCammon, R.B., 1969. *Models of Geologic Processes. An Introduction to Mathematical Geology*. AGI, Washington, D.C., 150 pp.

Krumbein, W.C. and Miller, R.L., 1953. Design of experiments for statistical analysis of geological data, *J. Geol.*, 61(6):510–532.

Krumbein, W.C. and Sloss, L.L., 1951. *Stratigraphy and Sedimentation*. San Francisco, Freeman, 497 pp.

Kryukov, P.A. and Zhuchkova, A.A., 1963. Physical-chemical phenomena associated with driving out od solution from rocks. In: *Present-Day Concept of Bound Water in Rocks. Izd. Akd. Nauk SSSR, Laboratory of Hydrogeological Problems of F.P. Savarenskiy*, Moscow, pp. 95–105.

Kullback, S., 1959. *Information Theory and Statistics*. New York, John Wiley and Sons, Inc., 395 pp.

Kusakov, M.M. and Gudok, N.S., 1958. Influence of confining pressure on the flow through oil-saturated rocks. *Oil Economy*, (6):40–48.

Kuster, G.T. and Toksöz, M.N., 1974. Velocity and attenuation of seismic waves in two-phase media, Part I: theoretical formulations. *Geophysics*, 39:587–606.

Kuz'min, A.A., Uriman, V.I. and Akent'ev, E.P. 1975. Expulsion of water from clays into reservoirs during process of exploitation of oil and gas deposits. *Geol. Nefti Gaza*, 6:40–43.

Langnes, G.L., Robertson, J.O., Jr. and Chilingar, G.V., 1972. *Secondary Recovery and Carbonate Reservoirs*. Elsevier Science Publishers, Amsterdam, 304 pp.

Larsen, G. and Chilingar, G.V., 1983. *Diagenesis of Sediments and Sedimentary Rocks.* Elsevier Science Publishers, Amsterdam, 579 pp.

Latyshova, M.G., Vendelshtein, B.Yu. and Tuzov, V.P., 1975. *Analysis and Interpretation of Well Logging Data.* Moscow, Nedra Press, 272 pp.

Leibenzon, L.S., 1947. *Flow of Natural Liquids and Gases through Porous Medium.* Moscow-Leningrad, Gostekhizdat., 244 pp.

Leverett, M.C., 1939. Flow of oil-water mixtures through unconsolidated sands. *Trans. AIME,* 132:149–152.

Leverett, M.C., 1941. Capillary behavior in porous solids. *Trans. AIME, TP 696,* 142:152–169.

Levorsen, A.I., 1967. *Geology of Petroleum,* 2nd edition: San Francisco, Freeman and Co., 724 pp.

Lo, K.Y., 1969. The pore pressure-strain relationship of normally consolidated undisturbed clays, I. Theoretical considerations. *Can. Geotechn. J.,* 6:383–394.

Lomtadze, V.D., 1972. *Laboratory Techniques for Investigation of Physical-Mechanical Properties of Rocks.* Leningrad, Nedra Press, 312 pp.

Lucia, F.J., 1983. Petrophysical parameters estimated from visual descriptions of carbonate rocks. A field classification of carbonate pore space. *J. Petrol Technol.,* 35(3):629–637.

Lucia, F.J., 1999. *Carbonate Reservoir Characterization.* Springer-Verlag, Berlin Heidelberg New York, 226 pp.

Lynch, E.J., 1962. *Formation Evaluation.* Harper's Geoscience Series, Harper & Row, 422 pp.

Magara, K., 1968. Compaction and migration of fluids in Miocene mudstone, Nagaoka Plain, Japan. *Am. Assoc. Petrol. Geologists Bull,* 52:2466–2501.

Magara, K., 1974. Compaction, ion filtration, and osmosis in shale and their significance in primary migration. *Am. Assoc. Petrol. Geologists Bull.,* 58:283–290.

Magara, K., 1978. *Compaction and Fluid Migration. Practical Petroleum Geology Development.* Elsevier Science Publishers, Amsterdam, 296 pp.

Mao Sisong, Ding Yuan, Zhou Juxiang, and Lu Naigang, 1981. *Regression Analysis and Experiment Design.* East China Normal University, 375 pp.

Mar'enko, Yu.I., 1978. *Oil and Gas –Bearing Carbonate Rocks.* Moscow, Izdatel'stvo "Nedra," 240 pp.

Marion, D., 1990. *Acoustical, Mechanical, and Transport Properties of Sediments and Granular Material.* Ph.D. dissertation, Stanford University, CA, 136 pp.

Marmorshtein, L.M., 1975. *Reservoir and Sealing Properties of Sedimentary Rocks under Various Temperature and Pressure Conditions.* Ministry of Geology, Scientific and Industrial Association "Sevmorgeo", Proc. Scientific Institute of Arctic Geology, Vol. 180, Leningrad, Nedra Press, 195 pp.

Marmorshtein, L.M., 1985. *Petrophysical Properties of Sedimentary Rocks at High Pressure and Temperature.* Moscow, Nedra Press, 190 pp.

Marsden, S.S. 1968. Wettability: The elusive key to water flooding, *J. Petrol. Eng.,* 1:82–87.

Mavko, G., 2000. *Rock Physics for Geophysical Reservoir Characterization and Recovery Monitoring.* Course Notes. Rock Physics Laboratory. Stanford University, CA.

Mavko, G., 2005. *Rock Physics.* Course notes. Available from World Wide Web: <http://pangea.stanford.edu/courses/gp262/ [Cited 2007-10-20].

Mavko, G., Mukerji, T. and Dvorkin, J., 1998. *The Rock Physics Handbook.*: Cambridge Univ. Press, 329 pp.

McBride, E.F., 1963. A classification of common sandstone. *J. Sediment. Petrology,* 33:664–669.

Meade, R.H., 1964. Removal of water and rearrangement of particles during the compaction of clayey sediments-review. *U.S. Geol. Surv., Prof. Pap.,* 497B, 1–23.

Meade, R.H., 1966. Factors influencing the early stages of compaction of clays and sands-review. *J. Sediment. Petrology,* 36:1085–1101.

Middleton, G.M., 1962. On sorting coefficients and the lognormality of the grain-size distribution of sandstones: A discussion. *J. Geol.,* 70:754–756.

Mikhailov, N.N., 1987. *Changes in Physical Parameters of Rocks in the Near Borehole Zones*: Moscow, Nedra Press.

Miller, R.L. and Kahn, J.S., 1962. *Statistical Analysis in the Geological Sciences.* New York, John Willey and Sons, Inc, 483 pp.

Mirchink, M.F., 1948. *Textbook of Oilfield Geology.* Gostoptekhizdat, Moscow.

Mobarek, S.A.M., 1971. *The Effect of Temperature on Wave Velocities in Porous Rocks.* Master's Thesis. University of California, Berkeley, CA, 83 pp.

Monicard, R.P., 1980. *Properties of Reservoir Rocks: Core Analysis.* Gulf Publ. Co., Houston, TX (also in French: Editions Technip, Paris), 168 pp.

Morita, N., Whitfill, D.L., Nygaad, O. and Bale, A., 1988. A quick method to determine subsidence, reservoir compaction, and in situ stress induced by reservoir depletion. *Proc. Soc. Petrol. Eng., Formation Damage Control Symposium, Bakersfield, Calif. SPE* 17150, pp. 73–84.

Morozovich, Ya.R., 1967. On the methodology of studying the dependence of physical properties of rocks on the confining pressure and stress condition. In: *Well Logging Problems*, Moscow, pp. 62–66.

Muravyov, I., Andriasov, R., Gimatudinov, SH., Govorova, G. and Polozkov, V., 1958. *Development and Exploitation of Oil and Gas Fields*, Peace (Mir) Publishers, Moscow 503 pp. (In English).

Muskat, M., 1937. *The Flow of Homogenous Fluids Through Porous Media.* New York, McGraw-Hill Book Co., 763 pp.

Muskat, M., 1949. *Physical Principles of Oil Production. New York*, New York, McGraw-Hill Book Co, 922 pp.

Myers, L.L., 1963. Dynamic Phenomena of sediment model and its application to Niigata area, Japan. *Am. Assoc. Petrol. Geologists Bull.,* 71:810–821.

Nalimov, V.V., 1971. *Theory of Experiment.* Moscow, Nauka Press, 207 pp.

Nalimov, V.V. and Chernova, N.A., 1965. *Statistical Methods of Design of Extreme Experiments.* Moscow, Nauka Press, 340 pp.

Nobes, D.C., Villingar, H., Davies, E.E. and Law, L.K., 1986. Estimation of marine sediments bulk physical properties at depth from sea-floor geophysical measurements. *J. Geophys. Res.,* 51:2193–2199.

Norris, A.N., 1985. A differential scheme for the effective moduli of composites. *Mechanics Materials,* 4:1–16.

Nur, A.M., 1969. *Effect of Stress and Fluid Inclusions on Wave Propagation in Rock.* Ph.D. Thesis, MIT, 164 pp.

Nur, A. and Simmons, G., 1969. Stress-induced velocity anisotropy in rock: An experimental study. *J. Geophys. Res.,* 74:6667–6674.

Odeh, A.S. and Yuster, S.T., 1953. The effect of viscosity on relative permeability. *Paper 264- G presented to Los Angeles Meeting of American Institute of Mining and Metallurgical Engineers,* Los Angeles, CA.

Ornatskiy, N.V., 1950. *Soil Mechanics*. Moscow University Publication, 420 pp.

Parkhomenko, E.I., 1965. *Electrical Properties of Rocks*. Nauka Press, 164 pp.

Paulsson, B., Fuller, B., Karrenbach, M., and Heuermann, P., 2001. Borehole data: Closer to the rocks, *Am. Assoc. Petrol. Geologists Explorer*, pp. 22–23.

Pavlova, N.N., 1975. *Deformation and Reservoir Properties of Rocks*. Nedra Press, 240 pp.

Perrier, R. and Quiblier, J., 1974. Thickness changes in sedimentary layers during compaction history; methods for quantitative evaluation. *Am. Assoc. Petrol. Geologists Bull.*, 58:507–520.

Petkevich, G.I. and Verbitskiy, T.Z., 1970. *Acoustic Investigations of Rocks in Oil Producing Wells*. Kiev, Naukova Dumka Press, 126 pp.

Pettijohn, F.J., 1957. *Sedimentary Rocks*, 2nd edition. New York, Harper and Row Publishers, 718 pp.

Pickett, G.R., 1970. Applications for borehole geophysics in geophysical exploration. *Geophysics*, 35:81–92.

Pirson, S.J., 1958. *Oil Reservoir Engineering*, 2nd edition. New York, McGraw-Hill Book Co., 441 pp.

Pirson, S.J., 1963. *Handbook of Well Log Analysis*. Prentice-Hall, 326 pp.

Pleskunin, V.I. and Voronina, E.D., 1979. *Theoretical Basis of Organizing and Analysis of Sampled Data in the Experimental Studies*. Leningrad State University Publication, 232 pp.

Podio, A.L., Gregory, A.R. and Gray, K.E., 1968. Dynamic properties of dry and water-saturated Green River Shale under stress. *J. Soc. Petrol. Eng.*, 8:389–404.

Ponikarov, V.P. (editor), 1969. *Geology and Mineral Resources of Foreign Countries (Syria Section)*. Moscow, Nedra Press, 216 pp.

Poskitt, T.J., 1969. Consolidation of saturated clay with variable permeability and porosity. *Geotechnique*, 234–252.

Potapov, I.I., 1954. *Apsheron Oil and Gas Region. Geological Description*. Baku, Azerbaijan Academy of Sciences Publication, 567 pp.

Potapov, I.I., 1964. *Geotectonics*. Rostov University Publication, 256 pp.

Powers, M.C., 1967. Fluid-release mechanisms in compacting marine mudrocks and their importance in oil exploration., *Am. Assoc. Petrol. Geologists Bull.*, 51:1240–1254.

Pray, L.C., 1960. Compaction in calcilutites. *Geol. Soc. Am., Bull.*, 71:1946.

Proshlyakov, B.K., 1974. *Secondary Changes in Oil and Gas Reservoir Rocks*. Moscow, Nedra Press, 232 pp.

Proshlyakov, B.K., Galyanova, T.I., and Pimenov, Yu.G., 1987. *Sedimentary Reservoir Rocks at Great Depths*. Moscow, Nedra Press, 200 pp.

Proshlyakov, B.K., 1960. Reservoir properties of rocks as a function of their depth and lithology. *Geol. Nefti Gaza*, 12:24–29.

Purcell, W.R., 1949. Capillary pressures – their measurements using mercury and the calculation of permeability therefrom. *Trans. AIME*, TP2603, 186:39–46.

Pustovalov, L.V., 1940. *Petrography of Sedimentary Rocks*. Vol. I, II: Moscow, Gostoptekhizdat, ??? pp.

Raghavan, R., 1974. Consolidation and rebound processes in one-dimensional porous columns. *J. Geophys. Res.*, 79:1687–1698.

Raghavan, R. and Miller, F.G., 1975. Mathematical analysis of sand compaction. In: G.V. Chilingar and K.H. Wolf, Eds., *Compaction of Coarse-Grained Sediments*, Elsevier, Amsterdam, pp. 403–524.

Rawlings, J.O., 1988. *Applied Regression Analysis: A Research Tool*. Wadsworth and Brooks/Cole Advanced Books and Software, Pacific Grove, CA, 553 pp.

Reuss, A., 1929. Berechnung der Fliessgrenze von Mischkristallen auf Grund der Plastizitätsbedingung für Einkristalle. *Z. Angew. Math. Mech.*, 9:49–54.

Ricken, W., 1987. The carbonate compaction law: a new tool. *Sedimentology*, 34:571–584.

Rieke, H.H., III, Chilingar, G.V. and Mannon, R.W., 1972. Application of petrography and statics to the study of some petrophysical properties of carbonate reservoir rocks. In: G.V. Chilingar, R.W. Mannon and H.H. Rieke III, Eds., *Oil and Gas Production from Carbonate Rocks*. Am. Elsevier, New York, NY, 408 pp.

Rieke, H.H., III, Chilingar, G.V. and Robertson, J.O., Jr., 1964. High-pressure (up to 500,000 psi) compaction studies on various clays. *Proceedings 22nd International Geology Congress, New Delhi, Sec.*15, pp. 22–38.

Rieke, H.H., III and Chilingarian, G.V., 1974. *Compaction of Argillaceous Sediments*. Elsevier, Amsterdam, 424 pp.

Rieke, H.H., III, Ghose, S.K., Fahhad, S.A. and Chilingar, G.V., 1969. Some data on compressibility of various clays, *Proc. Internat. Clay Conf.*, 1:817–828.

Robertson, E.C., 1955. Experimental study of the strength of rocks. *Geo. Soc. Am. Bull.*, 66:1294–1314.

Ryskal, O.E., Malinin, V.F. and Akhmetov, R.T., 1989. Porosity Types in complex carbonate reservoirs based on petrophysical studies. *J. Petrol. Sci. Eng.*, 2:27–30.

Romm, Ye.S., 1966. *Filtering Properties of Fractured Rocks*. Moscow, Nedra Press, 283 pp.

Romm, Ye.S., 1985. *Structural Models of Porous Space of Rocks*. Leningrad, Nedra Press, 240 pp.

Samedov, F I. and Buryakovsky, L.A., 1957. Relationships between petrophysical properties of the reservoir rocks of Neftyanye Kamni (Neft Dashlary) Field. *Azerbaijan Oil Economy*, 8:20–23.

Samuels, G., 1950. The effect of base exchange on the engineering properties of soils. *Build. Res. Stn. (G.B.) Note, C176*, 16 pp.

Sarkisyan, S.G., Politykina, M.A and Chilingarian, G.V., 1973. Effect of postsedimentation processes on carbonate reservoir rocks in Volga-Urals Region, USSR. *Am. Assoc. Petrol. Geologist* Bull., 57:1305–1313.

Sautin, S.N., 1975. *Experimental Design in Chemistry and Chemical Technology*. Leningrad, Chemistry Press, 48 pp.

Scherer, M., 1987. Parameters influencing porosity in sandstone: A model for sandstone porosity prediction. *Am. Assoc. Petrol. Geologists Bull.*, 71:485–491.

Scheidegger, A.E., 1957. *The Physics of Liquid Flow through Porous Media*. Toronto, Univ. of Toronto Press, 236 pp.

Schlumberger, 1972. *Log Interpretation, Vol. 1 – Principles*. Schlumberger Limited. 113 pp.

Schlumberger, 1987. *Log Interpretation Principles/Applications*. Schlumberger Educational Service, Houston, TX, 198 pp.

Scorer, J.D.T. and Miller, F.G., 1974. A review of reservoir rock compressibility and its relationship to oil and gas recovery. *Ins. Petrol.* (London), 74–003, 25 pp.

Shannon, C.E., 1948. *A Mathematical Theory of Communication. Bell System Tech. J.*, 27:379–423, 623–656.

Sharapov, I.P., 1965. *Applications of Statistical Analysis to Geological Data Processing*. Moscow, Nedra Press, 260 pp.

Sharma, S.K., Nohwar, U.S., Raj, H., Koithara, J. and Chilingar, G.V., 1980. Interrelationship among various petrophysical parameters of carbonate reservoir rocks in the Bombay High Oil Field, India. *Energy Sources*, 5(1):53–69.

Sheriff, R.E., 1984. *Encyclopedic Dictionary of Exploration Geophysics*, 2nd edition. Soc. Exploration Geophysicists, 323 pp.

Sheriff, R.E., 1985. Aspects of seismic resolution, *In:* (O.R. Berg and D. Woolverton, Eds., *Seismic Stratigraphy II: An integrated approach to hydrocarbon exploration.* Am. Assoc. Petrol. Geol. Memoir 39, Tulsa, Oklahoma, pp. 1–10.

Sheriff, R.E. and Geldart, L.P., 1983. *Exploration Seismology, 2, Data Processing and Interpretation.* Cambridge University Press, 130 pp.

Shinn, E.A., Halley, R.B., Hudson, J.H. and Lidz, B.H., 1977. Limestone compaction-an enigma. *Geology*, 5:21–24.

Shirkovskiy, A.I., 1969, Complex Field Investigation in Gas Condensate Deposits of Cambay, India, *Geology, Exploration and Development of Gas and Gas-Condensate Deposits*, Vniie- Gazprom, Moscow, 69–70.

Shirkovskiy, M.A., 1971. *Determination and Utilization of Physical parameters of Porous Medium During Development of Gas Condensate deposits.* VNIIZ-GAZPROM, Moscow, 48 pp.

Shreiner, L.A., Baidyuk, B.V., Pavlova, N.N. and Yakushev, V.P., 1968. *Deformation Properties of Rocks under High Pressures and Temperatures.* Moscow, Nedra Press, 358 pp.

Skempton, A.W., 1970. The consolidation of clays by gravitational compaction. *Q. J. Geol. Soc.*, London, 125:373–411.

Slobod, R.L., Chambers, A. and Prehn, W.L., 1951, Use of centrifuge for determining connate water, residual oil and capillary pressure curves of small core samples. *Trans. AIME*, TP 3074, 192:127–134.

Slobod, R.L. and Blum, H.A., 1952. Method for determining wettability of reservoir rocks. *Trans. AIME*, 195:1–4.

Smirnov, N.V. and Dunin-Barkovskiy, I.V., 1965. *Textbook for Probability Theory and Mathematical Statistics for Technical Applications.* Moscow, Nauka Press, 320 pp.

Smith, J.E., 1971a. Shale compaction. *Soc. Petol. Eng. J.*, 13:12–22.

Smith, J.E., 1971b. The dynamics of shale compaction and evolution of pore-fluid pressures. *Math. Geol.*, 3:239–263.

Smith, W.O., 1961. *Mechanism of Gravity Drainage and Its Relation to Specific Yield of Uniform Sands.*, U.S. Geological Survey. Professional Paper 402-a, 12 pp.

Stetyukha, Ye.I., 1964. *Equations of Correlation between Physical Properties of Rocks and Depth of Burial.* Moscow, Nedra Press, 134 pp.

Sawabini, C.T., Chilingar, G.V. and Allen, D.R., 1974. Compressibility of unconsolidated, arkosic oil sands. *Soc. Petrol. Eng. J.*, 14:132–138.

Taylor, J., 1950. Pore space reduction in sandstone. *Am. Assoc. Petrol. Geologists Bull*, 34:701–716.

Tegland, E.R., 1977. 3-D seismic techniques boost field development. *Oil Gas J.*, 75(37):79–82.

Teodorovich, G.I., 1965. Expanded classification of sandstones based upon composition. *Proc. USSR Academy of Sciences, Geologic Series* 6:75–95.

Terry, D. and Chilingar, G.V. 1955. Summary of concerning some additional aids in studying sedimentary formation by M.S. Shvetsov. *J. Sediment. Petrology.* 25(3):229–234.

Terzaghi, K. and Peck, R.P., 1948. *Soil Mechanics in Engineering Practice*. John Wiley and Sons, New York, N.Y., 729 pp.

Terzaghi, K., 1961. *Theory of Mechanics of Soils*. Gosstroyizdat, Moscow, 508 pp. (In Russian, translated from German).

Tiab, D. and Donaldson, E.C., 1996. *Petrophysics: Theory and Practice of Measuring Reservoir Rock and Fluid Transport Properties*. Gulf Publishing Company, Houston, TX, 706 pp.

Tixier, M.P., Alger, R.P., and Tanguy, D.R., 1960. New developments in induction and sonic Logging. *J. Petrol. Techn.*, 12 (5):79–87.

Thomsen, L., 1986. Weak elastic anisotropy. *Geophysics*, 151:1954–1966.

Tkhostov, B.A., 1963. *Initial Rock Pressure in Oil and Gas Deposits*. Pergamon, New York, NY, 118 pp.

Tkhostov, B.A., Vezirova, A.D., Vendel'shteyn, B.Yu. and Dolrynin, V.M., 1970. *Oil in Fractured Reservoirs*, Izd. Nedra, Leningrad, pp. 173–197.

Tosaya, C.A., 1982. *Acoustical Properties of Clay Bearing Rocks*, Ph.D. Dissertation, Stanford University, Stanford, CA, 136 pp.

Trask, P.D., 1942. Suggested classification of detrital particles, *Natl. Res. Council Rept. Comm., Sedimentation*. 1941–1942, Appendix G, 3 pp.

Trebin, F.A., 1945. *Permeability of Sands and Sandstones to Oil*. Moscow-Leningrad, Gostoptekhizdat, 141 pp.

Tsvetkova, M.A., 1954. Influence of mineralogic composition of sandy rocks on filtration capacity and oil production. *Tr. Inst. Nefti Akad. Nauk, USSR*, 3 pp.

Udden, J.A., 1914. Mechanical composition of clastic sediments. *Geol. Soc. Am., Bull.*, 24:655–744.

Van der Knaap, W., 1959. Nonlinear Behavior of Elastic Porous Media, *Trans. AIME*, 216:179–186.

Van der Knaap, W. and van der Vlis, A.C., 1967. On the cause of subsidence in oil producing areas, *Seventh World Petrol. Cong.*, Mexico City, Elsevier Publ. Co., 3:85–95.

Vassoevich, N.B. and Bronovitskiy, A.V., 1962. Towards studying density and porosity of sedimentary rocks. *Tr. VNIGRI*. Gostoptekhizdat, 190:478–484

Vendelshtein, B.Yu, 1966. *Analysis and Interpretation of Spontaneous Potential Data from Oil and Gas Wells*. Moscow, Nedra Press, 206 pp.

Vendelshtein, B.Yu. *et al.*, 1984. *Open Hole Testing in Oil and Gas Wells*. Moscow, Nedra Press, 230 pp.

Vennard, J.K., 1961. *Elementary Fluid Mechanics*, 4th ed., John Wiley, New York, 570 pp.

Vistelius, A.B., 1967. *Studies in Mathematical Geology*, New York, Consultants Bureau, 294 pp.

Voigt, W., 1910. *Lehrbuch der Kristallphysik*. Leipzig und Berlin, B.G. Teubner, 964 pp.

Volarovich, M.P., 1940. On the use of high pressure in experimental studies in geologic and geophysical sciences. *Proc. Academy of Sciences of the USSR, Chemical Science Division*, (6):985–996.

Volarovich, M.P., 1960. Studies of physical-mechanical properties of rocks at various pressures. In: *Problems of Tectonophysics*, Moscow: pp. 9–37.

Volarovich, M.P., 1974. *Stress-strain Properties of Rocks and Minerals at High Pressure and Temperature*. Moscow, Nauka Press.

Walton, G.G., 1972. Three-dimensional seismic method. *Geophysics*, 37:417–430.

Wallace, W.E., 1965. Application of electric log measured pressure to drilling problems and a new simplified chart for well site pressure computation. *Log Anal.*, 6:4–10.

Watt, J.P., 1980. Hashin-Shtrikman bounds on the effective elastic moduli of polycrystals with monoclinic symmetry. *J. Appl. Phy.*, 51:1520–1524.

Watt, J.P. and Peselnick, L. 1980. Clarification of the Hashin-Shtrikman bounds on the effective elastic moduli of polycrystals with hexagonal, trigonal, and tetragonal symmetries. *J. Appl. Phys.*, 51:1525–1530.

Waxman, M.H. and Smits, L.J.H., 1968. Electrical conductivities in oil-bearing shaly sands. *Soc. Petrol. Eng. J.*, (6):107–122.

Weaver, C.E., 1959, The clay petrology of sediments. *Proc. 6th Natl. Conf. Clays, Clay Minerals*, pp. 154–187.

Weaver, C.E. and Beck, K.C., 1969. *Changes in the Clay-Water System with Depth, Temperature and Time.* OWRR Proj. No. A-008-GA, Soc. Ceram. Eng. Georgia Institute of Technology, Atlanta, GA, WRC-0769, 95 pp.

Weller, F.A., 1959. Compaction of sediments, *Am. Assoc. Petrol. Geologists Bull.*, 43(2):273–310.

Weller, J.M., 1960. *Stratigraphic Principles and Practices*. Harper and Row, New York, NY, 725 pp.

Wentworth, C.K., 1922. A scale of grade and class terms for clastic sediments. *J. Geol.*, 30:377–392.

Wentworth, C.K., 1929. Method of computing mechanical composition types of sediments. *Am. Assoc. Petrol. Geologists Bull.*, 40:771–790.

Winsauer, W.O., Shearin, H.M., Jr., Masson, P.H., and Williams, M., 1952. Resistivity of brine saturated sands in relation to pore geometry. *Am. Assoc. Petrol. Geologists Bull.*, 36(2):253–278.

Winsauer, W.O. and Gaither, A.A., 1953. A study of porosity and grain relationship in experimental sands. *J. Sediment. Petrology.*, 23(3):147–251.

Wolf, K.H. and Chilingarian, G.V., 1976. Compaction digenesis of carbonate sediments and rocks. In: Chilingarian, G.V. and Wolf, K.H., Eds., *Compaction of Coarse-Grained Sediments.* Elsevier, Amsterdam, pp. 719–768.

Wolfe, M.J., 1968. Lithification of a carbonate mud: Senonian Chalk in Northern Ireland. *Sediment. Geol.*, 2:263–290.

Wyllie, M.R.J. and Gardner, G.H.F., 1958. The generalized Kozeny-Carman equation. *World Oil*, 146(4):210–213.

Wyllie, M.R.J. and Gregory, A.R., 1955. Effect of porosity and particle shape on Kozeny-Carman constants. *Ind. Eng. Chem.*, 47:1379–1388.

Wyllie, M.R.J., Gregory, A.R. and Gardner, G.H.F., 1956. Elastic wave velocities in heterogeneous and porous media, *Geophysics*, 21(1):41–70.

Wyllie, M.R.J., Gregory, A.R. and Gardner, G.H.F., 1958. An experimental investigation of factors affecting elastic wave velocities in porous media. *Geophysics*, 23(3):459–493.

Wyllie, M.R.J. and Rose, W.D., 1950. Application of the Kozeny equation to consolidated porous media. *Nature*, 155:972.

Wyllie, M.R.J. and Rose, W.D., 1950. Some theoretical considerations related to the quantitative evaluation of the physical characteristics of reservoir rocks from electrical log data, *J. Petrol. Techn.*, 2(4):105–118.

Wyllie, M.R.J. and Spangler, M.B., 1952. Application of electrical resistivity measurements to problem of fluid flow in porous media. *Am. Assoc. Petrol. Geologists Bull.*, 36:359–403.

Yin, H., 1992. *Acoustic Velocity and Attenuation of Rocks, Isotropy, Intrinsic Anisotropy, and Stress Induced Anisotropy*. Ph.D. dissertation, Stanford University, Stanford, CA, 227 pp.

Yuster, S.T., 1951. Theoretical considerations of multiphase flow in idealized capillary systems. *Proc. Third World Petroleum Congress*, The Hague, Section III, pp. 437–455.

Zimmerman, R.W., 1984. *The Effect of Pore Structure on the Pore and Bulk Compressibilities of Consolidated Sandstones*, Ph.D. Thesis, University of California, Berkeley, CA, 116 pp.

Subject Index

Also of Interest

Check out these other related titles from Scrivener Publishing

By the same author

Mechanics of Fluid Flow, by Basniev, Dmitriev, and Chilingar. Coming in August 2012, ISBN 9781118385067. The mechanics of fluid flow is one of the most important fundamental engineering disciplines explaining both natural phenomena and human-induced processes. A group of some of the best-known petroleum engineers in the world give a thorough understanding of this important discipline, central to the operations of the oil and gas industry.

Petroleum Accumulation Zones on Continental Margins, by Grigorenko, Chilingar, Sobolev, Andiyeva, and Zhukova. Coming in August 2012, ISBN 9781118385074. Some of the best-known petroleum engineers in the world have come together to produce one of the first comprehensive publications on the detailed (zonal) forecast of offshore petroleum potential, a must-have for any petroleum engineer or engineering student.

Other books by Scrivener Publishing

Zero-Waste Engineering, by Rafiqul Islam, ISBN 9780470626047. In this controversial new volume, the author explores the question of zero-waste engineering and how it can be done, efficiently and profitably. *NOW AVAILABLE!*

Sustainable Energy Pricing, by Gary Zatzman, ISBN 9780470901632. In this controversial new volume, the author explores a new science of energy pricing and how it can be done in a way that is sustainable for the world's economy and environment. *NOW AVAILABLE!*

Flow Assurance, by Boyun Guo and Rafiqul Islam, January 2013, ISBN 9780470626085. Comprehensive and state-of-the-art guide to flow assurance in the petroleum industry.

An Introduction to Petroleum Technology, Economics, and Politics, by James Speight, ISBN 9781118012994. The perfect primer for anyone wishing to learn about the petroleum industry, for the layperson or the engineer. *NOW AVAILABLE!*

Ethics in Engineering, by James Speight and Russell Foote, ISBN 9780470626023. Covers the most thought-provoking ethical questions in engineering. *NOW AVAILABLE!*

Formulas and Calculations for Drilling Engineers, by Robello Samuel, ISBN 9780470625996. The most comprehensive coverage of solutions for daily drilling problems ever published. *NOW AVAILABLE!*

Emergency Response Management for Offshore Oil Spills, by Nicholas P. Cheremisinoff, PhD, and Anton Davletshin, ISBN 9780470927120. The first book to examine the Deepwater Horizon disaster and offer processes for safety and environmental protection. *NOW AVAILABLE!*

Advanced Petroleum Reservoir Simulation, by M.R. Islam, S.H. Mousavizadegan, Shabbir Mustafiz, and Jamal H. Abou-Kassem, ISBN 9780470625811. The state of the art in petroleum reservoir simulation. *NOW AVAILABLE!*

Energy Storage: A New Approach, by Ralph Zito, ISBN 9780470625910. Exploring the potential of reversible concentrations cells, the author of this groundbreaking volume reveals new technologies to solve the global crisis of energy storage. *NOW AVAILABLE!*